Advanced Topics in Mathematical Analysis and Applications

Advanced Topics in Mathematical Analysis and Applications

Editor: Matt Ferrier

NY RESEARCH PRESS
New York

Published by NY Research Press
118-35 Queens Blvd., Suite 400,
Forest Hills, NY 11375, USA
www.nyresearchpress.com

Advanced Topics in Mathematical Analysis and Applications
Edited by Matt Ferrier

International Standard Book Number: 978-1-64725-442-1 (Hardback)

Cataloging-in-Publication Data

Advanced topics in mathematical analysis and applications / edited by Matt Ferrier.
p. cm.
Includes bibliographical references and index.
ISBN 978-1-64725-442-1
1. Mathematics. 2. Mathematical analysis. 3. Applied mathematics. I. Ferrier, Matt.
QA36 .A383 2023
510--dc23

Contents

Preface

Mathematical analysis is a mathematical discipline that deals with continuous functions, limits, and related theories such as differentiation, integration, infinite sequences, series and analytic functions. The field of mathematical analysis has multiple branches, which include real analysis, complex analysis, functional analysis, harmonics analysis and differential equations. Real analysis is a branch of mathematical analysis that involves the study of real numbers and real-valued functions of a real variable. Complex analysis examines the functions of complex numbers, and has several applications in physics and mathematics which include hydrodynamics, thermodynamics, mechanical engineering, electrical engineering, quantum theory, number theory, applied mathematics and algebraic geometry. Harmonic analysis is concerned with the representation of functions and signals as the superposition of basic waves and it finds applications in multiple areas including music theory, representation theory and tidal analysis. This book outlines the significance and applications of mathematical analysis in detail. Students, researchers, and experts associated with this field will benefit from it.

Significant researches are present in this book. Intensive efforts have been employed by authors to make this book an outstanding discourse. This book contains the enlightening chapters which have been written on the basis of significant researches done by the experts.

Finally, I would also like to thank all the members involved in this book for being a team and meeting all the deadlines for the submission of their respective works. I would also like to thank my friends and family for being supportive in my efforts.

Editor

1

Axiomatic Approach in the Analytic Theory of Singular Perturbations

Margarita Besova † **and Vasiliy Kachalov** *,†

Department of Mathematical Sciences, National Research University "MPEI", ul. Krasnokazarmennaya 14, 111250 Moscow, Russia; besova.margarita@yandex.ru
* Correspondence: vikachalov@rambler.ru
† These authors contributed equally to this work.

Abstract: Introduced by S.A. Lomov, the concept of a pseudoanalytic (pseudoholomorphic) solution laid the foundation for the development of the singular perturbation analytical theory. In order for this concept to work in case of linear problems, an apparatus for the theory of exponential type vector spaces was developed. When considering nonlinear singularly perturbed problems, an algebraic approach is currently used. This approval is based on the properties of algebra homomorphisms for holomorphic functions with various numbers of variables, as a result of which it is possible to obtain pseudoholomorphic solutions. In this paper, formally singularly perturbed equations are considered in topological algebras, which allows the authors to formulate the main concepts of the singular perturbation analytical theory from the standpoint of maximal generality.

Keywords: ε-regular function; invariants of equations and systems; ε-pseudoregular solution; essentially singular manifold

MSC: 34E15

---

1. Introduction

The basic concept of the singular perturbation analytic theory is the concept pseudoholomorphic solution, i.e., such a solution, which can be presented as a series in powers of a small parameter that converges in the usual sense (and not asymptotically). The nature of this convergence is determined by the topology of the spaces in which the investigated problems are considered. As a rule, spaces of holomorphic functions (of one or several variables) are used. In this regard, it was possible to formulate the main principles for the theory of singularly perturbed differential equations and systems—under fairly general assumptions that they possess holomorphics in small parameter first integrals [1,2]. Moreover, a connection between the first integrals and homomorphisms of algebras of holomorphic functions with various numbers of variables was established. The pseudoholomorphic solutions themselves are obtained as a result of applying the implicit function theorem. In the presented paper, all of these constructions will be carried out in topological algebras for formally singularly perturbed equations.

2. Algebraic and Analytic Aspects of the Theory of Singular Perturbations

Let $\mathcal{J}_a$ be a complete topological commutative algebra with unit e and let $X, Y_1, \ldots, Y_k, \ldots$ be a sequence of open sets $\mathcal{J}_a$. Let us denote by $\mathcal{A}_0, \mathcal{A}_1, \ldots, \mathcal{A}_k, \ldots$ the spaces of functions continuous on the sets $X, X \times Y_1, \ldots, X \times Y_1 \times \ldots \times Y_k, \ldots$ respectively with their values in $\mathcal{J}_a$. Let us formulate the block I of necessary conditions:

(1°) If the sequence $\{x_i\}_{i=0}^{\infty}$ is a bounded set [3] in $\mathcal{J}_a$, then the series $x_0 + \varepsilon x_1 + \ldots + \varepsilon^i x_i + \ldots$ converges at $|\varepsilon| < 1$.

(2°) If the sequence $\{h_{i,k}\}_{i=1}^{\infty} \subset \mathcal{A}_k$ is such that the series

$$\sum_{i=0}^{\infty} \varepsilon^i h_{i,k}(x, y_1, \ldots, y_k) \tag{1}$$

converges on each set $T \times T_1 \times \ldots \times T_k$, where T is an arbitrary compact from X; T_1 is an arbitrary compact from $Y_1, \cdots$; T_k is an arbitrary compact set from Y_k in some neighborhood of the value $\varepsilon = 0$, the function $\Phi \in \mathcal{A}_0$ and it can be extended to all $\mathcal{J}_a$, then we have

$$\Phi\left(\sum_{i=0}^{\infty} \varepsilon^i h_{i,k}\right) = \Phi(h_{0,k}) + \sum_{i=1}^{\infty} \varepsilon^i g_{i,k}$$

and the last row with coefficients from $\mathcal{A}_k$ is convergent.

Definition 1. *The function $f(x, y_1, \ldots, y_k, \varepsilon) \in \mathcal{A}_k$ represented by* (1), *is called ε-regular.*

(3°) If the system

$$\begin{cases} F_1(x, y_1, \ldots, y_k, \varepsilon) = q_1, \\ \ldots\ldots\ldots\ldots\ldots\ldots \\ F_k(x, y_1, \ldots, y_k, \varepsilon) = q_k \end{cases}$$

with ε-regular left-hand sides is uniquely solvable with respect to $\{y_1, \ldots, y_k\}$ for $\varepsilon = 0$ in some neighborhood of the point $x_0 \in X$, then it is also uniquely solvable in some neighborhood of the same point and thus functions $y_m(x, \varepsilon) \in \mathcal{A}_0$ $(m = \overline{1,k})$ are ε-regular.

Remark 1. *The conditions of the block I are satisfied if $\mathcal{J}_a = \mathbb{C}$, $X, Y_1, Y_2, \ldots$ are simply connected regions, $\mathcal{A}_0, \mathcal{A}_1, \ldots$ are spaces of holomorphic functions on $X, X \times Y_1, X \times Y_1 \times Y_2, \ldots$ respectively.*

In order to formulate the conditions of block II, we give some definitions.

Definition 2. *s-product of tuples $\boldsymbol{\varphi} = \{\varphi_1, \ldots, \varphi_k\}$ and $\boldsymbol{\psi} = \{\psi_1, \ldots, \psi_k\}$ is a function $\boldsymbol{\varphi} ⓢ \boldsymbol{\psi} = \varphi_1\psi_1 + \ldots + \varphi_k\psi_k$.*

Definition 3. *Let $f \in \mathcal{A}_k$, $\varphi_i(x) \in \mathcal{A}_0$, $i = \overline{1,k}$. The composition f and $\boldsymbol{\varphi} = \{\varphi_1, \ldots, \varphi_k\}$ is determined by the formula $f \circ \boldsymbol{\varphi} = f(x, \varphi_1(x), \ldots, \varphi_k(x))$ as usual.*

Block of conditions II:

(1°) All algebras $\mathcal{A}_0, \mathcal{A}_1, \ldots, \mathcal{A}_k, \ldots$ contain constant functions and linear functions. We consider the embeddings $\mathcal{A}_0 \subset \mathcal{A}_1 \subset \ldots \subset \mathcal{A}_k \subset \ldots$ together with topologies to be obvious.

(2°) On all spaces $\mathcal{A}_0, \mathcal{A}_1, \ldots, \mathcal{A}_k, \ldots$, a linear operation ∂_0 is defined such that $\partial_0 p = 0$, where p is a constant function, $\partial_0 x = e$ and $\partial_0 f = 0$ if $f \in \mathcal{A}_k$ and does not depend on x. On each space $\mathcal{A}_k$ $(k = 1, 2, \ldots)$, linear operations $\{\partial_i\}_{i=1}^{k}$ are defined and they comply with the following laws:

- (a) $\partial_i p = 0$, $i = \overline{1,k}$, where $p \in \mathcal{A}_k$ is a constant function;
- (b) $\partial_i y_i = e$, $i = \overline{1,k}$;
- (c) if the function $f \in \mathcal{A}_k$ does not depend on y_m, then $\partial_i f = 0$ for $i \neq m$.

(3°) The operations $\{\partial_i\}_{i=0}^{\infty}$ form a commutative ring.

(4°) An operation d is introduced, and it satisfies the following rules:

- (a) $d \equiv \partial_0$ on $\mathcal{A}_0$;
- (b) $d(f \circ g) = \partial_0 f \cdot \partial_0 g \;\; \forall f, g \in \mathcal{A}_0$;

(c) if $f \in \mathcal{A}_k$, $\varphi_i(x) \in \mathcal{A}_0$ $(i = \overline{1,k})$, $\boldsymbol{\varphi} = \{\varphi_1, \ldots, \varphi_k\}$, then $d(f \circ \boldsymbol{\varphi}) = \partial f \circledS \partial \boldsymbol{\varphi}$, where $\partial f \equiv \{\partial_0 f, \partial_1 f, \ldots, \partial_k f\}$, $\partial_0 \boldsymbol{\varphi} = \{e, \partial_0 \varphi_1, \ldots, \partial_0 \varphi_k\}$ are tuples of length $(k+1)$.

(5°) For every natural number k in the algebra $\mathcal{A}_k$, there exist a lot of tuples $\mathbf{f} = \{f_1(x, y_1, \ldots, y_k), \ldots, f_k(x, y_1, \ldots, y_k)\}$ such that the operator $D_k^{\mathbf{f}} = \mathbf{f} \circledS \boldsymbol{\partial}$, where $\boldsymbol{\partial} = \{\partial_1, \ldots, \partial_k\}$, with a specially defined domain $\mathcal{D}(D_k^{\mathbf{f}})$ is surjective and has the inverse $J_k^{\mathbf{f}}$, which has the following property: for arbitrary compact sets $T \subset X, T_1 \subset Y_1, \ldots, T_k \subset Y_k$ there is a number $C > 0$ such that, for an arbitrary function $\varphi(x) \in \mathcal{A}_0$, the set $\Gamma_k^{\varphi} = \{C^{-n}(J_k^{\mathbf{f}} \partial_0)^n \varphi(x, y_1, \ldots, y_k) \in T \times T_1 \times \ldots \times T_k\}_{n=1}^{\infty}$ is bounded in $\mathcal{J}_a$.

Let us consider the case $k = 1$. We investigate the following equation:

$$\varepsilon d y_1 = F(x, y_1), \tag{2}$$

in which $F \in \mathcal{A}_1$ and ε is a small complex parameter. The function $y_1(x) \in \mathcal{A}_0$ satisfying the initial condition

$$y_1(x^0, \varepsilon) = y_1^0, \tag{3}$$

where $x^0 \in X$, $y_1^0 \in Y_1$, is required to be found.

Definition 4. *The invariant of Equation* (2) *is the function $U(x, y_1, \varepsilon) \in \mathcal{A}_1$, which turns into a constant on the solution $y_1(x, \varepsilon)$ of this equation.*

Theorem 1. *When the blocks of conditions I and II are satisfied, then Equation* (2) *has ε-regular invariants.*

Proof of Theorem 1. If $U(x, y, \varepsilon)$ is an invariant of the Equation (2), then, as it follows from Definition 4, we have

$$\varepsilon \partial_0 U + D_1^F U = 0, \tag{4}$$

where $D_1^F = F \partial_1$.

We seek a solution of Equation (4) in the form of a series in powers of ε:

$$U(x, y_1, \varepsilon) = U_0(x, y_1) + \varepsilon U_1(x, y_1) + \ldots + \varepsilon^n U_n(x, y_1) + \ldots \tag{5}$$

for the coefficients of the equation above the following series of equations holds:

$$\begin{aligned} &D_1^F U_0 = 0, \\ &D_1^F U_1 = -\partial_0 U_0, \\ &\ldots\ldots\ldots\ldots\ldots \\ &D_1^F U_n = -\partial_0 U_{n-1}. \\ &\ldots\ldots\ldots\ldots\ldots \end{aligned} \tag{6}$$

As a solution to the first equation of this series, we take an arbitrary function $\varphi(x) \in \mathcal{A}_0$. To satisfy the condition (5°) of block II, we assume that the domain of the surjective operator D_1^F consists of functions from $\mathcal{A}_1$ that vanish when $y_1 = y_1^0$ $\forall x \in X$, and the inverse operator J_1^F is such that, for any compact sets $T \subset X$, $T_1 \subset Y_1$, there exists a number $C > 0$ such that, for an arbitrary function $\varphi(x) \in \mathcal{A}_0$ set $\Gamma_1^{\varphi} = \{C^{-n}(J_1^F \partial_0)^n \varphi, \ (x, y_1) \in T \times T_1\}_{n=1}^{\infty}$ is limited in $\mathcal{J}_a$.

As a result, all equations of the series (6), starting from the second, are uniquely solvable:

$$U(x, y_1, \varepsilon) = \varphi - \varepsilon (J_1^F \partial_0) \varphi + \ldots + (-1)^n \varepsilon^n (J_1^F \partial_0)^n \varphi + \ldots \tag{7}$$

and this series converges in some neighborhood of the value $\varepsilon = 0$ on the set $T \times T_1$. Theorem 1 is proved. □

Remark 2. *As it comes out from the form of series (7), we can consider $U(x, y_1, \varepsilon)$ for each fixed ε as the image of the linear operator $H_\varepsilon : \mathcal{A}_0 \to \mathcal{A}_1$ given by the formula*

$$H_\varepsilon = I - \varepsilon(J_1^F \partial_0) + \ldots + (-1)^n \varepsilon^n (J_1^F \partial_0)^n + \ldots,$$

where I is the identity operator. Thus, $U = H_\varepsilon[\varphi]$.

Theorem 2. *$\{H_\varepsilon\}$ forms a ε-regular family for homomorphisms of the algebra $\mathcal{A}_0$ into the algebra $\mathcal{A}_1$.*

Proof of Theorem 2. Let U and V be invariants of the Equation (2). Obviously, then there exists a function Φ such that $V = \Phi(U)$, and therefore $H_\varepsilon[\varphi(x)] = \Phi(H_\varepsilon[x])$. If in this equality we put $y_1 = y_1^0$, then $\varphi(x) = \Phi(x)\ \forall x \in X$, therefore

$$H_\varepsilon[\varphi(x)] = \varphi(H_\varepsilon[x]). \tag{8}$$

The equality (8) is called the commutation relation.

Now, let $\varphi_1(x), \varphi_2(x) \in \mathcal{A}_0$; then,

$$H_\varepsilon[\varphi_1\varphi_2] = (\varphi_1\varphi_2)(H_\varepsilon[x]) = \varphi_1(H_\varepsilon[x])\varphi_2(H_\varepsilon[x]) = H_\varepsilon[\varphi_1]H_\varepsilon[\varphi_2],$$

where $H_\varepsilon : \mathcal{A}_0 \to \mathcal{A}_1$ is a homomorphism. Theorem 2 is proved. □

For the concepts given below, we need a definition introduced by S.A. Lomov for the notion of the essentially singular manifold [4].

Definition 5. *Let $\varphi(x) \in \mathcal{A}_0$, $\varphi(x_0) = 0$, $\Phi \in \mathcal{A}_0$, let it allow continuation to all $\mathcal{J}_a$, and let T_0 be some compact from X containing the point x_0. The set $Q^+(\varphi, \Phi, T_0) = \{q : \Phi(\varphi(x)/\varepsilon),\ x \in T_0,\ \varepsilon > 0\}$ is called an essentially singular variety generated by the point $\varepsilon = 0$. Moreover, we say that it has the correct structure if*

$$Q^+ = \bigcup_{m=1}^{\infty} \Pi_m,$$

where $\Pi_1 \subset \Pi_2 \subset \ldots$ is an increasing compact system.

We introduce the concept of ε-pseudoregularity necessary for studying the analytic properties of a solution of $y(x, \varepsilon)$.

Definition 6. *The solution to the problems (2), (3) is called ε-pseudoregular if $y_1(x, \varepsilon) = \widetilde{Y}(x, \varphi(x)/\varepsilon, \varepsilon)$, in which $\varphi(x) \in \mathcal{A}_0$; the function $\widetilde{Y}(x, \eta, \varepsilon)$ is ε-regular for all $(x, \eta) \in T_0 \times G$ where T_0 is some compact set containing the point x_0, G is some unlimited set from $\mathcal{J}_a$.*

Theorem 3. *If the essentially singular manifold $Q^+(\varphi, \Phi, T_0)$ is a bounded set in $\mathcal{J}_a$ and the equation*

$$(J_1^F \partial_0)\varphi = \varphi(x)/\varepsilon \tag{9}$$

has a unique solution of the form $y_1 = Y_{1,0}(x, q)\big|_{q=\Phi(\varphi(x)/\varepsilon)}$ such that the function $Y_{1,0}(x, q)$ coincides with the contraction to the set $T_0 \times Q^+$ of some function from $\mathcal{A}_1$, then problems (2), (3) have a ε-pseudoregular solution.

Proof of Theorem 3. For the invariant represented by the Formula (7), we compose the equality

$$(J_1^F \partial_0)\varphi - \varepsilon(J_1^F \partial_0)^2\varphi + \ldots + (-1)^{n-1}\varepsilon^{n-1}(J_1^F \partial_0)^n \varphi + \ldots = \varphi(x)/\varepsilon,$$

which defines the solution to the problems (2), (3). We apply the function Φ to its left-hand and right-hand sides and, using the condition (2°) of block I, we obtain the following equality:

$$\Phi((J_1^F\partial_0)\varphi) + \varepsilon\Psi(x, y_1, \varepsilon) = q, \tag{10}$$

where $\Psi(x, y_1, \varepsilon)$ is some ε-regular function.

Let the small parameter $\varepsilon > 0$ in Equation (2) be such that the following expression holds:

$$\{q : q = \Phi(\varphi(x)/\varepsilon),\ x \in T_0\} = \Pi_m$$

for some natural number m (depending on ε). In accordance with the theorem conditions and the condition (3°) of block I, Equation (10) is solvable in some neighborhood σ_{xq} of each point $(x, q) \in T_0 \times \Pi_m$ and has a solution $y_1 = Y_1(x, q, \varepsilon)$ that is ε-regular in a neighborhood of $|\varepsilon| < \varepsilon_{xq}$, where $\varepsilon_{xq} > 0$ and is determined by this neighborhood. From the cover $\{\sigma_{xq}\}$ of the compact set $T_0 \times \Pi_m$, we choose the finite subcover $\{\sigma_{xq}\}_{i=1}^N$. Then, $y_1 = Y_1(x, q, \varepsilon)$ will be a ε-regular function in the smallest neighborhood of the point $\varepsilon = 0$ defined by a finite subcover; the function $y_1 = Y_1(x, \Phi(\varphi(x)/\varepsilon), \varepsilon)$ will give a ε-pseudoregular solution to the problem (2), (3) on the part $\widetilde{T}_0 \subset T_0$ such that the set $\{(x, q) : x \in \widetilde{T}_0,\ q = \Phi(\varphi(x)/\varepsilon)\} \subset T_0 \times \Pi_m$. The theorem is proved. □

3. Invariants and ε-Pseudoregular Solutions of Systems of Equations

We take into the consideration the system of equations

$$\begin{cases} \varepsilon dy_1 = F_1(x, y_1, \ldots, y_k), \\ \ldots\ldots\ldots\ldots\ldots\ldots\ldots \\ \varepsilon dy_k = F_k(x, y_1, \ldots, y_k), \end{cases} \tag{11}$$

the right-hand sides of which belong to the algebra $\mathcal{A}_k$. It is required to find its solution $\mathbf{y}(x, \varepsilon) = \{y_1(x, \varepsilon), \ldots, y_k(x, \varepsilon)\}$ satisfying the initial conditions

$$y_1(x^0, \varepsilon) = y_1^0,\ \ldots,\ y_k(x^0, \varepsilon) = y_k^0. \tag{12}$$

We rewrite system (11) by introducing the following denotation:

$$\begin{aligned} &\mathbb{F}(x, \mathbf{y}) = \{F_1(x, \mathbf{y}), \ldots, F_k(x, \mathbf{y})\}, \\ &\mathbf{y}^0 = \{y_1^0, \ldots, y_k^0\}. \end{aligned}$$

Thus, we have

$$\begin{aligned} &\varepsilon d\mathbf{y} = \mathbb{F}(x, \mathbf{y}), \\ &\mathbf{y}(x^0, \varepsilon) = \mathbf{y}^0 \end{aligned} \tag{13}$$

to be the initial investigated problem.

Definition 7. *The function $U(x, \mathbf{y}, \varepsilon) \in \mathcal{A}_k$ is called the invariant of the system* (11) *if it turns into a constant on the solution $\mathbf{y}(x, \varepsilon)$.*

We formulate a theorem similar to Theorem 1.

Theorem 4. *The system* (11) *has ε-regular invariants.*

Proof of Theorem 4. The proof is carried out according to the same scheme as in the case of a single equation. □

Definition 8. *The solution of the* $\mathbf{y}(x,\varepsilon)$ *problem* (13) *is called* ε*-pseudoregular if* $\mathbf{y}(x,\varepsilon)=\widetilde{\mathbb{Y}}(x,\boldsymbol{\varphi}(x)/\varepsilon,\varepsilon)$, *in which* $\boldsymbol{\varphi}(x)=\{\varphi_1(x),\ldots,\varphi_k(x)\}$, $\varphi_i(x)\in\mathcal{A}_0$ $(i=\overline{1,k})$ *and the function* $\widetilde{\mathbb{Y}}(x,\boldsymbol{\eta},\varepsilon)$ *in which* $\boldsymbol{\eta}=(\eta_1,\ldots,\eta_k)$, *is* ε*-regular for all* $(x,\eta_1,\ldots,\eta_k)\in T_0\times G_1\times\ldots\times G_k$ *where* T_0 *is some compact set from* X *containing* x_0 *and* G_i $(i=\overline{1,k})$ *are unbounded sets from* $\mathcal{J}_a$.

Theorem 5. *Let the following conditions be fulfilled:*

(1°) *The functions* Φ_i *and* φ_i *are such that the essentially singular manifolds* $Q_i^+(\varphi_i,\Phi_i,T_0)$ $(i=\overline{1,k})$ *are bounded sets in* $\mathcal{J}_a$.

(2°) *The equation* $D_k^{\mathbb{F}}V=e$ *has the solutions* $\{V_1(x,\mathbf{y}),\ldots,V_k(x,\mathbf{y})\}$ *such that system*

$$\begin{cases}\partial_0\varphi_1\cdot V_1(x,\mathbf{y}))=\varphi_1(x)/\varepsilon,\\ \cdots\cdots\cdots\cdots\cdots\cdots\\ \partial_0\varphi_k\cdot V_k(x,\mathbf{y}))=\varphi_k(x)/\varepsilon\end{cases}$$

has the only solution

$$\mathbf{y}=\mathbb{Y}_0(x,\mathbf{q})\Big|_{\substack{\mathbf{q}=(q_1,\ldots,q_k)\\ q_i=\Phi_i(\varphi_i(x)/\varepsilon)}},$$

and each component $Y_{0,i}(x,\mathbf{q})$ $(i=\overline{1,k})$ *of it coincides with the restriction to the set* $T_0\times Q_1^+\times\ldots\times Q_k^+$ *of some function from* $\mathcal{A}_k$.

Then, the solution $\mathbf{y}(x,\varepsilon)$ *of the problem* (13) *is* ε*-pseudoregular.*

Proof of Theorem 5. We write the equalities for the invariants of the system (11) in the following form:

$$\begin{cases}\partial_0\varphi_1\cdot V_1(x,\mathbf{y})-\varepsilon(J_k^{\mathbb{F}}\partial_0)^2\varphi_1+\varepsilon^2(J_k^{\mathbb{F}}\partial_0)^3\varphi_1-\ldots=\varphi_1(x)/\varepsilon,\\ \cdots\cdots\cdots\cdots\cdots\cdots\cdots\cdots\cdots\cdots\cdots\cdots\\ \partial_0\varphi_k\cdot V_k(x,\mathbf{y})-\varepsilon(J_k^{\mathbb{F}}\partial_0)^2\varphi_k+\varepsilon^2(J_k^{\mathbb{F}}\partial_0)^3\varphi_k-\ldots=\varphi_k(x)/\varepsilon.\end{cases}\tag{14}$$

In order for this system to determine the solution of the problem (11), (12) (or (13)), we assume (see condition (5°) of block II) that $\mathcal{D}(D_k^{\mathbb{F}})$ consists of functions that vanish when $\mathbf{y}=\mathbf{y}^0$ for any $x\in X$ and $V_i(x,\mathbf{y}^0)=0$, $i=\overline{1,k}$.

We apply the functions $\Phi_1,\ldots,\Phi_k$ to the equations of the system (14), respectively. Then, in accordance with the condition (2°) of block I, we obtain the system

$$\begin{cases}\Phi_1(\partial_0\varphi_1\cdot V_1(x,\mathbf{y}))+\varepsilon\Psi_1(x,\mathbf{y},\varepsilon)=q_1,\\ \cdots\cdots\cdots\cdots\cdots\cdots\cdots\cdots\\ \Phi_k(\partial_0\varphi_k\cdot V_k(x,\mathbf{y}))+\varepsilon\Psi_k(x,\mathbf{y},\varepsilon)=q_k.\end{cases}\tag{15}$$

Let the small parameter $\varepsilon>0$ in the system (13) be such that

$$\{q_i: q_i=\Phi_i(\varphi(x)/\varepsilon),\ x\in T_0\}=\Pi_{m_i},\ \ i=\overline{1,k}$$

for natural $m_1,\ldots,m_k$. By the condition of the (2°) Theorem 5, the system (15) for $\varepsilon=0$ has a unique solution $\mathbf{y}=\mathbb{Y}_0(x,\mathbf{q})$ and, therefore, in accordance with the condition (3°) of block I, this system is solvable in some neighborhood $\sigma_{x\mathbf{q}}$ of each point $(x,\mathbf{q})\in T_0\times\Pi_{m_1}\times\ldots\times\Pi_{m_k}$, and its solution $\mathbb{Y}(x,\mathbf{q},\varepsilon)$ is ε-regular there for $|\varepsilon|<\varepsilon_{x\mathbf{q}}$. After that, from the cover $\{\sigma_{x\mathbf{q}}\}$ of the compact set $T_0\times\Pi_{m_1}\times\ldots\times\Pi_{m_k}$, we choose a finite subcover and $\mathbb{Y}(x,\mathbf{q},\varepsilon)$ will be ε-regular in the minimal neighborhood from the neighborhood of $\varepsilon=0$ corresponding to a finite subcover. As in the proof of Theorem 3, we choose $\widetilde{T}_0\subset T_0$, a compact set on which there exists a ε-regular solution

$$\mathbf{y}(x,\varepsilon) = \mathbb{Y}(x,\mathbf{q},\varepsilon)\Big|_{\substack{q_i=\Phi_i(\varphi_i(x)/\varepsilon) \\ (i=\overline{1,k})}}.$$

The theorem is proved. □

4. Concrete Implementations of the Theory

In this section of the article, we assume that $\mathcal{J}_a = \mathbb{C}$, $X = P_0 \equiv \{z \in \mathbb{C} : |z - z^0| < r_0\}$, $Y_i = P_i \equiv \{w_i \in \mathbb{C} : |w_i - w_i^0| < r_i\}, i = \overline{1,k}$. We shall use the following denotations $\mathbf{w} = \{w_1, \dots, w_k\}$, $\mathbf{w}^0 = \{w_1^0, \dots, w_k^0\}$, $\mathbb{P}^k = P_1 \times \dots \times P_k$ a polycircle of $\mathbb{C}^k$.

Let $\mathcal{A}_0$ be the algebra of holomorphic functions in the P_0 circle of the variable z; let $\mathcal{A}_1$ be the algebra of holomorphic functions in the $P_0 \times P_1$ bicircle of the variables (z, w_1), …; let $\mathcal{A}_k$ be the algebra of holomorphic functions of the variables $(z, w_1, \dots, w_k)$ in the polycircle $P_0 \times \mathbb{P}^k$. It is clear that, if $\partial_0 = \partial_z, \partial_1 = \partial_{w_1}, \dots, \partial_k = \partial_{w_k}$, then all the conditions of block I and the conditions (1°)—(4°) of block II are satisfied. In the concepts given below, we show that the condition (5°) also holds under fairly general assumptions.

Thus, we investigate the Cauchy problem for $\varepsilon > 0$:

$$\begin{aligned} &\varepsilon\frac{d\mathbf{w}}{dz} = \mathbb{F}(z,\mathbf{w}), \;\; z \in \widetilde{P}_0 = \{z \in \mathbb{C} : |z - z^0| < \widetilde{r}_0, \; 0 < \widetilde{r}_0 < r_0\}, \\ &\mathbf{w}(z^0,\varepsilon) = \mathbf{w}^0, \end{aligned} \tag{16}$$

where $\mathbb{F}(z,\mathbf{w}) = \{F_1(z,\mathbf{w}), \dots, F_k(z,\mathbf{w})\}$, $F_i(z,\mathbf{w}) \in \mathcal{A}_k$ for $i = \overline{1,k}$.

From the nonlinear system (16), we come to the linear equation of its integrals (invariants):

$$\varepsilon\partial_z\mathbb{U} + D_k^{\mathbb{F}}\mathbb{U} = 0. \tag{17}$$

Here, $D_k^{\mathbb{F}} = F_1\partial_{w_1} + \dots + F_k\partial_{w_k}$ is the linear partial differential operator of the first order in partial derivatives: $\mathbb{U} = \{U^{[1]}, \dots, U^{[k]}\}$, where $\{U^{[i]}\}_{i=1}^k$ is the system of independent integrals.

First of all, we present an integral method for solving inhomogeneous linear differential equations of the first order with partial derivatives [5].

Let Λ be a holomorphically smooth surface in $\mathbb{C}^k$ and we need to solve the initial problem

$$\begin{aligned} &D_k^{\mathbb{F}}V = f, \;\; f \in \mathcal{A}_k, \\ &V\big|_{\mathbf{w}\in\Lambda} = 0. \end{aligned} \tag{18}$$

Let us suppose that the surface Λ is given by the coordinates $\widetilde{\mathbf{w}} = \{\widetilde{w}_1, \dots, \widetilde{w}_{k-1}\}$ and, namely, $\Lambda = \{\mathbf{w} \in \mathbb{C}^k : w_i = \lambda(\widetilde{\mathbf{w}}), i = \overline{1,k}\}$, where $\lambda_i(\widetilde{\mathbf{w}})$ are functions holomorphic in some region $\mathbb{C}^{k-1}$. Next, we compose the equation system for the characteristic equation

$$\frac{d\mathbf{w}}{ds} = \mathbb{F}(z,\mathbf{w}), \tag{19}$$

in which $s \in \mathbb{C}$ is an independent variable, and z acts as a parameter. Let $\mathbf{w} = \mathbf{g}(z, \widetilde{\mathbf{w}}, s)$ be a solution to the system (19) with the initial condition

$$\mathbf{w}\big|_{s=0} = \boldsymbol{\lambda}(\widetilde{\mathbf{w}}),$$

where $\boldsymbol{\lambda} = \{\lambda_1, \dots, \lambda_k\}$.

The existence and uniqueness theorem guarantees the unique solvability of the system $\mathbf{g}(z, \widetilde{\mathbf{w}}, s) = \mathbf{w}$ relative to $\widetilde{\mathbf{w}}$ and s: $s = S(z,\mathbf{w})$, $\widetilde{\mathbf{w}} = \widetilde{\mathbb{W}}(z,\mathbf{w})$. We denote the operator of replacing variables $(s, \widetilde{\mathbf{w}})$ by the variable $\mathbf{w}$ by $R(z)$ and the backward replacement operator is denoted by $R^{-1}(z,s)$:

$$R(z)[\chi(z,s,\widetilde{\mathbf{w}})] = \chi(z,S(z,\mathbf{w}),\widetilde{\mathbb{W}}(z,\mathbf{w})),$$
$$R^{-1}(z,s)[\theta(z,\mathbf{w})] = \theta(z,\mathbf{g}(z,\widetilde{\mathbf{w}},s)).$$

Then, as you know, if the phase trajectories in the system of characteristics are transversal (not tangent) to the surface Λ, then a solution to the Cauchy problem (17) exists, is unique, and is expressed by the following formula:

$$V(z,\mathbf{w}) = \int_0^s f(z,\mathbf{g}(z,\widetilde{\mathbf{w}},s_1))ds_1\Big|_{\substack{s=S(z,\mathbf{w}),\\ \widetilde{\mathbf{w}}=\widetilde{\mathbb{W}}(z,\mathbf{w})}}. \tag{20}$$

We return to Equation (18). We have

$$\mathbb{U}(z,\mathbf{w},\varepsilon) = \mathbb{U}_0(z,\mathbf{w}) + \varepsilon\mathbb{U}_1(z,\mathbf{w}) + \ldots + \varepsilon^n\mathbb{U}_n(z,\mathbf{w}) + \ldots, \tag{21}$$

and, as this takes place

$$\begin{aligned} &D_k^{\mathbb{F}}\mathbb{U}_0(z,\mathbf{w}) = 0,\\ &D_k^{\mathbb{F}}\mathbb{U}_1(z,\mathbf{w}) = -\partial_z\mathbb{U}_0(z,\mathbf{w}),\\ &\cdots\cdots\cdots\cdots\cdots\cdots\\ &D_k^{\mathbb{F}}\mathbb{U}_n(z,\mathbf{w}) = -\partial_z\mathbb{U}_{n-1}(z,\mathbf{w}).\\ &\cdots\cdots\cdots\cdots\cdots\cdots \end{aligned} \tag{22}$$

As a solution to the first equation of this series, we take the vector function $\mathbb{U}_0 = \{\varphi_1(z),\ldots,\varphi_k(z)\}$, $\varphi_i(z) \in \mathcal{A}_0$ for $i = \overline{1,k}$. The solution to the second equation of the series (22) is the vector function $\mathbb{U}_1 = \{-\partial_z\varphi_1 V^{[1]},\ldots,-\partial_z\varphi_k V^{[k]}\}$ where $\{V^{[1]},\ldots,V^{[k]}\}$ are functionally independent solutions of the equation $D_k^{\mathbb{F}}V = 1$ and such that $V^{[i]}(z,\mathbf{w}^0) = 0\ \forall z \in P_0$, $i = \overline{1,k}$. We find solutions to other equations using Formula (20), assuming that $\mathbf{w}^0 \in \Lambda$:

$$\begin{aligned} \mathbb{U}_2(z,\mathbf{w}) &= -R(z)\int_0^s R^{-1}(z,s_1)\partial_z\mathbb{U}_1(z,\mathbf{w})ds_1,\\ \mathbb{U}_3(z,\mathbf{w}) &= R(z)\int_0^s ds_1 R^{-1}(z,s_1)\partial_z R(z)\int_0^{s_1} R^{-1}(z,s_2)\partial_z\mathbb{U}_1(z,\mathbf{w})ds_2,\\ &\cdots\cdots\cdots\cdots\cdots\cdots\\ \mathbb{U}_n(z,\mathbf{w}) &= (-1)^{n-1}R(z)\int_0^s ds_1 R^{-1}(z,s_1)\partial_z R(z)\int_0^{s_1} ds_2 R^{-1}(z,s_2)\partial_z\ldots\\ &\quad\ldots R(z)\int_0^{s_{n-2}} R^{-1}(z,s_{n-1})\partial_z\mathbb{U}_1(z,\mathbf{w})ds_{n-1}.\\ &\cdots\cdots\cdots\cdots\cdots\cdots \end{aligned} \tag{23}$$

Next, to each natural $n \geq 2$, we associate $(n-1)$ concentric circles $C_m = \{z : |z - z^0| = t_m\}$ where

$$t_m = \widetilde{r}_0 + \frac{r - \widetilde{r}_0}{n-1}m, \quad m = \overline{1, n-1}$$

and $\widetilde{r}_0 < r < r_0$.

These circles are situated at the same distance from each other:

$$\rho = t_{n-1} - t_{n-2} = \ldots = t_2 - \widetilde{r}_0 = \frac{r - \widetilde{r}_0}{n-1}.$$

We use the equalities (23) with the Cauchy integral formula:

$$\begin{aligned}
\mathbb{U}_2(z,\mathbf{w}) &= -\tfrac{1}{2\pi i}R(z)\int_0^s R^{-1}(z,s_1)\oint_{C_1}\tfrac{dz_1}{(z_1-z)^2}\mathbb{U}_1(z_1,\mathbf{w})ds_1,\\
\mathbb{U}_3(z,\mathbf{w}) &= \tfrac{1}{(2\pi i)^2}R(z)\int_0^s ds_1R^{-1}(z,s_1)\oint_{C_1}\tfrac{dz_1}{(z_1-z)^2}R(z_1)\int_0^{s_1}R^{-1}(z_1,s_2)\oint_{C_2}\tfrac{dz_2}{(z_2-z_1)^2}\mathbb{U}_1(z_2,\mathbf{w})ds_2,\\
&\ldots\ldots\ldots\ldots\ldots\ldots\ldots\ldots\ldots\ldots\ldots\ldots\ldots\\
\mathbb{U}_n(z,\mathbf{w}) &= \tfrac{(-1)^{n-1}}{(2\pi i)^{n-1}}R(z)\int_0^s ds_1R^{-1}(z,s_1)\oint_{C_1}\tfrac{dz_1}{(z_1-z)^2}R(z_1)\int_0^{s_1}ds_2R^{-1}(z_1,s_2)\oint_{C_2}\tfrac{dz_2}{(z_2-z_1)^2}\cdots\\
&\quad\ldots R(z_{n-2})\int_0^{s_{n-2}}R^{-1}(z_{n-2},s_{n-1})\oint_{C_{n-1}}\tfrac{dz_{n-1}}{(z_{n-1}-z_{n-2})^2}\mathbb{U}_1(z_{n-1},\mathbf{w})ds_{n-1}.\\
&\ldots\ldots\ldots\ldots\ldots\ldots\ldots\ldots\ldots\ldots\ldots\ldots\ldots
\end{aligned}\tag{24}$$

We represent $\mathbb{U}_n(z,\mathbf{w})$ in the following form:

$$\mathbb{U}_n(z,\mathbf{w}) = \frac{(-1)^{n-1}}{(2\pi i)^{n-1}}\int_0^s ds_1\int_0^{s_1}ds_2\ldots\int_0^{s_{n-2}}ds_{n-1}\oint_{C_1}\frac{dz_1}{(z-z_1)^2}\cdots\oint_{C_{n-2}}\frac{dz_{n-2}}{(z_{n-3}-z_{n-2})^2}\cdot$$
$$\cdot\oint_{C_{n-1}}\frac{R(z)R^{-1}(z,s_1)R(z_1)R^{-1}(z_1,s_2)\ldots R(z_{n-2})R^{-1}(z_{n-2},s_{n-1})\mathbb{U}_1(z_{n-1},\mathbf{w})dz_{n-1}}{(z_{n-2}-z_{n-1})^2}.$$

Let $\|\cdot\|_k$ be the norm in $\mathbb{C}^k$; then, for all $z\in\widehat{P}_0=\{z\in\mathbb{C}:|z-z_0|<r_0\}$ and all $\mathbf{w}$ from some subregion $\widehat{\mathbb{P}}_k$ of the polycircle $\mathbb{P}_k$, the following inequality takes place:

$$\|\mathbb{U}_n(z,\mathbf{w})\|_k\le\left|\int_0^s ds_1\int_0^{s_1}ds_2\ldots\int_0^{s_{n-2}}ds_{n-1}\right|\frac{1}{(2\pi)^{n-1}}H_{n-1}\|\mathbb{U}_1(z,\mathbf{w})\|_k,$$

where

$$\begin{aligned}
H_{n-1} &= \oint_{C_1}\tfrac{|dz_1|}{|z-z_1|^2}\cdots\oint_{C_{n-1}}\tfrac{|dz_{n-1}|}{|z_{n-2}-z_{n-1}|^2}=\\
&=\int_0^{2\pi}\tfrac{t_1d\alpha}{t_1^2+|z|^2-2t_1|z|\cos\alpha}\cdots\int_0^{2\pi}\tfrac{t_{n-1}d\alpha}{t_{n-1}^2+t_{n-2}^2-2t_{n-1}t_{n-2}\cos\alpha}=\\
&=\tfrac{(2\pi)^{n-1}t_1t_2\ldots t_{n-1}}{(t_1-|z|^2)(t_2^2-t_1^2)\ldots(t_{n-1}^2-t_{n-2}^2)}\le\tfrac{(2\pi)^{n-1}r_0^{n-1}(n-1)^{n-1}}{2^{n-2}\widetilde{r}_0^{\,n-1}(r-\widetilde{r}_0)^{n-1}}.
\end{aligned}$$

As we have

$$\left|\int_0^s ds_1\int_0^{s_1}ds_2\ldots\int_0^{s_{n-2}}ds_{n-1}\right|=\frac{|s|^{n-1}}{(n-1)!},$$

then

$$\|\mathbb{U}_n(z,\mathbf{w})\|_k\le\frac{r_0^{n-1}(n-1)^{n-1}}{2^{n-2}\widetilde{r}_0^{\,n-1}(r-\widetilde{r}_0)^{n-1}(n-1)!}\|\mathbb{U}_1(z,\mathbf{w})\|_k,$$

and from that the convergence of the series (21) on any compact set from the set $\widehat{P}_0\times\widehat{\mathbb{P}}_k$ follows.

Thus, it is proved that the components of the vector $\mathbb{U}(z,\mathbf{w},\varepsilon)$ form an independent system of integrals (invariants) and are holomorphic (ε-regular) at the point $\varepsilon=0$. It is also clear that there is a statement similar to Theorem 5 on the existence of a pseudoholomorphic (ε-pseudoregular) solution of the Cauchy problem (16). Without loss of generality, we assume that $z_0=0$.

Theorem 6. *Let the entire functions* $\{\Phi_1,\ldots,\Phi_k\}$ *and the functions* $\{\varphi_1(z),\ldots,\varphi_k(z)\}$ *which are holomorphic in the circle* P_0 *be such that the essentially singular manifolds* $\{Q_1^+(\varphi_1,\Phi_1,T_0),\ldots,Q_k^+(\varphi_k,\Phi_k,T_0)\}$ *created by the functions described above where* T_0 *is some segment of the real axis, the left end of which coincides with the origin and belongs to the circle* $\widehat{P}_0$*, are sets bounded in* $\mathbb{C}$*; and the system of equations*

$$\begin{cases} \varphi_1'(z)V_1(z,\mathbf{w}) = \varphi_1(z)/\varepsilon, \\ \cdots\cdots\cdots\cdots\cdots\cdots\cdots \\ \varphi_k'(z)V_k(z,\mathbf{w}) = \varphi_k(z)/\varepsilon, \end{cases}$$

in which $\{V_1(z,\mathbf{w}),\ldots,V_k(z,\mathbf{w})\}$ *are independent solutions of the equation* $D_k^{\mathbb{F}}V = 1$*, has a solution of the form*

$$\mathbf{w} = \mathbb{W}_0(z,\mathbf{q})\Big|_{\mathbf{q}=\{\Phi_1(\varphi_1(z)/\varepsilon),\ldots,\Phi_k(\varphi_k(z)/\varepsilon)\}},$$

each component $W_{0,i}(z,\mathbf{q})$ $(i=\overline{1,k})$ *of it is holomorphic on the set* $T_0 \times Q_1^+ \times \ldots \times Q_k^+$*. Then, the solution* $\mathbf{w}(z,\varepsilon)$ *of the initial problem* (16) *is pseudoholomorphic at the point* $\varepsilon = 0$ *(*ε*-pseudoregular).*

It should be noted [6] that this solution can be continued in a pseudoholomorphic way for a fixed $\varepsilon > 0$ from some segment $[0,\widetilde{T}_0]$ (see the end of the proof of Theorem 5) by segment $[0,T_0]$.

5. Conclusions

Further development of the axiomatic approach in the analytical singular perturbation theory will allow us to consider a more general class of equations with a small parameter, in particular, an analogue of nonlinear differential equations in partial derivatives (for example, equations of the Navier–Stokes type, etc.). This is very urgent since the range of problems leading to singularly perturbed problems is constantly expanding. In this sense, the "Dyson argument" that appeared in theoretical physics is quite indicative—the solutions of the equations arising in astrophysics can depend holomorphically on the gravitational constant only after isolating a revealing the essentially singular manifold [7].

As for the current state of the singular perturbation of theory, the asymptotic approach prevails there. In our opinion, when solving most singularly perturbed equations, the following methods are used: the Vasilieva–Butuzov–Nefedov boundary function method [8,9], the Maslov method [10], the Lomov regularization method [4,11], and the Bogolyubov–Krylov–Mitropolsky average method [12,13]. In the case of more specific situations, these methods are combined and new approaches to the asymptotic integration are proposed [14].

Author Contributions: Conceptualization and writing—review and editing by M.B. and V.K. All authors have read and agreed to the published version of the manuscript.

References

1. Kachalov, V.I. On a method of holomorphic regularization of singularly perturbed problems. *Russ. Math. (Izv. VUZov)* **2017**, *61*, 44–50. [CrossRef]
2. Kachalov, V.I. On holomorphic regularization of strongly nonlinear singularly perturbed problems. *Ufa Math. J.* **2018**, *10*, 35–42. [CrossRef]
3. Rudin, W. *Functional Analysis*; McGraw-Hill Book Company: New York, NY, USA, 1973.
4. Lomov, S.A.; Lomov, I.S. *Fundamentals of the Mathematical Theory of Boundary Layer*; Moscow State University: Moscow, Russia, 2011.
5. Samoilenko, A.M.; Krivosheya, S.A.; Perestyuk, N.A. *Differential Equations*; Vysshaya Shkola: Moscow, Russia, 1989.
6. Bobodzhanov, A.A.; Safonov, V.F.; Kachalov, V.I. Asymptotic and Pseudoholomorphic Solutions of Singularly Perturbed Differential and Integral Equations in the Lomov's Regularization Method. *Axioms* **2019**, *8*, 27. [CrossRef]

7. Krivoruchenko, M.I.; Nadyozhin, D.K.; Yudin, A.V. Hydrostatic Equilibrium of Stars without Electroneutrality constraint. *Phys. Rev. D* **2018**, *97*, 083016. [CrossRef]
8. Vasilieva, A.B.; Butuzov, V.F. *Asymptotic Expansions of Solutions of Singularly Perturbed Equations*; Nauka: Moscow, Russia, 1989.
9. Vasilieva, A.B.; Butuzov, V.F.; Nefedov, N.N. Contrast structures in singularly perturbed problems. *Fundam. Appl. Math.* **1998**, *4*, 799–851.
10. Maslov, V.P. *Asymptotic Methods and Perturbation Theory*; Nauka: Moscow, Russia, 1988.
11. Lomov, S.A. *Introduction to the General Theory of Singular Perturbations*; Nauka: Moscow, Russia, 1989.
12. Bogolyubov, N.N; Mitropolsky, Yu.A. *Asymptotic Methods in the Theory of Nonlinear Oscillations*; Publ. H. of the Academy of Sciences of the USSR: Moscow, Russia, 1963.
13. Mitropolsky, Y.A. *Averaging Method in Nonlinear Mechanics*; Naukova Dumka: Kiev, Ukraine, 1971.
14. Glizer, V.Y. Conditions of Functional Null Controllability for Some Types of Singularly Perturbed Nonlineal Systems with Delays. *Axioms* **2019**, *8*, 80. [CrossRef]

2

On Fractional q-Extensions of some q-Orthogonal Polynomials

P. Njionou Sadjang [1,*] and S. Mboutngam [2]

[1] Department of Common Core, National Higher Polytechnic School Douala, Douala, Cameroon
[2] Higher Teachers' Training College, University of Maroua, Maroua, Cameroon; mbsalif@gmail.com
* Correspondence: pnjionou@yahoo.fr

Abstract: In this paper, we introduce a fractional q-extension of the q-differential operator $D_{q^{-1}}$ and prove some of its main properties. Next, fractional q-extensions of some classical q-orthogonal polynomials are introduced and some of the main properties of the newly-defined functions are given. Finally, a fractional q-difference equation of Gaussian type is introduced and solved by means of the power series method.

Keywords: Q-fractional calculus; Q-hypergeometric functions; Q-orthogonal polynomials; fractional q-difference equations

---

1. Introduction

Q-fractional calculus is the field of mathematical analysis which deals with the investigation and applications of derivatives and integrals of arbitrary (real or complex) order (see [1–8] and the references therein). It is an interesting topic having interconnections with various problems in function theory, integral and differential equations, and other branches of analysis. It has been continually developed, stimulated by ideas and results in various fields of mathematical analysis. This is demonstrated by the many publications—hundreds of papers in the past years—and by the many conferences devoted to the problems of fractional calculus.

A family $\{P_n(x)\}$, $(n \in \mathbb{N} := \{0,1,2\dots\}, k_n \neq 0)$ of polynomials of degree exactly n is a family of classical q-orthogonal polynomials of the q-Hahn class if it is the solution of a q-differential equation of the type

$$\sigma(x)D_qD_{1/q}P_n(x) + \tau(x)D_qP_n(x) + \lambda_nP_n(x) = 0, \tag{1}$$

where $\sigma(x) = ax^2 + bx + c$ is a polynomial of at most second order and $\tau(x) = dx + e$ is a polynomial of first order. Here, The q-difference operator D_q is defined by

$$D_qf(x) = \frac{f(x) - f(qx)}{(1-q)x}, \quad x \neq 0,\ q \neq 1,$$

and $D_qf(0) := f'(0)$ by continuity, provided $f'(0)$ exists.

The polynomial systems that are a solution of (1) form the q-Hahn tableau (see [9,10] and the references therein). These systems are contained in the so-called Askey–Wilson scheme [11]. The following systems are members of the q-Hahn tableau: the Big q-Jacobi polynomials, the q-Hahn polynomials, the Big q-Laguerre polynomials, the Little q-Jacobi polynomials, the q-Meixner polynomials, the Quantum q-Krawtchouk polynomials, the q-Krawtchouk polynomials, the Affine q-Krawtchouk polynomials, the Little q-Laguerre polynomials, the q-Laguerre polynomials, the Alternative q-Charlier (also called q-Bessel) polynomials, the q-Charlier polynomials, the Al-Salam–Carlitz I polynomials, the Al-Salam–Carlitz II polynomials, the Stieltjes–Wigert polynomials, the Discrete q-Hermite I polynomials, and the Discrete q-Hermite II polynomials.

In [12], the authors have defined some fractional extensions of the Jacobi polynomials from their Rodrigues representation and provided several properties of these new functions. They introduced a fractional version of the Gauss hypergeometric differential equation and used the modified power series method to provide some of its solutions. Note that, very recently, these Jacobi functions were used in [3] to provide new connection formulas for Jacobi polynomials. Note also that the previous Jacobi functions contain ultraspherical, Chebyshev of first, second, third and fourth kinds and Legendre functions as special cases.

In [13], the authors defined the C-Laguerre functions from the Rodrigues representation of the Laguerre polynomials by replacing the ordinary derivative by a fractional type derivative, then they gave several properties of the new defined functions.

In this work, we introduce a new q-differential operator of fractional order and use it to introduce the Little q-Jacobi, the Little q-Laguerre and the q-Laguerre functions. The hypergeometric representations of the new defined functions are given and the limit transitions are provided. Note that we obtained the results for the Big q-Jacobi, the Big q-Laguerre, the Big q-Legendre, the Al-Salam Carlitz-I and II and the Stieltjes–Wigert polynomials but did not include them because they are not of nice form.

The paper is organised as follows:

1. In Section 2, we present the preliminary results and definitions that are useful for a better reading of this manuscript.
2. In Section 3, we introduce the fractional q-calculus
3. In Section 4, we introduce a new fractional q-differential operator $D^{\alpha}_{q^{-1}}$ and apply it to some functions,
4. In Section 5, fractional q-extensions of some q-orthogonal polynomials are given and their basic hypergeometric representation provided. We prove for some of these new defined functions some limit transitions.
5. In Section 6, we introduce a fractional q-extension of the q-hypergeometric q-difference equation and provide some of its solution.

2. Preliminary Definitions and Results

This section contains some preliminary definitions and results that are useful for a better reading of the manuscript. The q-hypergeometric series, a fractional q-derivative and fractional q-integral are defined. The reader will consult the References [4,6,11,14] for more informations about these concepts and some applications.

Definition 1 (See [11]). *The basic hypergeometric or q-hypergeometric series ${}_r\phi_s$ is defined by the series*

$${}_r\phi_s\left(\begin{matrix} a_1,\cdots,a_r \\ b_1,\cdots,b_s\end{matrix}\middle| q;z\right) := \sum_{n=0}^{\infty}\frac{(a_1,\cdots,a_r;q)_n}{(b_1,\cdots,b_s;q)_n}\left((-1)^n q^{\binom{k}{2}}\right)^{1+s-r}\frac{z^n}{(q;q)_n},$$

where

$$(a_1,\cdots,a_r)_n := (a_1;q)_n\cdots(a_r;q)_n,$$

with

$$(a_i;q)_n = \begin{cases} \prod\limits_{j=0}^{n-1}(1-a_iq^j) & \text{if } n=1,2,3,\cdots \\ 1 & \text{if } n=0 \end{cases}.$$

For $n=\infty$ we set

$$(a;q)_\infty = \prod_{n=0}^{\infty}(1-aq^n),\ |q|<1.$$

The notation $(a;q)_n$ is the so-called q-Pochhammer symbol.

From the definition of $(a;q)_\infty$, it follows that for $0 < |q| < 1$, and for a nonnegative integer n, we have

$$(a;q)_n = \frac{(a;q)_\infty}{(aq^n;q)_\infty}.$$

Definition 2 (See [11]). *For any complex number λ,*

$$(a;q)_\lambda = \frac{(a;q)_\infty}{(aq^\lambda;q)_\infty}, 0 < |q| < 1,$$

where the principal value of q^λ is taken.

We will also use the following common notations

$$[a]_q = \frac{1-q^a}{1-q}, a \in \mathbb{C}, q \neq 1,$$

$$\begin{bmatrix} n \\ m \end{bmatrix}_q = \frac{(q;q)_n}{(q;q)_m (q;q)_{n-m}}, 0 \leq m \leq n,$$

and

$$(x \ominus y)_q^n = (x-y)(x-qy)\cdots(x-q^{n-1}y).$$

called the q-bracket and the q-binomial coefficients and the q-power respectively.

Proposition 1 ([11], Page 16). *The basic hypergeometric series fulfil the following identities*

$$ {}_1\phi_0\left(\begin{matrix} a \\ - \end{matrix}\middle| q; z\right) = \sum_{n=0}^{\infty} \frac{(a;q)_n}{(q;q)_n} z^n = \frac{(az;q)_\infty}{(z;q)_\infty}, \quad 0 < |q| < 1, \ |z| < 1, \tag{2}$$

$$ {}_1\phi_1\left(\begin{matrix} a \\ c \end{matrix}\middle| \frac{c}{a}\right) = \frac{(c/a;q)_\infty}{(c;q)_\infty}, \quad 0 < |q| < 1. \tag{3}$$

Relation (2) is the so-called q-binomial theorem.

The next proposition gives some important Heine transformation formulas for basic hypergeometric series.

Proposition 2 ([15], p. 10). *The following transformation formulas hold*

$$ {}_2\phi_1\left(\begin{matrix} a, b \\ c \end{matrix}\middle| z\right) = \frac{(az, b;q)_\infty}{(c, z;q)_\infty} {}_2\phi_1\left(\begin{matrix} c/b, z \\ az \end{matrix}\middle| q; b\right), \quad |z| < 1, \tag{4}$$

$$ {}_2\phi_1\left(\begin{matrix} a, b \\ c \end{matrix}\middle| q; z\right) = \frac{(abz/c;q)_\infty}{(z;q)_\infty} {}_2\phi_1\left(\begin{matrix} c/a, b/a \\ c \end{matrix}\middle| q; abz/c\right), \tag{5}$$

$$ {}_2\phi_1\left(\begin{matrix} a, b \\ c \end{matrix}\middle| z\right) = \frac{(az;q)_\infty}{(z;q)_\infty} {}_2\phi_2\left(\begin{matrix} a, c/b \\ c, az \end{matrix}\middle| bz\right), \quad |z| < 1,$$

$$ {}_2\phi_1\left(\begin{matrix} a, b \\ 0 \end{matrix}\middle| z\right) = \frac{(az;q)_\infty}{(z;q)_\infty} {}_1\phi_1\left(\begin{matrix} a \\ az \end{matrix}\middle| q; bz\right), \quad |z| < 1. \tag{6}$$

Proposition 3 ([16]). *The q-hypergeometric q-difference equation*

$$z(q^c - q^{a+b+1}z)(D_q^2 u)(z) \ + \left([c]_q - (q^b[a]_q + q^a[b+1]_q)z\right)(D_q u)(z) - [a]_q[b]_q u(z) = 0 \tag{7}$$

has particular solutions

$$u_1(z) = {}_2\phi_1\left(\begin{matrix} q^a, q^b \\ q^c \end{matrix}\middle| q; z\right), \quad \textit{and} \quad u_2(z) = z^{1-c}{}_2\phi_1\left(\begin{matrix} q^{1+a-c}, q^{1+b-c} \\ q^{2-c} \end{matrix}\middle| q; z\right).$$

Definition 3 (See [11]). *The q-Gamma function is defined by*

$$\Gamma_q(x) := \frac{(q;q)_\infty}{(q^x;q)_\infty}(1-q)^{1-x}, \quad 0 < q < 1. \tag{8}$$

Remark 1. *From Definition 2, the q-Gamma function is also represented by*

$$\Gamma_q(x) = (1-q)^{1-x}(q;q)_{x-1}.$$

Note also that the q-Gamma function satisfies the functional equation

$$\Gamma_q(x+1) = [x]_q\Gamma_q(x), \quad \textit{with} \quad \Gamma_q(1) = 1.$$

Note that for arbitrary complex α,

$$\begin{bmatrix} \alpha \\ k \end{bmatrix}_q = \frac{(q^{-\alpha};q)_k}{(q;q)_k}(-1)^k q^{\alpha k - \binom{k}{2}} = \frac{\Gamma_q(\alpha+1)}{\Gamma_q(k+1)\Gamma_q(\alpha-1)}. \tag{9}$$

The exponential function has two different natural q-extensions, denoted by $e_q(z)$ and $E_q(z)$, which can be defined by

$$e_q(z) := {}_1\phi_0\left(\begin{matrix} 0 \\ - \end{matrix}\middle| q; z\right) = \sum_{n=0}^{\infty}\frac{z^n}{(q;q)_n} = \frac{1}{(z;q)_\infty}, \quad 0 < |q| < 1, \ |z| < 1, \tag{10}$$

and

$$E_q(z) := {}_0\phi_0\left(\begin{matrix} - \\ - \end{matrix}\middle| q, -z\right) = \sum_{n=0}^{\infty}\frac{q^{\binom{n}{2}}}{(q;q)_n}z^n = (-z;q)_\infty, \quad 0 < |q| < 1. \tag{11}$$

These q-analogues of the exponential function are clearly related by

$$e_q(z)E_q(-z) = 1.$$

Corollary 1. *The following expansions apply.*

$$\frac{e_q(\alpha z)}{e_q(\beta z)} = \sum_{n=0}^{\infty}\frac{(\beta\alpha^{-1};q)_n}{(q;q)_n}(\alpha z)^n = {}_1\phi_0\left(\begin{matrix} \beta\alpha^{-1} \\ - \end{matrix}\middle| q; \alpha z\right);$$

$$\frac{e_q(\alpha z)}{E_q(\beta z)} = \sum_{n=0}^{\infty}\left[\sum_{k=0}^{n}\begin{bmatrix} n \\ k \end{bmatrix}_q\left(-\frac{\alpha}{\beta}\right)^k\right]\frac{(-\beta z)^n}{(q;q)_n}.$$

Next, we recall some basic knowledge about fractional q-calculus. The usual starting point for a definition of fractional operators in q-calculus taken in [1,2,7,8,17], is the q-analogue of the Riemann–Liouville fractional integral

$$I_q^\alpha f(z) = \frac{z^{\alpha-1}}{\Gamma_q(\alpha)}\int_0^z (tq/z;q)_{\alpha-1}f(t)d_qt = \frac{1}{\Gamma_q(\alpha)}\int_0^z (z \ominus qt)_q^{\alpha-1}f(t)d_qt. \tag{12}$$

This q-integral was motivated from the q-analogue of the Cauchy formula for a repeated q-integral

$$\begin{aligned} I^n_{q,a}f(z) &= \int_a^z d_qt \int_a^t d_qt_{n-1} \int_a^{t_{n-1}} d_qt_{n-2} \cdots \int_a^{t_2} f(t_1)d_qt_1 \\ &= \frac{z^{n-1}}{[n-1]_q!]} \int_0^z (tq/z;q)_{n-1} f(t)d_qt. \end{aligned} \tag{13}$$

The reduction of the multiple q-integral to a single one was considered by Al-Salam in [18]. In [8], the authors allow the lower parameter in (12) to be any real number $a \in (0, z)$. There are several definitions of the fractional q-integral and fractional q-derivatives. We adopt in this work the definition of the fractional q-integral given in [7].

Definition 4 (See [7]). *The fractional q-integral is*

$$(I^\alpha_{q,c}f)(x) = \frac{x^{\alpha-1}}{\Gamma_q(\alpha)} \int_c^x (qt/x;q)_{\alpha-1} f(t)d_qt = \frac{1}{\Gamma_q(\alpha)} \int_c^x (x \ominus qt)_q^{\alpha-1} f(t)d_qt, \ \ (\alpha \in \mathbb{R}^+).$$

In [7], it is proved that for $\alpha \in \mathbb{R}^+, \lambda, \lambda+\alpha \in \mathbb{R} \setminus \{-1,-2,\cdots\}$, equation is valid:

$$I^\alpha_{q,c}(x \ominus c)_q^\lambda = \frac{\Gamma_q(\lambda+1)}{\Gamma_q(\alpha+\lambda+1)} (x \ominus c)_q^{\alpha+\lambda}, \quad (0 < c < x). \tag{14}$$

Definition 5 (See [7]). *The fractional q-derivative of Riemann–Liouville type of order $\alpha \in \mathbb{R}^+$ is*

$$(D^\alpha_{q,c}f)(x) = (D_q^{\lceil \alpha \rceil} I_{q,c}^{\lceil \alpha \rceil - \alpha} f)(x), \tag{15}$$

where $\lceil \alpha \rceil$ denotes the smallest integer greater or equal to α.

Mahmoud Annaby and Zeinab Mansour ([17], p. 148) prove that the Riemann–Liouville fractional operator $D^\alpha_{q,0}$ coincides with a q-analogue of the Grünwald Letnikov fractional operator defined by

$$\begin{aligned} (\mathcal{D}^\alpha_q f)(x) &= \frac{1}{x^\alpha(1-q)^\alpha} \sum_{n=0}^\infty (-1)^n \begin{bmatrix} \alpha \\ n \end{bmatrix}_q \frac{f(q^n x)}{q^{\frac{n(n-1)}{2}+n(\alpha-n)}} \\ &= \frac{1}{x^\alpha(1-q)^\alpha} \sum_{n=0}^\infty \frac{(q^{-\alpha};q)_n}{(q;q)_n} q^n f(q^n x). \end{aligned} \tag{16}$$

3. More q-Fractional Operator

Since many Rodrigues-type formulas for some of the orthogonal polynomials of the q-Hahn class are expressed in terms of the q-operator $D_{q^{-1}}$ instead of D_q, and since our new functions are defined using the Rodrigues-type formula of each family, there is a need to develop a fractional calculus for $D_{q^{-1}}$. The more natural way to do it is to start by the power derivative of $D_{q^{-1}}$. The following proposition (see [19]) gives the result.

Proposition 4 (See [19]). *Let $n \in \mathbb{N}_0$ and f a given function defined on $\{q^k,\ k \in \mathbb{Z}\}$. Then the following power derivative rule for $D_{q^{-1}}$ applies*

$$D^n_{q^{-1}}f(x) = \frac{q^{\binom{n+1}{2}}}{(1-q)^n x^n} \sum_{k=0}^n (-1)^k \begin{bmatrix} n \\ k \end{bmatrix}_q q^{\binom{k}{2}-(n-1)k} f(q^{k-n}x). \tag{17}$$

Proof. This proof is also from [19]. The result is clear for $n = 0$. Assume the assertion is true for $n \geq 0$, then:

$$
\begin{aligned}
D_{q^{-1}}^{n+1} f(x) &= \frac{q^{\binom{n+1}{2}}}{(1-q)^n} D_{q^{-1}} \left(x^{-n} \sum_{k=0}^{n} (-1)^k \begin{bmatrix} n \\ k \end{bmatrix}_q q^{\binom{k}{2}-(n-1)k} f(q^{k-n}x) \right) \\
&= \frac{q^{\binom{n+1}{2}}}{(1-q)^n} \frac{q}{(1-q)x} \left(x^{-n} \sum_{k=0}^{n} (-1)^k \begin{bmatrix} n \\ k \end{bmatrix}_q q^{\binom{k}{2}-(n-1)k} f(q^{k-n}x) \right. \\
&\quad \left. -q^n x^{-n} \sum_{k=0}^{n} (-1)^k \begin{bmatrix} n \\ k \end{bmatrix}_q q^{\binom{k}{2}-(n-1)k} f(q^{k-n-1}x) \right) \\
&= \frac{-q^{\binom{n+1}{2}} q^{n+1}}{(1-q)^{n+1} x^{n+1}} \left(\sum_{k=1}^{n+1} (-1)^{k+1} \begin{bmatrix} n \\ k-1 \end{bmatrix}_q q^{\binom{k-1}{2}-(n-1)(k-1)-n} f(q^{k-n-1}x) \right. \\
&\quad \left. - \sum_{k=0}^{n} (-1)^k \begin{bmatrix} n \\ k \end{bmatrix}_q q^{\binom{k}{2}-(n-1)k} f(q^{k-n-1}x) \right) \\
&= \frac{q^{\binom{n+2}{2}}}{(1-q)^{n+1} x^{n+1}} \left(\sum_{k=1}^{n+1} (-1)^{k} \begin{bmatrix} n \\ k-1 \end{bmatrix}_q q^{\binom{k-1}{2}-(n-1)(k-1)-n} f(q^{k-n-1}x) \right. \\
&\quad \left. - \sum_{k=0}^{n} (-1)^k \begin{bmatrix} n \\ k \end{bmatrix}_q q^{\binom{k}{2}-(n-1)k} f(q^{k-n-1}x) \right) \\
&= \frac{q^{\binom{n+2}{2}}}{(1-q)^{n+1} x^{n+1}} \sum_{k=0}^{n} (-1)^k \begin{bmatrix} n+1 \\ k \end{bmatrix}_q q^{\binom{k}{2}-nk} f(q^{k-n-1}x).
\end{aligned}
$$

□

Note that, using the obvious relation $\begin{bmatrix} n \\ k \end{bmatrix}_q = \begin{bmatrix} n \\ n-k \end{bmatrix}_q$, and reversing the order of summation, (17) reads

$$
\begin{aligned}
D_{q^{-1}}^{n} f(x) &= \frac{q^{\binom{n+1}{2}}}{(1-q)^n x^n} \sum_{k=0}^{n} (-1)^{n-k} \begin{bmatrix} n \\ n-k \end{bmatrix}_q q^{\binom{n-k}{2}-(n-1)(n-k)} f(xq^{-k}) \\
&= \frac{q^{\binom{n+1}{2}}}{(q-1)^n x^n} \sum_{k=0}^{n} (-1)^{k} \begin{bmatrix} n \\ k \end{bmatrix}_q q^{\binom{n-k}{2}-(n-1)(n-k)} f(xq^{-k}).
\end{aligned}
$$

Next, using the fact that

$$
\binom{n-k}{2} - (n-1)(n-k) = \frac{(n-k)(n-k-1) - 2(n-1)(n-k)}{2} = -\binom{n}{2} + \binom{k}{2},
$$

we get

$$
\begin{aligned}
D_{q^{-1}}^{n} f(x) &= \frac{q^{\binom{n+1}{2}}}{(q-1)^n x^n} \sum_{k=0}^{n} (-1)^{k} \begin{bmatrix} n \\ k \end{bmatrix}_q q^{-\binom{n}{2}+\binom{k}{2}} f(xq^{-k}) \\
&= \frac{q^n}{(q-1)^n x^n} \sum_{k=0}^{n} (-1)^{k} \begin{bmatrix} n \\ k \end{bmatrix}_q q^{\binom{k}{2}} f(xq^{-k}). \qquad (18)
\end{aligned}
$$

Remark 2 (See [19,20]). *It is known that*

$$
D_q^n f(x) = \frac{1}{(1-q)^n x^n} \sum_{k=0}^{n} (-1)^k \begin{bmatrix} n \\ k \end{bmatrix}_q q^{\binom{k}{2}-(n-1)k} f(q^k x). \qquad (19)
$$

Note that replacing q by q^{-1} in (19), *it follows that*

$$\begin{aligned} D^n_{q^{-1}} f(x) &= \frac{1}{(1-q^{-1})^n x^n} \sum_{k=0}^{n} (-1)^k \begin{bmatrix} n \\ k \end{bmatrix}_{q^{-1}} q^{-\binom{k}{2}-(n-1)k} f(q^{-k}x) \\ &= \frac{q^n}{(q-1)^n x^n} \sum_{k=0}^{n} (-1)^k \begin{bmatrix} n \\ k \end{bmatrix}_{q^{-1}} q^{-\binom{k}{2}-(n-1)k} f(q^{-k}x). \end{aligned}$$

Taking care that

$$[n]_{q^{-1}} = \frac{1-q^{-n}}{1-q^{-1}} = \frac{1}{q^{n-1}} [n]_q,$$

it follows that

$$[n]_{q^{-1}}! = q^{-\binom{n}{2}} [n]_q!,$$

and so we get

$$\begin{bmatrix} n \\ k \end{bmatrix}_{q^{-1}} = \frac{q^{-\binom{n}{2}} [n]_q!}{q^{-\binom{k}{2}} [k]_q! q^{-\binom{n-k}{2}} [n-k]_q!} = q^{\binom{k}{2}+\binom{n-k}{2}-\binom{n}{2}} \begin{bmatrix} n \\ k \end{bmatrix}_q.$$

Now, we can write

$$\begin{aligned} D^n_{q^{-1}} f(x) &= \frac{1}{(1-q^{-1})^n x^n} \sum_{k=0}^{n} (-1)^k q^{\binom{k}{2}+\binom{n-k}{2}-\binom{n}{2}} \begin{bmatrix} n \\ k \end{bmatrix}_q q^{-\binom{k}{2}-(n-1)k} f(q^{-k}x) \\ &= \frac{q^n}{(q-1)^n x^n} \sum_{k=0}^{n} (-1)^k \begin{bmatrix} n \\ k \end{bmatrix}_q q^{\binom{n-k}{2}-\binom{n}{2}-(n-1)k} f(q^{-k}x) \\ &= \frac{q^n}{(q-1)^n x^n} \sum_{k=0}^{n} (-1)^k \begin{bmatrix} n \\ k \end{bmatrix}_q q^{\binom{k}{2}-2\binom{n}{2}+(n-1)(n-k)-(n-1)k} f(q^{-k}x) \\ &= \frac{q^n}{(q-1)^n x^n} \sum_{k=0}^{n} (-1)^k \begin{bmatrix} n \\ k \end{bmatrix}_q q^{\binom{k}{2}} f(xq^{-k}). \end{aligned}$$

This is exactely another way to write the result (17) *obtained in [19] thanks to* (18).

We are about to define a fractional extension of $D_{q^{-1}}$. Since $\begin{bmatrix} n \\ k \end{bmatrix}_q = 0$ for $k > n$, we can write (18) as

$$D^n_{q^{-1}} f(x) = \frac{q^n}{(q-1)^n x^n} \sum_{k=0}^{\infty} (-1)^k \begin{bmatrix} n \\ k \end{bmatrix}_q q^{\binom{k}{2}} f(xq^{-k}). \tag{20}$$

We extend Equation (20) to any arbitrary complex number α, $D^{\alpha}_{q^{-1}}$ in the following way:

$$D^{\alpha}_{q^{-1}} f(x) = \frac{q^{\alpha}}{(q-1)^{\alpha} x^{\alpha}} \sum_{k=0}^{\infty} (-1)^k \begin{bmatrix} \alpha \\ k \end{bmatrix}_q q^{\binom{k}{2}} f(xq^{-k}),$$

provided that the infinite series of the right hand side converges. Now, using Equation (9), we obtain

$$\begin{aligned} D^{\alpha}_{q^{-1}} f(x) &= \frac{q^{\alpha}}{(q-1)^{\alpha} x^{\alpha}} \sum_{k=0}^{\infty} (-1)^k \frac{(q^{-\alpha};q)_k}{(q;q)_k} (-1)^k q^{\alpha k-\binom{k}{2}} q^{\binom{k}{2}} f(xq^{-k}) \\ &= \frac{q^{\alpha}}{(q-1)^{\alpha} x^{\alpha}} \sum_{k=0}^{\infty} \frac{(q^{-\alpha};q)_k}{(q;q)_k} q^{\alpha k} f(xq^{-k}). \end{aligned}$$

We then set the following definition.

Definition 6. *For any complex number α, we define the fractional operator $D^{\alpha}_{q^{-1}}$ by*

$$D^{\alpha}_{q^{-1}}f(x) = \frac{q^{\alpha}}{(q-1)^{\alpha}x^{\alpha}}\sum_{k=0}^{\infty}\frac{(q^{-\alpha};q)_k}{(q;q)_k}q^{\alpha k}f(xq^{-k}), \tag{21}$$

provided that the right hand side of (21) *converges.*

Note also that we could use directly (17) to write

$$\begin{aligned}
{}_{\star}D^{\alpha}_{q^{-1}}f(x) &= \frac{q^{\frac{\alpha(\alpha-1)}{2}}}{(1-q)^{\alpha}x^{\alpha}}\sum_{k=0}^{\infty}(-1)^k\begin{bmatrix}\alpha\\k\end{bmatrix}_q q^{\binom{k}{2}-(\alpha-1)k}f(q^{k-\alpha}x)\\
&= \frac{q^{\frac{\alpha(\alpha-1)}{2}}}{(1-q)^{\alpha}x^{\alpha}}\sum_{k=0}^{\infty}(-1)^k\frac{(q^{-\alpha};q)_k}{(q;q)_k}(-1)^k q^{\alpha k-\binom{k}{2}}q^{\binom{k}{2}-(\alpha-1)k}f(q^{k-\alpha}x)\\
&= \frac{q^{\frac{\alpha(\alpha-1)}{2}}}{(1-q)^{\alpha}x^{\alpha}}\sum_{k=0}^{\infty}\frac{(q^{-\alpha};q)_k}{(q;q)_k}q^k f(q^{k-\alpha}x),
\end{aligned}$$

which may be looked as another fractional extension of $D_{q^{-1}}$.

Proposition 5. *For $\alpha \in \mathbb{R}\setminus\{1,2,3,\cdots\}$, the following derivative rule applies*

$$D^{\lambda}_{q^{-1}}x^{\alpha} = \begin{cases}\dfrac{q^{\lambda}}{(q-1)^{\lambda}}\dfrac{(q^{-\alpha};q)_{\infty}}{(q^{\lambda-\alpha};q)_{\infty}}x^{\alpha-\lambda}, & \text{if } \alpha-\lambda\in\mathbb{R}\setminus\mathbb{N}\\ 0, & \text{if } \alpha-\lambda\in\mathbb{N}\end{cases}. \tag{22}$$

Proof. From (21), we get

$$\begin{aligned}
D^{\lambda}_{q^{-1}}x^{\alpha} &= \frac{q^{\lambda}}{(q-1)^{\lambda}x^{\lambda}}\sum_{k=0}^{\infty}\frac{(q^{-\lambda};q)_k}{(q;q)_k}q^{\lambda k}(xq^{-k})^{\alpha}\\
&= \frac{q^{\lambda}x^{\alpha}}{(q-1)^{\lambda}x^{\lambda}}\sum_{k=0}^{\infty}\frac{(q^{-\lambda};q)_k}{(q;q)_k}q^{(\lambda-\alpha)k}\\
&= \frac{q^{\lambda}}{(q-1)^{\lambda}}x^{\alpha-\lambda}{}_1\phi_0\left(\begin{matrix}q^{-\lambda}\\-\end{matrix}\middle| q;q^{\lambda-\alpha}\right)\\
&= \frac{q^{\lambda}}{(q-1)^{\lambda}}\frac{(q^{-\alpha};q)_{\infty}}{(q^{\lambda-\alpha};q)_{\infty}}x^{\alpha-\lambda}.
\end{aligned}$$

□

Before we state the semi-group property of the operator $D^{\alpha}_{q^{-1}}$, we state the following summation results that will be useful for the proof.

Lemma 1. (See [14], Lemma 10) *The following relation applies:*

$$\sum_{n=0}^{\infty}\sum_{k=0}^{\infty}A(k,n) = \sum_{n=0}^{\infty}\sum_{k=0}^{n}A(k,n-k). \tag{23}$$

Lemma 2. (See [15], Exercise 1.3) *The following multiplication formula holds true*

$$(ab;q)_n = \sum_{k=0}^{n}\begin{bmatrix}n\\k\end{bmatrix}_q b^k(a;q)_k(b;q)_{n-k} = \sum_{k=0}^{n}\begin{bmatrix}n\\k\end{bmatrix}_q a^{n-k}(a;q)_k(b;q)_{n-k}.$$

Using Lemmas 1 and 2 and the definition of $D^{\lambda}_{q^{-1}}$, we prove the following proposition.

Proposition 6 (Semi-group property). *The following equation applies*

$$D^{\alpha}_{q^{-1}} D^{\beta}_{q^{-1}} f(x) = D^{\alpha+\beta}_{q^{-1}} f(x). \tag{24}$$

Proof. We have

$$\begin{aligned}
D^{\beta}_{q^{-1}} D^{\alpha}_{q^{-1}} f(x) &= D^{\beta}_{q^{-1}} \left[D^{\alpha}_{q^{-1}} f(x) \right] \\
&= \frac{q^{\alpha}}{(q-1)^{\alpha}} D^{\beta}_{q^{-1}} \left[\frac{1}{x^{\alpha}} \sum_{k=0}^{\infty} \frac{(q^{-\alpha};q)_k}{(q;q)_k} q^{\alpha k} f(xq^{-k}) \right] \\
&= \frac{q^{\alpha}}{(q-1)^{\alpha}} \frac{q^{\beta}}{(q-1)^{\beta} x^{\beta}} \sum_{n=0}^{\infty} \frac{(q^{-\beta};q)_n}{(q;q)_n} q^{\beta n} \left(\frac{1}{(xq^{-n})^{\alpha}} \sum_{k=0}^{\infty} \frac{(q^{-\alpha};q)_k}{(q;q)_k} q^{\alpha k} f(xq^{-n-k}) \right) \\
&= \frac{q^{\alpha+\beta}}{(q-1)^{\alpha+\beta} x^{\alpha+\beta}} \sum_{n=0}^{\infty} \sum_{k=0}^{\infty} \frac{(q^{-\beta};q)_n}{(q;q)_n} \frac{(q^{-\alpha};q)_k}{(q;q)_k} q^{\alpha(n+k)} q^{\beta n} f(xq^{-n-k}) \\
&= \frac{q^{\alpha+\beta}}{(q-1)^{\alpha+\beta} x^{\alpha+\beta}} \sum_{n=0}^{\infty} \left(\sum_{k=0}^{n} \frac{(q^{-\beta};q)_{n-k}}{(q;q)_{n-k}} \frac{(q^{-\alpha};q)_k}{(q;q)_k} q^{-\beta k} \right) q^{(\alpha+\beta)n} f(xq^{-n}) \\
&= \frac{q^{\alpha+\beta}}{(q-1)^{\alpha+\beta} x^{\alpha+\beta}} \sum_{n=0}^{\infty} \frac{(q^{-(\alpha+\beta)};q)_n}{(q;q)_n} q^{(\alpha+\beta)n} f(xq^{-n}) = D^{\alpha+\beta}_{q^{-1}} f(x).
\end{aligned}$$

This proves the proposition. □

4. Fractional q-Extensions of Some q-Orthogonal Polynomials

In this section we introduce some fractional q-extensions of some orthogonal polynomials of the q-Hahn class. The families that are of interest here are those which use D_q and $D_{q^{-1}}$ in their Rodrigues representations.

4.1. *The Little q-Jacobi Functions*

The Little q-Jacobi polynomials have the q-hypergeometric representation ([11], p. 482)

$$p_n(x;a,b|q) = {}_2\phi_1 \left(\begin{matrix} q^{-n}, abq^{n+1} \\ aq \end{matrix} \middle| q; qx \right).$$

They can also be represented by the Redrigues-type formula

$$w(x;\alpha,\beta;q) p_n(x;q^{\alpha},q^{\beta}|q) = \frac{q^{n\alpha+\binom{n}{2}}(1-q)^n}{(q^{\alpha+1};q)_n} D^{n}_{q^{-1}} [w(x;\alpha+n,\beta+n;q)],$$

where

$$w(x;\alpha,\beta;q) = \frac{(qx;q)_{\infty}}{(q^{\beta+1}x;q)_{\infty}} x^{\alpha}.$$

Definition 7. *Let $\lambda \in \mathbb{R}$, we define the fractional Little q-Jacobi functions by*

$$P_{\lambda}(x;q^{\alpha},q^{\beta}|q) = \frac{q^{\lambda\alpha+\frac{\lambda(\lambda-1)}{2}}(1-q)^{\lambda}}{w(x;\alpha,\beta;q)(q^{\alpha+1};q)_{\lambda}} D^{\lambda}_{q^{-1}} [w(x;\alpha+\lambda,\beta+\lambda;q)]. \tag{25}$$

Proposition 7. *The fractional Little q-Jacobi functions defined by relation (25) have the following basic hypergeometric representation*

$$P_\lambda(x;q^\alpha,q^\beta|q) = (-1)^\lambda q^{\lambda\alpha+\frac{\lambda(\lambda+1)}{2}} \frac{(q^{-(\alpha+\lambda)};q)_\infty}{(q^{\alpha+1};q)_\lambda (q^{-\alpha};q)_\infty} {}_2\phi_1\left(\left.\begin{matrix} q^{-\lambda}, q^{\alpha+\beta+\lambda+1} \\ q^{\alpha+1}\end{matrix}\right| q;qx\right).$$

Proof. Applying the q-binomial theorem (2), we have

$$\begin{aligned} w(x;\alpha,\beta;q) &= x^\alpha \frac{(qx;q)_\infty}{(q^{\beta+1}x;q)_\infty} \\ &= \sum_{n=0}^{\infty} \frac{(q^{-\beta};q)_n}{(q;q)_n} q^{n(\beta+1)} x^{n+\alpha}. \end{aligned}$$

Thus,

$$w(x;\alpha+\lambda,\beta+\lambda;q) = \sum_{n=0}^{\infty} \frac{(q^{-(\beta+\lambda)};q)_n}{(q;q)_n} q^{n(\beta+\lambda+1)} x^{n+\alpha+\lambda}.$$

Hence,

$$\begin{aligned} &D_{q^{-1}}^{\lambda}[w(x;\alpha+\lambda,\beta+\lambda;q)] \\ &= \sum_{n=0}^{\infty} \frac{(q^{-(\beta+\lambda)};q)_n}{(q;q)_n} q^{n(\beta+\lambda+1)} D_{q^{-1}}^{\lambda}\left[x^{n+\alpha+\lambda}\right] \\ &= \frac{q^\lambda}{(q-1)^\lambda} \sum_{n=0}^{\infty} \frac{(q^{-(\beta+\lambda)};q)_n}{(q;q)_n} \frac{(q^{-n}q^{-(\alpha+\lambda)};q)_\infty}{(q^{-n}q^{-\alpha};q)_\infty} q^{n(\beta+\lambda+1)} x^{n+\alpha}. \end{aligned}$$

Using the fact that

$$(aq^{-n};q)_\infty = (-a)^n (a^{-1}q;q)_n q^{-\binom{n+1}{2}} (a;q)_\infty,$$

we have

$$\frac{(q^{-n}q^{-(\alpha+\lambda)};q)_\infty}{(q^{-n}q^{-\alpha};q)_\infty} = \frac{(q^{-(\alpha+\lambda)};q)_\infty}{(q^{-\alpha};q)_\infty} \frac{(q^{\alpha+\lambda+1};q)_n}{(q^{\alpha+1};q)_n} q^{-\lambda n}.$$

Whence

$$\begin{aligned} &D_{q^{-1}}^{\lambda}[w(x;\alpha+\lambda,\beta+\lambda;q)] \\ &= \frac{q^\lambda x^\alpha}{(q-1)^\lambda} \frac{(q^{-(\alpha+\lambda)};q)_\infty}{(q^{-\alpha};q)_\infty} \sum_{n=0}^{\infty} \frac{(q^{-(\beta+\lambda)};q)_n}{(q;q)_n} \frac{(q^{\alpha+\lambda+1};q)_n}{(q^{\alpha+1};q)_n} q^{n(\beta+1)} x^n \\ &= \frac{q^\lambda x^\alpha}{(q-1)^\lambda} \frac{(q^{-(\alpha+\lambda)};q)_\infty}{(q^{-\alpha};q)_\infty} {}_2\phi_1\left(\left.\begin{matrix} q^{-(\beta+\lambda)}, q^{\alpha+\lambda+1} \\ q^{\alpha+1}\end{matrix}\right| q; xq^{\beta+1}\right). \end{aligned}$$

Now, applying the Heine–Euler transformation formula (5), we have

$${}_2\phi_1\left(\left.\begin{matrix} q^{-(\beta+\lambda)}, q^{\alpha+\lambda+1} \\ q^{\alpha+1}\end{matrix}\right| q; xq^{\beta+1}\right) = \frac{(qx;q)_\infty}{(q^{\beta+1};q)_\infty} {}_2\phi_1\left(\left.\begin{matrix} q^{-\lambda}, q^{\alpha+\beta+\lambda+1} \\ q^{\alpha+1}\end{matrix}\right| q;qx\right).$$

So we have

$$\frac{D_{q^{-1}}^{\lambda}[w(x;\alpha+\lambda,\beta+\lambda;q)]}{w(x;\alpha,\beta;q)} = \frac{q^\lambda}{(q-1)^\lambda} \frac{(q^{-(\alpha+\lambda)};q)_\infty}{(q^{-\alpha};q)_\infty} {}_2\phi_1\left(\left.\begin{matrix} q^{-\lambda}, q^{\alpha+\beta+\lambda+1} \\ q^{\alpha+1}\end{matrix}\right| q;qx\right).$$

We finally obtain

$$P_\lambda(x;q^\alpha,q^\beta;q) = (-1)^\lambda q^{\lambda\alpha+\frac{\lambda(\lambda+1)}{2}} \frac{(q^{-(\alpha+\lambda)};q)_\infty}{(q^{\alpha+1};q)_\lambda (q^{-\alpha};q)_\infty} {}_2\phi_1\left(\begin{matrix} q^{-\lambda}, q^{\alpha+\beta+\lambda+1} \\ q^{\alpha+1} \end{matrix} \middle| q;qx\right).$$

□

Proposition 8. *The following limit transition holds:*

$$\lim_{\lambda\to n} P_\lambda(x;q^\alpha,q^\beta;q) = p_n(x;q^\alpha,q^\beta;q),$$

where n is a nonnegative integer.

Proof. It is not difficult to see that

$$\begin{aligned}
&\lim_{\lambda\to n}\left((-1)^\lambda q^{\lambda\alpha+\frac{\lambda(\lambda+1)}{2}} \frac{(q^{-(\alpha+\lambda)};q)_\infty}{(q^{\alpha+1};q)_\lambda (q^{-\alpha};q)_\infty}\right) \\
&\qquad = (-1)^n q^{n\alpha+\binom{n+1}{2}} \frac{(q^{-(\alpha+n)};q)_\infty}{(q^{\alpha+1};q)_n (q^{-\alpha};q)_\infty} \\
&\qquad = (-1)^n q^{n\alpha+\binom{n+1}{2}} \frac{(-1)^n q^{-\alpha n} q^{-\binom{n+1}{2}} (q^{\alpha+1};q)_n (q^{-\alpha};q)_\infty}{(q^{\alpha+1};q)_n (q^{-\alpha};q)_\infty} = 1,
\end{aligned}$$

so,

$$\lim_{\lambda\to n} P_\lambda(x;q^\alpha,q^\beta;q) = p_n(x;q^\alpha,q^\beta;q).$$

□

4.2. The Little q-Laguerre Functions

The Little q-Laguerre polynomials have the q-hypergeometric representation ([11], p. 518)

$$p_n(x,a|q) = {}_2\phi_1\left(\begin{matrix} q^{-n}, 0 \\ aq \end{matrix} \middle| q;qx\right).$$

They can also be represented by the Rodrigues-type formula ([11], p. 520)

$$w(x;\alpha;q)p_n(x;q^\alpha;q) = \frac{q^{n\alpha+\binom{n}{2}}(1-q)^n}{(q^{\alpha+1};q)_n} D^n_{q^{-1}}[w(x;\alpha+n;q)],$$

with

$$w(x;\alpha;q) = x^\alpha (qx;q)_\infty.$$

Definition 8. *Let $\lambda \in \mathbb{R}$, we define the fractional Little q-Laguerre functions by*

$$P_\lambda(x;q^\alpha;q) = \frac{q^{\lambda\alpha+\binom{\lambda}{2}}(1-q)^\lambda}{w(x;\alpha;q)(q^{\alpha+1};q)_\lambda} D^\lambda_{q^{-1}}[w(x;\alpha+\lambda;q)]. \tag{26}$$

Proposition 9. *The fractional Little q-Laguerre functions defined by relation (26) have the following representation*

$$P_\lambda(x;q^\alpha|q) = (-1)^\lambda q^{\lambda\alpha+\frac{\lambda(\lambda+1)}{2}} \frac{(q^{-(\alpha+\lambda)};q)_\infty}{(q^{\alpha+1};q)_\lambda (q^{-\alpha};q)_\infty} {}_2\phi_1\left(\begin{matrix} q^{-\lambda}, 0 \\ q^{\alpha+1} \end{matrix} \middle| qx\right).$$

Proof. It is easy to see that

$$w(x;\alpha;q) = x^{\alpha}(qx;q)_{\infty} = \sum_{n=0}^{\infty} \frac{q^{\binom{n+1}{2}}}{(q;q)_n}(-1)^n x^{n+\alpha}.$$

So,

$$w(x;\alpha+\lambda;q) = \sum_{n=0}^{\infty} \frac{q^{\binom{n+1}{2}}}{(q;q)_n}(-1)^n x^{n+\alpha+\lambda}.$$

Thus, using the definitions of the fractional Little q-Laguerre functions and the fractional q-derivative (21), combined with the transformations (4) and (6) we have:

$$\begin{aligned}
D_{q^{-1}}^{\lambda}[w(x;\alpha+\lambda;q)] &= \sum_{n=0}^{\infty} \frac{q^{\binom{n+1}{2}}}{(q;q)_n}(-1)^n D_{q^{-1}}^{\lambda}[x^{n+\alpha+\lambda}] \\
&= \sum_{n=0}^{\infty} \frac{q^{\binom{n+1}{2}}}{(q;q)_n}(-1)^n \frac{q^{\lambda}}{(q-1)^{\lambda}} \frac{(q^{-n}q^{-(\alpha+\lambda)};q)_{\infty}}{(q^{-n}q^{-\alpha};q)_{\infty}} x^{n+\alpha} \\
&= \frac{x^{\alpha}q^{\lambda}}{(q-1)^{\lambda}} \frac{(q^{-(\alpha+\lambda)};q)_{\infty}}{(q^{-\alpha};q)_{\infty}} \sum_{n=0}^{\infty} \frac{(q^{\alpha+\lambda+1};q)_n}{(q^{\alpha+1};q)_n} \left((-1)^n q^{\binom{n}{2}}\right) \frac{(q^{1-\lambda}x)^n}{(q;q)_n} \\
&= \frac{x^{\alpha}q^{\lambda}}{(q-1)^{\lambda}} \frac{(q^{-(\alpha+\lambda)};q)_{\infty}}{(q^{-\alpha};q)_{\infty}} {}_1\phi_1\left(\begin{matrix} q^{\alpha+\lambda+1} \\ q^{\alpha+1} \end{matrix} \middle| q; xq^{-\lambda+1}\right) \\
&\overset{(6)}{=} \frac{x^{\alpha}q^{\lambda}}{(q-1)^{\lambda}} \frac{(q^{-(\alpha+\lambda)};q)_{\infty}}{(q^{-\alpha};q)_{\infty}} \frac{(q^{-\lambda};q)_{\infty}}{(q^{\alpha+1};q)_{\infty}} {}_2\phi_0\left(\begin{matrix} q^{\alpha+\lambda+1}, qx \\ 0 \end{matrix} \middle| q^{-\lambda}\right) \\
&\overset{(4)}{=} \frac{x^{\alpha}(qx;q)_{\infty}q^{\lambda}}{(q-1)^{\lambda}} \frac{(q^{-(\alpha+\lambda)};q)_{\infty}}{(q^{-\alpha};q)_{\infty}} {}_2\phi_1\left(\begin{matrix} q^{-\lambda}, 0 \\ q^{\alpha+1} \end{matrix} \middle| qx\right) \\
&= (-1)^{\lambda} q^{\lambda\alpha+\frac{\lambda(\lambda+1)}{2}} \frac{(q^{-(\alpha+\lambda)};q)_{\infty}}{(q^{\alpha+1};q)_{\lambda}(q^{-\alpha};q)_{\infty}} {}_2\phi_1\left(\begin{matrix} q^{-\lambda}, 0 \\ q^{\alpha+1} \end{matrix} \middle| qx\right).
\end{aligned}$$

□

Proposition 10. *Let n be a nonnegative integer, then the following limit transition holds true*

$$\lim_{\lambda\to n} P_{\lambda}(x;q^{\alpha};q) = p_n(x;q^{\alpha};q).$$

Proof. Since

$$\begin{aligned}
&\lim_{\lambda\to n}\left((-1)^{\lambda} q^{\lambda\alpha+\frac{\lambda(\lambda+1)}{2}} \frac{(q^{-(\alpha+\lambda)};q)_{\infty}}{(q^{\alpha+1};q)_{\lambda}(q^{-\alpha};q)_{\infty}}\right) \\
&\qquad = (-1)^n q^{n\alpha+\binom{n+1}{2}} \frac{(q^{-(\alpha+n)};q)_{\infty}}{(q^{\alpha+1};q)_n(q^{-\alpha};q)_{\infty}} \\
&\qquad = (-1)^n q^{n\alpha+\binom{n+1}{2}} \frac{(-1)^n q^{-\alpha n} q^{-\binom{n+1}{2}} (q^{\alpha+1};q)_n (q^{-\alpha};q)_{\infty}}{(q^{\alpha+1};q)_n(q^{-\alpha};q)_{\infty}} = 1,
\end{aligned}$$

we have, $\lim_{\lambda\to n} P_{\lambda}(x;q^{\alpha};q) = p_n(x;q^{\alpha};q)$. □

4.3. The q-Laguerre Functions

The q-Laguerre polynomials have the q-hypergeometric representation ([11], p. 522)

$$L_n^{(\alpha)}(x;q) = \frac{(q^{\alpha+1};q)_n}{(q;q)_n} {}_1\phi_1\left(\begin{matrix} q^{-n} \\ q^{\alpha+1} \end{matrix} \middle| q; -q^{n+\alpha+1}x\right).$$

They can also be represented by the Rodrigues-type formula ([11], p. 524)

$$w(x;\alpha;q)L_n^{(\alpha)}(x;q) = \frac{(1-q)^n}{(q;q)_n} D_q^n[w(x;\alpha+n;q)],$$

with

$$w(x;\alpha;q) = \frac{x^\alpha}{(-x;q)_\infty}.$$

Definition 9. *Let $\lambda \in \mathbb{R}$, we define the fractional q-Laguerre functions by*

$$L_\lambda^\alpha(x;q) = \frac{(1-q)^\lambda\,(-x,;q)_\infty}{(q;q)_\lambda x^\alpha} D_q^\lambda\left[\frac{x^{\alpha+\lambda}}{(-x;q)_\infty}\right]. \tag{27}$$

Proposition 11. *The fractional q-Laguerre functions defined by relation (27) have the following basic hypergeometric representation*

$$\begin{aligned} L_\lambda^\alpha(x;q) &= \frac{1}{(q;q)_\lambda} {}_2\phi_1\left(\begin{matrix} q^{-\lambda}, -x \\ 0 \end{matrix} \middle| q; q^{\alpha+\lambda+1}\right) \\ &= \frac{(q^{\alpha+1};q)_\lambda}{(q;q)_\lambda} {}_1\phi_1\left(\begin{matrix} q^{-\lambda} \\ q^{\alpha+1} \end{matrix} \middle| q; -xq^{\alpha+\lambda+1}\right). \end{aligned}$$

Proof. From the definition of the q-exponential (10), it follows that

$$\frac{x^{\alpha+\lambda}}{(-x;q)_\infty} = x^{\alpha+\lambda}e_q(-x) = \sum_{n=0}^{\infty} \frac{(-1)^n x^{\alpha+\lambda+n}}{(q;q)_n}.$$

Then,

$$\begin{aligned} D_{q,0}^\lambda\left[\frac{x^{\alpha+\lambda}}{(-x;q)_\infty}\right] &= \sum_{n=0}^{\infty} \frac{(-1)^n}{(q;q)_n} \frac{\Gamma_q(\alpha+\lambda+n+1)}{\Gamma_q(\alpha+n+1)} x^{\alpha+n} \\ &= x^\alpha \frac{\Gamma_q(\alpha+\lambda+1)}{\Gamma_q(\alpha+1)} \sum_{n=0}^{\infty} \frac{(q^{\alpha+\lambda+1};q)_n}{(q;q)_n(q^{\alpha+1};q)_n}(-x)^n \\ &= x^\alpha \frac{\Gamma_q(\alpha+\lambda+1)}{\Gamma_q(\alpha+1)} {}_2\phi_1\left(\begin{matrix} 0, q^{\alpha+\lambda+1} \\ q^{\alpha+1} \end{matrix} \middle| q; -x\right) \\ &= \frac{x^\alpha}{(1-q)^\lambda} \frac{(q^{\alpha+1};q)_\infty}{(q^{\alpha+\lambda+1};q)_\infty} {}_2\phi_1\left(\begin{matrix} 0, q^{\alpha+\lambda+1} \\ q^{\alpha+1} \end{matrix} \middle| q; -x\right). \end{aligned}$$

Using the Heine transformation (4), we have

$${}_2\phi_1\left(\begin{matrix} 0, q^{\alpha+\lambda+1} \\ q^{\alpha+1} \end{matrix} \middle| q; -x\right) = \frac{(q^{\alpha+\lambda+1};q)_\infty}{(q^{\alpha+1};q)_\infty(-x;q)_\infty} {}_2\phi_1\left(\begin{matrix} q^{-\lambda}; -x \\ 0 \end{matrix} \middle| q; q^{\alpha+\lambda+1}\right).$$

Hence, we have

$$D_{q,0}^{\lambda}\left[\frac{x^{\alpha+\lambda}}{(-x;q)_\infty}\right]=\frac{x^{\alpha}}{(1-q)^{\lambda}(-x;q)_\infty}{}_2\phi_1\left(\begin{matrix}q^{-\lambda};-x\\0\end{matrix}\middle|q;q^{\alpha+\lambda+1}\right).$$

Finally it follows that

$$L_{\lambda}^{(\alpha)}(x;q)=\frac{1}{(q;q)_\lambda}{}_2\phi_1\left(\begin{matrix}q^{-\lambda};-x\\0\end{matrix}\middle|q;q^{\alpha+\lambda+1}\right).$$

Next, using the transformation formula (6) we get

$$\begin{aligned}\frac{1}{(q;q)_\lambda}{}_2\phi_1\left(\begin{matrix}q^{-\lambda};-x\\0\end{matrix}\middle|q;q^{\alpha+\lambda+1}\right) &= \frac{1}{(q;q)_\lambda}\frac{(q^{\alpha+1};q)_\infty}{(q^{\alpha+\lambda+1};q)_\infty}{}_1\phi_1\left(\begin{matrix}q^{-\lambda}\\q^{\alpha+1}\end{matrix}\middle|q;-xq^{\alpha+\lambda+1}\right)\\ &= \frac{(q^{\alpha+1};q)_\lambda}{(q;q)_\lambda}{}_1\phi_1\left(\begin{matrix}q^{-\lambda}\\q^{\alpha+1}\end{matrix}\middle|q;-xq^{\alpha+\lambda+1}\right).\end{aligned}$$

This ends the proof of the proposition. □

5. Fractional q-Gauss Differential Equation

In this section, we give a fractional version of the q-hypergeometric q-difference equation given by Koorwinder in [16]. Next, we solve this fractional q-difference equation by means of modified power series.

Definition 10. *The fractional q-hypergeometric q-difference equation is defined for $0<\lambda\le 1$ by*

$$\begin{aligned}&z^{\lambda}(q^{c}-q^{a+b+1}z^{\lambda})(D_q^{2\lambda}u)(z)\\&\quad+\left([c]_q-(q^{b}[a]_q+q^{a}[b+1]_q)z^{\lambda}\right)(D_q^{\lambda}u)(z)-[a]_q[b]_qu(z)=0.\end{aligned}\tag{28}$$

Definition 11. *The fractional q-Gauss function is defined as the series*

$${}_2\phi_1^{\mu}\left(\begin{matrix}q^{a},q^{b}\\q^{c}\end{matrix}\middle|q;z\right)=u_0z^{\rho}\sum_{n=0}^{\infty}\prod_{k=0}^{n}\frac{g_{q,k}(\rho)}{f_{q,k+1}(\rho)}z^{n\lambda},\quad 0<\lambda\le 1,\tag{29}$$

where

$$f_{q,k}(\rho)=q^{c}\frac{\Gamma_q(1+\rho+k\lambda)}{\Gamma_q(1+\rho+(k-2)\lambda)}+[c]_q\frac{\Gamma_q(1+\rho+k\lambda)}{\Gamma_q(1+\rho+(k-1)\lambda)},\tag{30}$$

$$\begin{aligned}g_{q,k}(\rho) &= q^{a+b+1}\frac{\Gamma_q(1+\rho+k\lambda)}{\Gamma_q(1+\rho+(k-2)\lambda)}\\ &\quad+(q^{b}[a]_q+q^{a}[b+1]_q)\frac{\Gamma_q(1+\rho+k\lambda)}{\Gamma_q(1+\rho+(k-1)\lambda)}+[a]_q[b]_q,\end{aligned}\tag{31}$$

and $\rho>-1$ satisfies the equation

$$f_{q,0}(\rho)=\frac{\Gamma_q(1+\rho)}{\Gamma_q(1+\rho-2\lambda)}+[c]_q\frac{\Gamma_q(1+\rho)}{\Gamma_q(1+\rho-\lambda)}=0.\tag{32}$$

The following assertion is valid.

Theorem 1. *The fractional q-Gauss function* (29) *is a solution of the fractonal q-Gauss hypergeometric Equation* (28) *where* $f_{q,k}(\rho)$ *and* $g_{q,k}(\rho)$ *are given by* (30) *and* (31), *repectively and* ρ *satisfies the condition* (32).

Proof. We look for the solution under the following modified formal power series form

$$u(z) = \sum_{n=0}^{\infty} d_n z^{n\lambda+\rho}.$$

Then,

$$D_q^{\lambda} u(z) = \sum_{n=0}^{\infty} d_n \frac{\Gamma_q(n\lambda+\rho+1)}{\Gamma_q((n-1)\lambda+\rho+1)} z^{(n-1)\lambda+\rho},$$

and

$$D_q^{2\lambda} u(z) = \sum_{n=0}^{\infty} d_n \frac{\Gamma_q(n\lambda+\rho+1)}{\Gamma_q((n-2)\lambda+\rho+1)} z^{(n-2)\lambda+\rho}.$$

Inserting these fractional q-derivatives in (28), we obtain the following recurrence relation for the coefficients a_n,

$$f_{q,n}(\rho) d_{n+1} - g_{q,n+1}(\rho) d_n = 0,$$

with $f_{q,0}(\rho) = 0$, where $f_{q,n}(\rho)$ and $g_{q,n}(\rho)$ are given by (30) and (31) respectively. The theorem follows easily. □

6. Conclusions and Further Perspectives

In this work we have introduced a new fractional q-differential operator $D^{\alpha}_{q^{-1}}$ and have proved some of its main important properties. Then, we have used it to extend some families of classical q-orthogonal polynomials. We have also defined a fractional q-Gauss differential equation, extending the one introduced by Koorwinder in [16], and solve them by means of the power series method. It should be noted that we obtained the results the Big q-Jacobi, the Big q-Laguerre, the Big q-Legendre, the Al-Salam Carlitz-I and II and the Stieltjes–Wigert polynomials but did not include them because they are not of nice form.

As future works, we plan to provide similar extensions to classical orthogonal polynomials on quadratic and q-quadratic lattices. To do it, it will be necessary to introduce new differential operators of fractional order, namely, the Wilson operator [11,21]

$$\mathbf{D} f(x) = \frac{f(x+i/2) - f(x-i/2)}{2ix}$$

or the Askey–Wilson divided difference operator [11]

$$\mathcal{D}_q f(x) = \frac{\hat{f}(q^{1/2}e^{i\theta}) - \hat{f}(q^{-1/2}e^{i\theta})}{i(q^{1/2} - q^{-1/2}) \sin\theta}$$

with

$$\hat{f}(e^{i\theta}) = f(x), \quad x = \cos\theta.$$

For these operators, the paper of Cooper [22] will help to define their fractional extensions. This work is ongoing.

Author Contributions: Both authors equally contributed to this work. All authors have read and agreed to the published version of the manuscript.

References

1. Al-Salam, W.A. Some fractional q-integrals and q-derivatives. *Proc. Edinb. Math. Soc.* **1966**, *15*, 135–140. [CrossRef]
2. Argawal, R.P. Certain fractional q-integrals ans q-derivatives. *Proc. Camb. Philos. Soc.* **1975**, *66*, 365–370.
3. Abd-Elhameed, W.M. New formulae between Jacobi polynomials and some fractional Jacobi functions generalizing some connection formulae. *Anal. Math. Phys.* **2017**, *9*, 73–98. [CrossRef]
4. Cao, J.; Srivastava, H.M.; Liu, Z.G. Some iterated fractional q-integrals and their applications. *Fract. Calc. Appl. Anal.* **2018**, *21*, 672–695. [CrossRef]
5. Ernst, T. On the Triple Lauricella-Horn-Karlsson q-Hypergeometric Functions. *Axioms* **2020**, *9*, 93. [CrossRef]
6. Purohit, S.D.; Kalla, S.L. On the fractional q-calculus of a general class of q-polynomials. *Algebr. Groups Geom.* **2009**, *26*, 1–13.
7. Rajković, P.M.; Marinković, S.D.; Stanković, M.S. On q-analogues of Caputo derivative and Mittag-Leffler function. *Fract. Calc. Appl. Anal.* **2007**, *10*, 359–373.
8. Rajković, P.M.; Marinković, S.D.; Stanković, M.S. Fractional integrals and derivatives in q-calculus. *Appl. Anal. Discret. Math.* **2007**, *1*, 311–323.
9. Area, I.; Godoy, E.; Ronveaux, A.; Zarzo, A. Inversion problems in the q-Hahn tableau. *J. Symb. Comput.* **1998**, *136*, 1–10.
10. Foupouagnigni, M.; Koepf, W.; Tcheutia, D.D.; Njionou Sadjang, P. Representations of q-orthogonal polynomials. *J. Symb. Comput.* **2012**, *47*, 1347–1371. [CrossRef]
11. Koekoek, R.; Lesky, P.A.; Swarttouw, R.F. *Hypergeometric Orthogonal Polynomials and Their q-Analogues*; Springer: Berlin, Germany, 2010.
12. Gogovcheva, E.; Boyadjiev, L. Fractional extensions of Jacobi polynomials and Gauss hypergeometric function. *Fract. Calc. Appl. Anal.* **2005**, *8*, 432–438
13. Ishteva, M.; Scherer, R.; Boyadjiev, L. On the Caputo operator of fractional calculus and C-Laguerre functions. *Math. Sci. Res.* **2005**, *9*, 161–170.
14. Rainville, E.D. *Special Functions*; The Macmillan Company: New York, NY, USA, 1960.
15. Gasper, G.; Rahman, M. *Basic Hypergeometric Series*; Cambridge University Press: Cambridge, UK, 1990; Volume 35.
16. Koorwinder, T.H. q-Special Functions, a Tutorial. Available online: math/9403216.pdf (accessed on 21 March 2013).
17. Annaby, M.H.; Mansour, Z.S. *q-Fractional Calculus and Equations*; Springer: Berlin/Heidelberg, Germany, 2012.
18. Al-Salam, W.A. q-Analogues of Cauchy's Formulas. *Proc. Am. Math. Soc.* **1966**, *17*, 616–621.
19. Fischer, K.K. Identifikation Spezieller Funktionen, die Durch Rodriguesformeln Gegeben Sind. Ph.D. Thesis, Universität Kassel, Kassel, Germany 2016. Available online: http://kobra.bibliothek.uni-kassel.de/handle/urn:nbn:de:hebis:34-2016030349960 (accessed on 3 March 2016).
20. Annaby, M.H.; Mansour, Z.S. q-Taylor and interpolation series for Jackson q-difference operators. *J. Math. Anal. Appl.* **2008**, *344*, 472–483. [CrossRef]
21. Njionou Sadjang, P.; Koepf, W.; Foupouagnigni, M. On structure formulas for Wilson polynomials. *Integral Transform. Spec. Funct.* **2015**, *26*, 1000–1014. [CrossRef]
22. Cooper, S. *The Askey-Wilson Operator and the* ${}_6\phi_5$ *Summation Formula*; Massey University: Palmerston North, New Zealand, 2012.

3

Some Identities on Degenerate Bernstein and Degenerate Euler Polynomials

Taekyun Kim [1,*] and Dae San Kim [2]

[1] Department of Mathematics, Kwangwoon University, Seoul 139-701, Korea
[2] Department of Mathematics, Sogang University, Seoul 121-742, Korea; dskim@sogang.ac.kr
* Correspondence: tkkim@kw.ac.kr

Abstract: In recent years, intensive studies on degenerate versions of various special numbers and polynomials have been done by means of generating functions, combinatorial methods, umbral calculus, p-adic analysis and differential equations. The degenerate Bernstein polynomials and operators were recently introduced as degenerate versions of the classical Bernstein polynomials and operators. Herein, we firstly derive some of their basic properties. Secondly, we explore some properties of the degenerate Euler numbers and polynomials and also their relations with the degenerate Bernstein polynomials.

Keywords: degenerate Bernstein polynomials; degenerate Bernstein operators; degenerate Euler polynomials

---

1. Introduction

Let us denote the space of continuous functions on $[0,1]$ by $C[0,1]$, and the space of polynomials of degree $\leq n$ by $\mathbb{P}_n$. The Bernstein operator $\mathbb{B}_n$ of order n, $(n \geq 1)$, associates to each $f \in C[0,1]$ the polynomial $\mathbb{B}_n(f|x) \in \mathbb{P}_n$, and was introduced by Bernstein as (see [1,2]):

$$\begin{aligned}\mathbb{B}_n(f|x) &= \sum_{k=0}^{n} f(\frac{k}{n})\binom{n}{k}x^k(1-x)^{n-k} \\ &= \sum_{k=0}^{n} f(\frac{k}{n})B_{k,n}(x),\end{aligned} \tag{1}$$

(see [1–14]) where

$$B_{k,n}(x) = \binom{n}{k}x^k(1-x)^{n-k}, \ (n,k \in \mathbb{Z}_{\geq 0}) \tag{2}$$

are called either Bernstein polynomials of degree n or Bernstein basis polynomials of degree n.

The Bernstein polynomials of degree n can be defined in terms of two such polynomials of degree $n-1$. That is, the k-th Bernstein polynomial of degree n can be written as

$$B_{k,n}(x) = (1-x)B_{k,n-1}(x) + xB_{k-1,n-1}(x), \ (k,n \in \mathbb{N}). \tag{3}$$

From (2), the first few Bernstein polynomials $B_{k,n}(x)$ are given by

$$\begin{aligned}&B_{0,1}(x) = 1-x, \ B_{1,1}(x) = x, \ B_{0,2}(x) = (1-x)^2, \ B_{1,2}(x) = 2x(1-x), \\ &B_{2,2}(x) = x^2, \ B_{0,3}(x) = (1-x)^3, \ B_{1,3}(x) = 3x(1-x)^2, \\ &B_{2,3}(x) = 3x^2(1-x), \ B_{3,3}(x) = x^3, \cdots.\end{aligned}$$

Thus, we note that

$$x^k = x(x^{k-1}) = x\sum_{i=k-1}^{n} \frac{\binom{i}{k-1}}{\binom{n}{k-1}} B_{i,n}(x)$$
$$= \sum_{i=k-1}^{n} \frac{\binom{i}{k-1}}{\binom{n}{k-1}} \frac{i+1}{n+1} B_{i+1,n+1}(x)$$
$$= \sum_{i=k-1}^{n} \frac{\binom{i+1}{k}}{\binom{n+1}{k}} B_{i+1,n+1}(x).$$

For $\lambda \in \mathbb{R}$, L. Carlitz introduced the degenerate Euler poynomials given by the generating function (see [15,16])

$$\frac{2}{(1+\lambda t)^{\frac{1}{\lambda}}+1}(1+\lambda t)^{\frac{x}{\lambda}} = \sum_{n=0}^{\infty} \mathcal{E}_{n,\lambda}(x)\frac{t^n}{n!}, \tag{4}$$

When $x = 0$, $\mathcal{E}_{n,\lambda} = \mathcal{E}_{n,\lambda}(0)$ are called the degenerate Euler numbers. It is easy to show that $\lim_{\lambda\to 0} \mathcal{E}_{n,\lambda}(x) = E_n(x)$, where $E_n(x)$ are the Euler polynomials given by (see [15–18])

$$\frac{2}{e^t+1}e^{xt} = \sum_{n=0}^{\infty} E_n(x)\frac{t^n}{n!} \tag{5}$$

For $n \geq 0$, we define the λ-product as follows (see [8]):

$$(x)_{0,\lambda} = 1,\ (x)_{n,\lambda} = x(x-\lambda)(x-2\lambda)\cdots(x-(n-1)\lambda),\ (n \geq 1), \tag{6}$$

Observe here that $\lim_{\lambda\to 0}(x)_{n,\lambda} = x^n,\ (n \geq 1)$.

Recently, the degenerate Bernstein polynomials of degree n are introduced as (see [8])

$$B_{k,n}(x|\lambda) = \binom{n}{k}(x)_{k,\lambda}(1-x)_{n-k,\lambda},\ (x \in [0,1],\ n,k \geq 0), \tag{7}$$

From (7), it is not difficult to show that the generating function for $B_{k,n}(x|\lambda)$ is given by (see [8])

$$\frac{1}{k!}(x)_{k,\lambda}t^k(1+\lambda t)^{\frac{1-x}{\lambda}} = \sum_{n=k}^{\infty} B_{k,n}(x|\lambda)\frac{t^n}{n!}, \tag{8}$$

By (8), we easily get $\lim_{\lambda\to 0} B_{k,n}(x|\lambda) = B_{k,n}(x),\ (n,k \geq 0)$.

The Bernstein polynomials are the mathematical basis for Bézier curves which are frequently used in computer graphics and related fields. In this paper, we investigate the degenerate Bernstein polynomials and operators. We study their elementary properties (see also [8]) and then their further properties in association with the degenerate Euler numbers and polynomials.

Finally, we would like to briefly go over some of the recent works related with Bernstein polynomials and operators.

Kim-Kim in Ref. [19] gave identities for degenerate Bernoulli polynomials and Korobov polynomials of the first kind. The authors in Ref. [20] introduced a generalization of the Bernstein polynomials associated with Frobenius–Euler polynomials. The paper [21] deals with some identities of q-Euler numbers and polynomials associated with q-Bernstein polynomials. In Ref. [22], the authors studied a space-time fractional diffusion equation with initial boundary conditions and presented a numerical solution for that. Both normalized Bernstein polynomials with collocation and Galerkin methods are applied to turn the problem into an algebraic system. Kim in Ref. [23] introduced some identities on the q-integral representation of the product of the several q-Bernstein type polynomials. Grouped data are commonly encountered in applications. In Ref. [24], Kim-Kim studied some properties on degenerate Eulerian numbers and polynomials. The authors in Ref. [25] give an overview of several results related to partially degenerate poly-Bernoulli polynomials associated with Hermit polynomials.

2. Degenerate Bernstein Polynomials and Operators

The degenerate Bernstein operator of order n is defined, for $f \in C[0,1]$, as

$$\mathbb{B}_{n,\lambda}(f|x) = \sum_{k=0}^{n} f(\frac{k}{n})\binom{n}{k}(x)_{k,\lambda}(1-x)_{n-k,\lambda} = \sum_{k=0}^{n} f(\frac{k}{n})B_{k,n}(x|\lambda), \tag{9}$$

where $x \in [0,1]$ and $n,k \in \mathbb{Z}_{\geq 0}$.

Theorem 1. *For $n \geq 0$, we have*

$$\mathbb{B}_{n,\lambda}(f|0) = f(0)(1)_{n,\lambda},\ \mathbb{B}_{n,\lambda}(f|1) = f(1)(1)_{n,\lambda},$$

and

$$\mathbb{B}_{n,\lambda}(1|x) = (1)_{n,\lambda},\ \mathbb{B}_{n,\lambda}(x|x) = x\sum_{k=0}^{n-1}(-1)^k\lambda^k(n-1)_k(1)_{n-1-k,\lambda},\ (n \geq 1),$$

where $(x)_k = x(x-1)\cdots(x-k+1)$, $(k \geq 1)$, $(x)_0 = 1$.

Proof. From (9), we clearly have

$$\mathbb{B}_{n,\lambda}(1|x) = \sum_{k=0}^{n}\binom{n}{k}(x)_{k,\lambda}(1-x)_{n-k,\lambda}. \tag{10}$$

Now, we observe that

$$\begin{aligned}\sum_{n=0}^{\infty}(1)_{n,\lambda}\frac{t^n}{n!} &= (1+\lambda t)^{\frac{1}{\lambda}} = (1+\lambda t)^{\frac{x}{\lambda}}(1+\lambda t)^{\frac{1-x}{\lambda}}\\ &= \left(\sum_{l=0}^{\infty}(x)_{l,\lambda}\frac{t^l}{l!}\right)\left(\sum_{m=0}^{\infty}(1-x)_{m,\lambda}\frac{t^m}{m!}\right)\\ &= \sum_{n=0}^{\infty}\left(\sum_{l=0}^{n}\binom{n}{l}(x)_{l,\lambda}(1-x)_{n-l,\lambda}\right)\frac{t^n}{n!}.\end{aligned} \tag{11}$$

Comparing the coefficients on both sides of (11), we derive

$$(1)_{n,\lambda} = \sum_{l=0}^{n}\binom{n}{l}(x)_{l,\lambda}(1-x)_{n-l,\lambda}. \tag{12}$$

Combining (10) with (12), we have

$$\mathbb{B}_{n,\lambda}(1|x) = \sum_{k=0}^{n}\binom{n}{k}(x)_{k,\lambda}(1-x)_{n-k,\lambda} = (1)_{n,\lambda},\ (n \geq 0). \tag{13}$$

Furthermore, we get from (9) that for $f(x) = x$,

$$\begin{aligned}\mathbb{B}_{n,\lambda}(x|x) &= \sum_{k=0}^{n} \frac{k}{n}\binom{n}{k}(x)_{k,\lambda}(1-x)_{n-k,\lambda} \\ &= \sum_{k=1}^{n}\binom{n-1}{k-1}(x)_{k,\lambda}(1-x)_{n-k,\lambda} \\ &= \sum_{k=0}^{n-1}\binom{n-1}{k}(x)_{k,\lambda}(1-x)_{n-1-k,\lambda}(x-k\lambda) \\ &= x(1)_{n-1,\lambda} - (n-1)\lambda\sum_{k=0}^{n-1}\binom{n-1}{k}(x)_{k,\lambda}(1-x)_{n-1-k,\lambda}\frac{k}{n-1} \\ &= x(1)_{n-1,\lambda} - (n-1)\lambda B_{n-1,\lambda}(x|x).\end{aligned} \tag{14}$$

From (14), we can easily deduce the following Equation (15):

$$\begin{aligned}\mathbb{B}_{n,\lambda}(x|x) &= x(1)_{n-1,\lambda} - (n-1)\lambda\{x(1)_{n-2,\lambda} - (n-2)\lambda B_{n-2,\lambda}(x|x)\} \\ &= x(1)_{n-1,\lambda} - x(n-1)\lambda(1)_{n-2,\lambda} + (-1)^2(n-1)(n-2)\lambda^2 B_{n-2,\lambda}(x|x) \\ &= x(1)_{n-1,\lambda} - x(n-1)\lambda(1)_{n-2,\lambda} \\ &\quad + (-1)^2(n-1)(n-2)\lambda^2\{x(1)_{n-3,\lambda} - (n-3)\lambda B_{n-3,\lambda}(x|x)\} \\ &= x(1)_{n-1,\lambda} - x(n-1)\lambda(1)_{n-2,\lambda} + (-1)^2(n-1)(n-2)\lambda^2 x(1)_{n-3,\lambda} \\ &\quad + (-1)^3(n-1)(n-2)(n-3)\lambda^3 B_{n-3,\lambda}(x|x) \\ &= \cdots \\ &= x\sum_{k=0}^{n-1}(-1)^k\lambda^k(n-1)_k(1)_{n-1-k,\lambda}.\end{aligned} \tag{15}$$

□

Let f, g be continuous functions defined on $[0,1]$. Then, we clearly have

$$\mathbb{B}_{n,\lambda}(\alpha f + \beta g|x) = \alpha\mathbb{B}_{n,\lambda}(f|x) + \beta\mathbb{B}_{n,\lambda}(g|x),\ (n \geq 0), \tag{16}$$

where α, β are constants.

So, the degenerate Bernstein operator is linear. From (7), we note that

$$\mathbb{B}_{0,1}(x|\lambda) = 1-x,\ \mathbb{B}_{1,1}(x|\lambda) = x,\ \mathbb{B}_{0,2}(x|\lambda) = (1-x)^2 - \lambda(1-x),$$
$$\mathbb{B}_{1,2}(x|\lambda) = 2x(1-x),\ \mathbb{B}_{2,2}(x|\lambda) = x^2 - \lambda x.$$

It is not hard to see that

$$\begin{aligned}\sum_{n=0}^{\infty}(1-x)_{n,\lambda}\frac{t^n}{n!} &= (1+\lambda t)^{\frac{1-x}{\lambda}} = (1+\lambda t)^{\frac{1}{\lambda}}(1+\lambda t)^{-\frac{x}{\lambda}} \\ &= \left(\sum_{l=0}^{\infty}(1)_{l,\lambda}\frac{t^l}{l!}\right)\left(\sum_{m=0}^{\infty}(-x)_{m,\lambda}\frac{t^m}{m!}\right) \\ &= \sum_{n=0}^{\infty}\left(\sum_{l=0}^{n}\binom{n}{l}(1)_{n-l,\lambda}(-1)^l(x)_{l,-\lambda}\right)\frac{t^n}{n!}.\end{aligned}$$

This shows that we have

$$(1-x)_{n,\lambda} = \sum_{l=0}^{n}\binom{n}{l}(1)_{n-l,\lambda}(-1)^l(x)_{l,-\lambda},\ (n \geq 0). \tag{17}$$

Theorem 2. *For $f \in C[0,1]$ and $n \in \mathbb{Z}_{\geq 0}$, we have*

$$\mathbb{B}_{n,\lambda}(f|x) = \sum_{m=0}^{n} \binom{n}{m} (x)_{m,-\lambda} \sum_{k=0}^{m} \binom{m}{k} (-1)^{m-k} (1)_{n-m,\lambda} \frac{(x)_{k,\lambda}}{(x+(m-1)\lambda)_{k,\lambda}} f(\frac{k}{n}).$$

Proof. From (9), it is immediate to see that

$$\begin{aligned}\mathbb{B}_{n,\lambda}(f|x) &= \sum_{k=0}^{n} f(\frac{k}{n}) B_{k,n}(x|\lambda) = \sum_{k=0}^{n} f(\frac{k}{n}) \binom{n}{k} (x)_{k,\lambda}(1-x)_{n-k,\lambda} \\ &= \sum_{k=0}^{n} f(\frac{k}{n}) \binom{n}{k} (x)_{k,\lambda} \sum_{j=0}^{n-k} \binom{n-k}{j} (-1)^j (1)_{n-k-j,\lambda} (x)_{j,-\lambda}.\end{aligned} \tag{18}$$

We need to note the following:

$$\binom{n}{k}\binom{n-k}{j} = \binom{n}{k+j}\binom{k+j}{k}, \ (n,k \geq 0), \tag{19}$$

and

$$(x)_{j,-\lambda} = \frac{(x)_{k+j,-\lambda}}{(x+(j+k-1)\lambda)_{k,\lambda}}. \tag{20}$$

Let $k+j=m$. Then, by (19), we obviously have

$$\binom{n}{k}\binom{n-k}{j} = \binom{n}{m}\binom{m}{k}. \tag{21}$$

Combining (18) with (19)–(21) gives the following result:

$$\mathbb{B}_{n,\lambda}(f|x) = \sum_{m=0}^{n} \binom{n}{m} (x)_{m,-\lambda} \sum_{k=0}^{m} \binom{m}{k} (-1)^{m-k} (1)_{n-m,\lambda} \frac{(x)_{k,\lambda}}{(x+(m-1)\lambda)_{k,\lambda}} f(\frac{k}{n}).$$

□

Theorem 3. *For $n,k \in \mathbb{Z}_{\geq 0}$ and $x \in [0,1]$, we have*

$$B_{k,n}(x|\lambda) = \sum_{i=k}^{n} (-1)^{i-k} \binom{n}{i}\binom{i}{k} (x)_{i,-\lambda} \frac{(x)_{k,\lambda}}{(x+(i-1)\lambda)_{k,\lambda}} (1)_{n-i,\lambda}.$$

Proof. From (7), (17), and (20), we observe that

$$\begin{aligned}B_{k,n}(x|\lambda) &= \binom{n}{k} (x)_{k,\lambda}(1-x)_{n-k,\lambda} \\ &= \binom{n}{k} (x)_{k,\lambda} \sum_{i=0}^{n-k} \binom{n-k}{i} (-1)^i (x)_{i,-\lambda} (1)_{n-k-i,\lambda} \\ &= \sum_{i=0}^{n-k} (-1)^i \binom{n}{k}\binom{n-k}{i} (x)_{k+i,-\lambda} \frac{(x)_{k,\lambda}}{(x+(k+i-1)\lambda)_{k,\lambda}} (1)_{n-k-i,\lambda} \\ &= \sum_{i=k}^{n} (-1)^{i-k} \binom{n}{k}\binom{n-k}{i-k} (x)_{i,-\lambda} \frac{(x)_{k,\lambda}}{(x+(i-1)\lambda)_{k,\lambda}} (1)_{n-i,\lambda} \\ &= \sum_{i=k}^{n} (-1)^{i-k} \binom{n}{i}\binom{i}{k} (x)_{i,-\lambda} \frac{(x)_{k,\lambda}}{(x+(i-1)\lambda)_{k,\lambda}} (1)_{n-i,\lambda}.\end{aligned} \tag{22}$$

□

3. Degenerate Euler Polynomials Associated with Degenerate Bernstein Polynomials

Theorem 4. *For $n \geq 0$, the following holds true:*

$$\sum_{l=0}^{n}\binom{n}{l}(1)_{n-l,\lambda}\mathcal{E}_{l,\lambda}+\mathcal{E}_{n,\lambda}=\begin{cases}2, & \text{if } n=0,\\ 0, & \text{if } n>0.\end{cases}$$

Proof. From (4), we remark that

$$\begin{aligned}2&=\left(\sum_{l=0}^{\infty}\mathcal{E}_{l,\lambda}\frac{t^l}{l!}\right)\left((1+\lambda t)^{\frac{1}{\lambda}}+1\right)=\left(\sum_{l=0}^{\infty}\mathcal{E}_{l,\lambda}\frac{t^l}{l!}\right)\left(\sum_{m=0}^{\infty}(1)_{m,\lambda}\frac{t^m}{m!}+1\right)\\&=\sum_{n=0}^{\infty}\left(\sum_{l=0}^{n}\binom{n}{l}\mathcal{E}_{l,\lambda}(1)_{n-l,\lambda}\right)\frac{t^n}{n!}+\sum_{n=0}^{\infty}\mathcal{E}_{n,\lambda}\frac{t^n}{n!}\\&=\sum_{n=0}^{\infty}\left(\sum_{l=0}^{n}\binom{n}{l}(1)_{n-l,\lambda}\mathcal{E}_{l,\lambda}+\mathcal{E}_{n,\lambda}\right)\frac{t^n}{n!}.\end{aligned}\tag{23}$$

The result follows by comparing the coefficients on both sides of (23).
□

From Theorem 4, we note that

$$\mathcal{E}_{0,\lambda}=1,\ \mathcal{E}_{n,\lambda}=-\sum_{l=0}^{n}\binom{n}{l}(1)_{n-l,\lambda}\mathcal{E}_{l,\lambda},\ (n>0).$$

From these recurrence relations, we note that the first few degenerate Euler numbers are given by

$$\begin{aligned}\mathcal{E}_{0,\lambda}=1,\ \mathcal{E}_{1,\lambda}=-\frac{1}{2},\ \mathcal{E}_{2,\lambda}=\frac{1}{2}\lambda,\ \mathcal{E}_{3,\lambda}=\frac{1}{4}-\lambda^2,\ \mathcal{E}_{4,\lambda}=-\frac{3}{2}\lambda+3\lambda^3,\\ \mathcal{E}_{5,\lambda}=-\frac{1}{2}+\frac{35}{4}\lambda^2-12\lambda^4,\ \mathcal{E}_{6,\lambda}=\frac{15}{2}\lambda-\frac{225}{4}\lambda^3+60\lambda^5.\end{aligned}$$

Theorem 5. *For $n \geq 0$, we have*

$$\mathcal{E}_{n,\lambda}(1-x)=(-1)^n\mathcal{E}_{n,-\lambda}(x).$$

Especially, we have

$$\mathcal{E}_{n,\lambda}(2)=(-1)^n\mathcal{E}_{n,-\lambda}(-1),\ (n\geq 0).$$

Proof. By (4), we get

$$\begin{aligned}\sum_{n=0}^{\infty}\mathcal{E}_{n,\lambda}(1-x)\frac{t^n}{n!}&=\frac{2}{(1+\lambda t)^{\frac{1}{\lambda}}+1}(1+\lambda t)^{\frac{1-x}{\lambda}}=\frac{2}{(1+\lambda t)^{-\frac{1}{\lambda}}+1}(1+\lambda t)^{-\frac{x}{\lambda}}\\&=\sum_{n=0}^{\infty}\mathcal{E}_{n,-\lambda}(x)(-1)^n\frac{t^n}{n!}.\end{aligned}\tag{24}$$

Comparing the coefficients on both sides of (24), we have

$$\mathcal{E}_{n,\lambda}(1-x)=(-1)^n\mathcal{E}_{n,-\lambda}(x),\ (n\geq 0).\tag{25}$$

In particular, if we take $x = -1$, we get

$$\mathcal{E}_{n,\lambda}(2) = (-1)^n \mathcal{E}_{n,-\lambda}(-1), \ (n \geq 0). \tag{26}$$

□

Corollary 1. *For $n \geq 0$, we have*

$$\mathcal{E}_{n,\lambda}(2) = 2(1)_{n,\lambda} + \mathcal{E}_{n,\lambda}.$$

Proof. From (4), we have

$$\begin{aligned}
\sum_{n=0}^{\infty} \mathcal{E}_{n,\lambda}(2)\frac{t^n}{n!} &= \frac{2}{(1+\lambda t)^{\frac{1}{\lambda}}+1}(1+\lambda t)^{\frac{2}{\lambda}} \\
&= \frac{2}{(1+\lambda t)^{\frac{1}{\lambda}}+1}(1+\lambda t)^{\frac{1}{\lambda}}((1+\lambda t)^{\frac{1}{\lambda}}+1-1) \\
&= 2(1+\lambda t)^{\frac{1}{\lambda}} - \frac{2}{(1+\lambda t)^{\frac{1}{\lambda}}+1}(1+\lambda t)^{\frac{1}{\lambda}} \\
&= \sum_{n=0}^{\infty}\left(2(1)_{n,\lambda} - \sum_{l=0}^{n}\binom{n}{l}(1)_{n-l,\lambda}\mathcal{E}_{l,\lambda}\right)\frac{t^n}{n!}.
\end{aligned} \tag{27}$$

Thus, by (27), we get

$$\mathcal{E}_{n,\lambda}(2) = 2(1)_{n,\lambda} - \sum_{l=0}^{n}\binom{n}{l}(1)_{n-l,\lambda}\mathcal{E}_{l,\lambda} = 2(1)_{n,\lambda} + \mathcal{E}_{n,\lambda}, \ (n \geq 0).$$

□

Theorem 6. *For $n \geq 0, k \geq 1$, we have*

$$\mathcal{E}_{n,\lambda}(x) = 2\sum_{i=1}^{k}(-1)^{i-1}(x-i)_{n,\lambda} + (-1)^k\mathcal{E}_{n,\lambda}(x-k).$$

Proof. From (4), we easily see that

$$\mathcal{E}_{n,\lambda}(x) = \sum_{l=0}^{n}\binom{n}{l}\mathcal{E}_{l,\lambda}(x)_{n-l,\lambda}, \ (n \geq 0).$$

Now, we observe that

$$\begin{aligned}\frac{2}{(1+\lambda t)^{\frac{1}{\lambda}}+1}(1+\lambda t)^{\frac{x}{\lambda}} &= \frac{2}{(1+\lambda t)^{\frac{1}{\lambda}}+1}((1+\lambda t)^{\frac{1}{\lambda}}+1-1)(1+\lambda t)^{\frac{x-1}{\lambda}}\\ &= 2(1+\lambda t)^{\frac{x-1}{\lambda}} - \frac{2}{(1+\lambda t)^{\frac{1}{\lambda}}+1}(1+\lambda t)^{\frac{x-1}{\lambda}}\\ &= 2(1+\lambda t)^{\frac{x-1}{\lambda}} - 2(1+\lambda t)^{\frac{x-2}{\lambda}} + \frac{2}{(1+\lambda t)^{\frac{1}{\lambda}}+1}(1+\lambda t)^{\frac{x-2}{\lambda}}\\ &= 2(1+\lambda t)^{\frac{x-1}{\lambda}} - 2(1+\lambda t)^{\frac{x-2}{\lambda}} + 2(1+\lambda t)^{\frac{x-3}{\lambda}} - \frac{2}{(1+\lambda t)^{\frac{1}{\lambda}}+1}(1+\lambda t)^{\frac{x-3}{\lambda}}\\ &= 2(1+\lambda t)^{\frac{x-1}{\lambda}} - 2(1+\lambda t)^{\frac{x-2}{\lambda}} + 2(1+\lambda t)^{\frac{x-3}{\lambda}} - 2(1+\lambda t)^{\frac{x-4}{\lambda}}\\ &\quad + \frac{2}{(1+\lambda t)^{\frac{1}{\lambda}}+1}(1+\lambda t)^{\frac{x-4}{\lambda}}.\end{aligned} \tag{28}$$

Continuing the process in (28) gives the following result:

$$\begin{aligned}\sum_{n=0}^{\infty}\mathcal{E}_{n,\lambda}(x)\frac{t^n}{n!} &= 2\sum_{i=1}^{k}(-1)^{i-1}(1+\lambda t)^{\frac{x-i}{\lambda}} + (-1)^k\frac{2}{(1+\lambda t)^{\frac{1}{\lambda}}+1}(1+\lambda t)^{\frac{x-k}{\lambda}}\\ &= \sum_{n=0}^{\infty}\left(2\sum_{i=1}^{k}(-1)^{i-1}(x-i)_{n,\lambda}\right)\frac{t^n}{n!} + (-1)^k\frac{2}{(1+\lambda t)^{\frac{1}{\lambda}}+1}(1+\lambda t)^{\frac{x-k}{\lambda}}.\end{aligned} \tag{29}$$

The desired result now follows from (4) and (29).

□

Theorem 7. *For $n, k \geq 0$, we have*

$$B_{k,n+k}(x|\lambda) = \frac{1}{2}(x)_{k,\lambda}\binom{n+k}{k}(\mathcal{E}_{n,\lambda}(2-x)+\mathcal{E}_{n,\lambda}(1-x)).$$

Proof. In view of (8), we have

$$(x)_{k,\lambda}(1+\lambda t)^{\frac{1-x}{\lambda}} = \frac{k!}{t^k}\sum_{n=k}^{\infty}B_{k,n}(x|\lambda)\frac{t^n}{n!} = \sum_{n=0}^{\infty}B_{k,n+k}(x|\lambda)\frac{1}{\binom{n+k}{n}}\frac{t^n}{n!}. \tag{30}$$

On the other hand, (30) is also given by

$$(x)_{k,\lambda}(1+\lambda t)^{\frac{1-x}{\lambda}} = \sum_{n=0}^{\infty}(x)_{k,\lambda}(1-x)_{n,\lambda}\frac{t^n}{n!}. \tag{31}$$

From (30) and (31), we have

$$(x)_{k,\lambda}(1-x)_{n,\lambda} = \frac{1}{\binom{n+k}{n}}B_{k,n+k}(x|\lambda),\ (n,k\geq 0). \tag{32}$$

Now, we observe that

$$\begin{aligned}(x)_{k,\lambda}(1+\lambda t)^{\frac{1-x}{\lambda}} &= \frac{(x)_{k,\lambda}}{2}\frac{2}{(1+\lambda t)^{\frac{1}{\lambda}}+1}(1+\lambda t)^{\frac{1-x}{\lambda}}((1+\lambda t)^{\frac{1}{\lambda}}+1)\\ &= \frac{(x)_{k,\lambda}}{2}\left(\frac{2}{(1+\lambda t)^{\frac{1}{\lambda}}+1}(1+\lambda t)^{\frac{2-x}{\lambda}}+\frac{2}{(1+\lambda t)^{\frac{1}{\lambda}}+1}(1+\lambda t)^{\frac{1-x}{\lambda}}\right)\\ &= \frac{(x)_{k,\lambda}}{2}\left(\sum_{n=0}^{\infty}(\mathcal{E}_{n,\lambda}(2-x)+\mathcal{E}_{n,\lambda}(1-x))\frac{t^n}{n!}\right).\end{aligned} \tag{33}$$

By (31) and (33), we get

$$(x)_{k,\lambda}(1-x)_{n,\lambda} = \frac{(x)_{k,\lambda}}{2}(\mathcal{E}_{n,\lambda}(2-x)+\mathcal{E}_{n,\lambda}(1-x)),\ (n \geq 0). \tag{34}$$

Therefore, from (32) and (34), we have the result. □

4. Conclusions

In 1912, Bernstein first used Bernstein polynomials to give a constructive proof for the Stone–Weierstrass approximation theorem. The convergence of the Bernstein approximation of a function f to f is of order $1/n$, even for smooth functions, and hence the related approximation process is not used for computational purposes. However, by combining Bernstein approximations and the use of ad hoc extrapolation algorithms, fast techniques were designed (see the recent review [14] and paper [13]). Furthermore, about half a century later, they were used to design automobile bodies at Renault by Pierre Bézier. The Bernstein polynomials are the mathematical basis for Bézier curves, which are frequently used in computer graphics and related fields such as animation, modeling, CAD, and CAGD.

The study of degenerate versions of special numbers and polynomials began with the papers by Carlitz in Refs. [15,16]. Kim and his colleagues have been studying various degenerate numbers and polynomials by means of generating functions, combinatorial methods, umbral calculus, p-adic analysis, and differential equations. This line of study led even to the introduction to degenerate gamma functions and degenerate Laplace transforms (see [26]). These already demonstrate that studying degenerate versions of known special numbers and polynomials can be very promising and rewarding. Furthermore, we can hope that many applications will be found not only in mathematics but also in sciences and engineering. As we mentioned in the above, it was not until about fifty years later that Bernstein polynomials found their applications in real-world problems.

With this hope in mind, here we investigated the degenerate Bernstein polynomials and operators which were recently introduced as degenerate versions of the classical Bernstein polynomials and operators. We derived some of their basic properties. In addition, we studied some further properties of the degenerate Bernstein polynomials related to the degenerate Euler numbers and polynomials.

Author Contributions: T.K. and D.S.K. conceived the framework and structured the whole paper; T.K. wrote the paper; D.S.K. checked the results of the paper; D.S.K. and T. K. completed the revision of the article.

Acknowledgments: The first author's work in this paper was conducted during the sabbatical year of Kwangwoon University in 2018. The authors would like to express their sincere gratitude to the referees for their valuable comments, which have improved the presentation of this paper.

References

1. Kurt, V. Some relation between the Bernstein polynomials and second kind Bernoulli polynomials. *Adv. Stud. Contemp. Math. (Kyungshang)* **2013**, *23*, 43–48.

2. Kim, T. Identities on the weighted q–Euler numbers and q-Bernstein polynomials. *Adv. Stud. Contemp. Math. (Kyungshang)* **2012**, *22* , 7–12. [CrossRef]
3. Kim, T. A note on q–Bernstein polynomials. *Russ. J. Math. Phys.* **2011**, *18* , 73–82. [CrossRef]
4. Kim, T. A study on the q–Euler numbers and the fermionic q–integral of the product of several type q–Bernstein polynomials on $\mathbb{Z}_p$. *Adv. Stud. Contemp. Math. (Kyungshang)* **2013**, *23*, 5–11.
5. Kim, T.; Bayad, A.; Kim, Y.H. A study on the p–adic q–integral representation on $\mathbb{Z}_p$ associated with the weighted q–Bernstein and q-Bernoulli polynomials. *J. Inequal. Appl.* **2011**, *2011*, 513821. [CrossRef]
6. Kim, T.; Choi, J.; Kim, Y.-H. On the k–dimensional generalization of q–Bernstein polynomials. *Proc. Jangjeon Math. Soc.* **2011**, *14*, 199–207.
7. Kim, T.; Choi, J.; Kim, Y.H. Some identities on the q–Bernstein polynomials, q–Stirling numbers and q–Bernoulli numbers. *Adv. Stud. Contemp. Math. (Kyungshang)* **2010**, *20*, 335–341.
8. Kim, T.; Kim, D.S. Degenerate Bernstein Polynomials. *RACSAM* **2018**. [CrossRef]
9. Kim, T.; Lee, B.; Choi, J.; Kim, Y.H.; Rim, S.H. On the q–Euler numbers and weighted q–Bernstein polynomials. *Adv. Stud. Contemp. Math. (Kyungshang)* **2011**, *21*, 13–18.
10. Rim, S.-H.; Joung, J.; Jin, J.-H.; Lee, S.-J. A note on the weighted Carlitz's type q=Euler numbers and q-Bernstein polynomials. *Proc. Jangjeon Math. Soc.* **2012**, *15*, 195–201.
11. Simsek, Y. Identities and relations related to combinatorial numbers and polynomials. *Proc. Jangjeon Math. Soc.* **2017**, *20*, 127–135.
12. Siddiqui, M.A.; Agarwal, R.R.; Gupta, N. On a class of modified new Bernstein operators. *Adv. Stud. Contemp. Math. (Kyungshang)* **2014**, *24*, 97–107.
13. Costabile, F.; Gualtieri, M.; Serra, S. Asymptotic expansion and extrapolation for Bernstein polynomials with applications. *BIT Numer. Math.* **1996**, *36*, 676–687. [CrossRef]
14. Khosravian, H.; Dehghan, M.; Eslahchi, M.R. A new approach to improve the order of approximation of the Bernstein operators: Theory and applications. *Numer. Algorithms* **2018**, *77*, 111–150. [CrossRef]
15. Carlitz, L. Degenerate Stirling, Bernoulli and Eulerian numbers. *Utilitas Math.* **1979**, *15*, 51–88.
16. Carlitz, L. A degenerate Staudt-Clausen theorem. *Arch. Math.* **1956**, *7*, 28–33. [CrossRef]
17. Bayad, A.; Kim, T.; Lee, B.; Rim, S.-H. Some identities on Bernstein polynomials associated with q-Euler polynomials. *Abstr. Appl. Anal.* **2011**, *2011*, 294715. [CrossRef]
18. Bernstein, S. Demonstration du theoreme de Weierstrass basee sur le calcul des probabilites. *Commun. Soc. Math. Kharkow* **1912**, *13*, 1–2.
19. Kim, T.; Kim, D.S. Identities for degenerate Bernoulli polynomials and Korobov polynomials of the first kind. *Sci. China Math.* **2018**. [CrossRef]
20. Araci, S.; Acikgoz, M. A note on the Frobenius-Euler numbers and polynomials associated with Bernstein polynomials. *Adv. Stud. Contemp. Math.* **2012**, *22*, 399–406.
21. Ryoo, C.S. Some identities of the twisted q-Euler numbers and polynomials associated with q-Bernstein polynomials. *Proc. Jangjeon Math. Soc.* **2011**, *14*, 239–248.
22. Bayad, A.; Kim, T. Identities involving values of Bernstein, q–Bernoulli, and q–Euler polynomials. *Russ. J. Math. Phys.* **2011**, *18*, 133–143. [CrossRef]
23. Kim, T. Some identities on the q-integral representation of the product of several q-Bernstein-type polynomials. *Abstr. Appl. Anal.* **2011**, *2011*, 634675. [CrossRef]
24. Kim, D.S.; Kim, T. A note on degenerate Eulerian numbers and polynomials. *Adv. Stud. Contemp. Math. (Kyungshang)* **2017**, *27*, 431–440.
25. Khan, W.A.; Ahmad, M. Partially degenerate poly-Bernoulli polynomials associated with Hermite polynomials. *Adv. Stud. Contemp. Math. (Kyungshang)* **2018**, *28*, 487–496.
26. Kim, T.; Kim, D.S. Degenerate Laplace transform and degenerate gamma function. *Russ. J. Math. Phys.* **2017**, *24*, 241–248. [CrossRef]

4

Umbral Methods and Harmonic Numbers

Giuseppe Dattoli [1,†], Bruna Germano [2,†], Silvia Licciardi [1,*,†], and Maria Renata Martinelli [2,†]

[1] ENEA—Frascati Research Center, Via Enrico Fermi 45, 00044 Rome, Italy; giuseppe.dattoli@enea.it
[2] Department of Methods and Mathematic Models for Applied Sciences, University of Rome, La Sapienza, Via A. Scarpa, 14, 00161 Rome, Italy; bruna.germano@sbai.uniroma1.it (B.G.); martinelli@dmmm.uniroma1.it (M.R.M.)
[*] Correspondence: silviakant@gmail.com;
[†] These authors contributed equally to this work.

Abstract: The theory of harmonic-based functions is discussed here within the framework of umbral operational methods. We derive a number of results based on elementary notions relying on the properties of Gaussian integrals.

Keywords: harmonic numbers 11K99; operators 44A99, 47B99; umbral methods 05A40; special functions 33C52, 33C65, 33C99, 33B10, 33B15; Hermite polynomials 33C45

---

1. Introduction

Methods employing the concepts and the formalism of umbral calculus have been exploited in [1] to guess the existence of generating functions involving harmonic numbers [2]. The conjectures put forward in [1] have been proven in [3,4] and further elaborated in [5], and these were extended to hyper-harmonic numbers in [6].

In this note, we use the same point of view as [1], by discussing the possibility of exploiting the formalism developed therein in a wider context.

The umbral methods we are going to describe have certain advantages with respect to the ordinary techniques. The key idea is that of exploiting the harmonic number index as a power exponent; such a "promotion" allows the possibility of reducing the associated computational technicalities to elementary algebraic manipulations. Series involving harmonic numbers can, e.g., be treated as formal series of known functions (exponential, Gaussian, rational, etc.), and the relevant properties can be exploited to carry out computations, which are significantly more cumbersome and involved when conventional methods are employed.

2. Harmonic Numbers and Generating Functions

The harmonic numbers are defined by means of the following partial sum [2]:

$$h_n := \sum_{r=1}^{n} \frac{1}{r}, \quad \forall n \in \mathbb{N}_0 \,. \tag{1}$$

The integral representation for this family of numbers can be derived using a standard procedure, tracing back to Euler, which is sketched below.

Proposition 1. *The use of elementary integral transform yields, for the finite sum in Equation* (1), *the identity:*

$$h_n = \sum_{r=1}^{n} \int_0^{\infty} e^{-s\,r} ds, \quad \forall n \in \mathbb{N}_0 \,, \tag{2}$$

thereby getting the n-th harmonic number through Euler's integral [7–9]:

$$h_n = \int_0^1 \frac{1-x^n}{1-x} dx. \tag{3}$$

Proof. $\forall n \in \mathbb{N}_0$, by applying the Laplace transform, the theorem of uniform convergence and the sum of a geometric series, we obtain:

$$\begin{aligned} h_n &= \sum_{r=1}^{n} \int_0^\infty e^{-sr} ds = \int_0^\infty \left[\left(\sum_{r=0}^{n} e^{-sr} \right) - 1 \right] ds \\ &= \int_0^\infty \frac{1-(e^{-s})^{n+1}}{1-e^{-s}} - 1\, ds = \int_{-\infty}^0 \frac{1-(e^{s})^{n+1}}{1-e^{s}} - 1\, ds \\ &= \int_{-\infty}^0 \frac{e^{(n+1)s}-e^s}{e^s-1} ds \end{aligned}$$

and by applying the change of variables $e^s \to x$, we eventually end up with:

$$h_n = \int_0^1 \frac{1-x^n}{1-x} dx.$$

□

According to [8], from this point onwards, the definition in Equation (3) can be so extended to any real value of n, and therefore, it can be exploited as an alternative definition holding for n a positive real.

Definition 1. *The function:*

$$\varphi_h(z) := \int_0^1 \frac{1-x^z}{1-x} dx, \quad \forall z \in \mathbb{R}^+, \tag{4}$$

is called the harmonic number umbral vacuum, or simply the vacuum.

Definition 2. *The operator:*

$$\hat{h} := e^{\partial_z} \tag{5}$$

realizes the vacuum shift operator, zbeing the domain's variable of the function on which the operator acts For a deeper introduction to umbral calculus, see [10,11]

Theorem 1. *The umbral operator, $\hat{h}^n$, $\forall n \in \mathbb{R}^+$, defines the harmonic numbers, h_n, as the action of the shift operator* (5) *on the vacuum* (4)*:*

$$\hat{h}^n \varphi_h(z)\Big|_{z=0} := \hat{h}^n \varphi_{h_z}\Big|_{z=0} = h_n \tag{6}$$

or simply:

$$\begin{aligned} \hat{h}^n &\equiv h_n, \\ h_0 &= 0. \end{aligned} \tag{7}$$

Proof. $\forall n \in \mathbb{R}^+$, by applying the shift operator (5) on the vacuum (4), we obtain:

$$\begin{aligned} \hat{h}^n \varphi_{h_0} &= \hat{h}^n \varphi_{h_z}\Big|_{z=0} = e^{n\partial_z} \varphi_{h_z}\Big|_{z=0} = \varphi_{h_{z+n}}\big|_{z=0} = \int_0^1 \frac{1-x^{z+n}}{1-x} dx\Big|_{z=0} \\ &= \int_0^1 \frac{1-x^n}{1-x} dx = h_n. \end{aligned}$$

□

Properties 1. *$\forall n, m \in \mathbb{R}^+$, we get:*

$$\begin{aligned} &(i) \quad \hat{h}^n \hat{h}^m = \hat{h}^{n+m}\,, \\ &(ii) \left(\hat{h}^n\right)^m = \hat{h}^{nm}\,. \end{aligned} \tag{8}$$

The proof is a fairly direct consequence of the realization given in Equation (5).

Definition 3. *We call the Harmonic-Based Exponential Function (HBEF) the series:*

$$_he(x) := e^{\hat{h}\,x}\varphi_{h_0} = 1 + \sum_{n=1}^{\infty} \frac{h_n}{n!}\,x^n. \tag{9}$$

This function, as already discussed in [1], has quite remarkable properties.

The relevant derivatives can accordingly be expressed as (see the concluding part of the paper for further comments):

$$\begin{aligned} &_he(x,m) := \left(\frac{d}{dx}\right)^m {}_he(x) = \hat{h}^m e^{\hat{h}\,x}\varphi_{h_0} = h_m + \sum_{n=1}^{\infty} \frac{h_{n+m}}{n!}\,x^n, \quad \forall x \in \mathbb{R}, \forall m \in \mathbb{N} \\ &_he(x,k+m) = \left(\frac{d}{dx}\right)^m {}_he(x,k), \quad \forall k \in \mathbb{N}, \end{aligned} \tag{10}$$

and according to Equation (9), we also find that:

$$\int_0^{\infty} {}_he(-\alpha\,x)\,e^{-x}dx = \int_0^{\infty} e^{-(\alpha\,\hat{h}+1)\,x}dx = \frac{1}{\alpha\,\hat{h}+1}, \quad |\,\alpha\,| < 1. \tag{11}$$

Corollary 1. *By expanding the umbral function on the r.h.s. of Equation* (11)*, we obtain:*

$$\frac{1}{\alpha\,\hat{h}+1} = 1 + \sum_{n=1}^{\infty}(-1)^n\alpha^n h_n, \quad |\,\alpha\,| < 1. \tag{12}$$

Proof. By using the Taylor expansion and Equation (7), for $|\,\alpha\,| < 1$, we have:

$$\frac{1}{\alpha\,\hat{h}+1} = \sum_{n=0}^{\infty}(-\alpha\hat{h})^n = 1 + \sum_{n=1}^{\infty}(-1)^n\alpha^n\hat{h}^n = 1 + \sum_{n=1}^{\infty}(-1)^n\alpha^n h_n,$$

□

This is an expected conclusion, achievable by direct integration, underscored here to stress the consistency of the procedure.

A further interesting example comes from the following "Gaussian" integral.

$$\int_{-\infty}^{\infty} {}_he(-\alpha\,x)\,e^{-x^2}dx = \int_{-\infty}^{\infty} e^{-(\alpha\,\hat{h}\,x+x^2)}\,dx = \sqrt{\pi}e^{\frac{\alpha^2\,\hat{h}^2}{4}} \;\forall \alpha \in \mathbb{R}. \tag{13}$$

The last term in Equation (13) has been obtained by treating $\hat{h}$ as an ordinary algebraic quantity and then by applying the standard rules of the Gaussian integration.

We notice that, using Equation (9), we obtain:

$$_{h^2}e\left(\frac{\alpha^2}{4}\right) := e^{\frac{\hat{h}^2\alpha^2}{4}}\varphi_{h_0} = 1 + \sum_{r=1}^{\infty}\frac{h_{2r}}{r!}\left(\frac{\alpha}{2}\right)^{2r}. \tag{14}$$

Let us now consider the following slightly more elaborate example, involving the integration of two "Gaussians", namely the ordinary case and its analogous HBEF.

Example 1.

$$\int_{-\infty}^{\infty} {}_{h}e(-\alpha x^2)\, e^{-x^2} dx = \int_{-\infty}^{\infty} e^{-(\hat{h}\alpha+1)x^2} dx\, \varphi_{h_0} = \sqrt{\frac{\pi}{1+\alpha\hat{h}}}\varphi_{h_0}, \quad |\alpha| < 1. \tag{15}$$

This last result, obtained after applying elementary rules, can be worded as follows: the integral in Equation (15) depends on the operator function on its r.h.s., for which we should provide a computational meaning. The use of the Newton binomial yields:

$$\begin{aligned}\sqrt{\frac{\pi}{1+\alpha\hat{h}}}\varphi_{h_0} &= \sqrt{\pi}\sum_{r=0}^{\infty}\binom{-\frac{1}{2}}{r}\left(\alpha\hat{h}\right)^r \varphi_{h_0} = \sqrt{\pi}\left(1+\sqrt{\pi}\sum_{r=1}^{\infty}\frac{\alpha^r h_r}{\Gamma\left(\frac{1}{2}-r\right) r!}\right)\\ &= \sqrt{\pi}\left(1+\sum_{r=1}^{\infty}\binom{2r}{r}\frac{(-\alpha)^r h_r}{2^{2r}}\right),\\ |\alpha| &< 1\,.\end{aligned} \tag{16}$$

The correctness of this conclusion has been confirmed by the numerical check, as well.

It is evident that the examples we have provided show that the use of concepts borrowed from umbral theory offers a fairly powerful tool to deal with the "harmonic-based" functions.

3. Harmonic-Based Functions and Differential Equations

In the following, we will further push the formalism to stress the associated flexibility.

We note indeed that the function:

$${}_{\sqrt{h}}e(x) := e^{\hat{h}^{\frac{1}{2}}x}\varphi_{h_0} = 1+\sum_{n=1}^{\infty}\frac{\left(\sqrt{\hat{h}}\,x\right)^n}{n!}\varphi_{h_0} = 1+\sum_{n=1}^{\infty}\frac{h_{n/2}}{n!}x^n, \forall x\in\mathbb{R}, \tag{17}$$

defines, $\forall\alpha\in\mathbb{R}$, an HBEF through the following Gauss transform:

$$\int_{-\infty}^{+\infty} {}_{\sqrt{h}}e(\alpha x)\, e^{-x^2} dx = \int_{-\infty}^{+\infty} e^{\hat{h}^{\frac{1}{2}}\alpha x - x^2} dx\, \varphi_{h_0} = \sqrt{\pi} e^{\hat{h}\left(\frac{\alpha}{2}\right)^2}\varphi_{h_0} = \sqrt{\pi}\, {}_{h}e\left(\left(\frac{\alpha}{2}\right)^2\right). \tag{18}$$

On the other side, Equation (17) can be expressed in terms of the HBEF, ${}_{h}e(x)$, using appropriate integral transform methods [12].

Definition 4. *If:*

$$g_{\frac{1}{2}}(\eta) = \frac{1}{2\sqrt{\pi\eta^3}}e^{-\frac{1}{4\eta}}, \quad \forall\eta\in\mathbb{R}^+, \tag{19}$$

is the Levy distribution of order $\frac{1}{2}$, *then [12]:*

$$e^{-p^{\frac{1}{2}}x} = \int_0^{\infty} e^{-p\eta x^2} g_{\frac{1}{2}}(\eta)\, d\eta\,, \quad \forall p\in\mathbb{R}^+, \tag{20}$$

is the associated Levy integral transform.

The use of Equations (17) and (19) allows us to write the following identity.

Corollary 2.

$$_{\sqrt{h}}e(-x) = \int_0^\infty {}_he(-\eta\, x^2)\, g_{\frac{1}{2}}(\eta)\, d\eta. \tag{21}$$

Proof.

$$_{\sqrt{h}}e(-x) = e^{-\hat{h}^{\frac{1}{2}} x}\varphi_{h_0} = \int_0^\infty e^{-\hat{h}\eta x^2} g_{\frac{1}{2}}(\eta)\, d\eta\ \varphi_{h_0} = \int_0^\infty {}_he(-\eta\, x^2)\, g_{\frac{1}{2}}(\eta)\, d\eta\ .$$

□

The possibility of defining $_{\sqrt[k]{h}}e(x)$ will be discussed elsewhere.

Theorem 2. *The function $_he(x)$ satisfies the first order non-homogeneous differential equation:*

$$\begin{cases} {}_he'(x) = \dfrac{d}{dx}{}_he(x) = {}_he(x) + \dfrac{e^x - x - 1}{x}, & \forall x \in \mathbb{R}_0, \\ {}_he(0) = 1. \end{cases} \tag{22}$$

Proof. Equation (10), for $m = 1$, yields:

$$_he'(x) = {}_h\, e(x, 1) = 1 + \sum_{n=1}^{\infty} \frac{h_{n+1}}{n!}\, x^n\ . \tag{23}$$

Since $h_{n+1} = h_n + \frac{1}{n+1}$, we find:

$$1 + \sum_{n=1}^{\infty} \frac{h_{n+1}}{n!}\, x^n = {}_he(x) + \frac{1}{x}\,(e^x - x - 1) \tag{24}$$

and hence, Equation (22) follows. □

Corollary 3. *The solution of Equation (22) yields for the HBEF the explicit expression in terms of ordinary special functions $\forall x \in \mathbb{R}^+$:*

$$\begin{aligned} &{}_he(x) = 1 + e^x\,(\ln(x) + E_1(x) + \gamma)\,, \\ &E_1(x) = \int_x^\infty \frac{e^{-t}}{t} dt, \\ &(\ln(x) + E_1(x) + \gamma) = -\sum_{n=1}^{\infty} \frac{(-x)^n}{n\, n!}, \end{aligned} \tag{25}$$

where γ is the Euler–Mascheroni–constant .

The previous expression is the generating function of harmonic numbers originally derived by Gosper (see [2,13]).

By iterating the previous procedure, we find the following general recurrence.

Corollary 4.

$$_he(x, m) = {}_he(x) + \sum_{r=0}^{m-1} \left(\frac{d}{dx}\right)^r \frac{e^x - 1 - x}{x}\ . \tag{26}$$

Definition 5. *The binomial expansion:*

$$h_n(x) := (x + \hat{h})^n \varphi_{h_0} = x^n + \sum_{s=1}^{n} \binom{n}{s}\, x^{n-s}\, h_s, \quad \forall x \in \mathbb{R}, \forall n \in \mathbb{N}_0, \tag{27}$$

specifies the harmonic polynomials.

They are easily shown to be linked to the HBEF by means of the generating function as follows.

Corollary 5.

$$\sum_{n=0}^{\infty} \frac{t^n}{n!} h_n(x) = e^{x\,t}{}_h e(t), \quad \forall x, t \in \mathbb{R}. \tag{28}$$

Proof. It is readily checked that:

$$\sum_{n=0}^{\infty} \frac{t^n}{n!} h_n(x) = \sum_{n=0}^{\infty} \frac{t^n}{n!} (x+\hat{h})^n \varphi_{h_0} = e^{t(x+\hat{h})} \varphi_{h_0} = e^{x\,t}{}_h e(t)\,.$$

□

According to Equation (28), $h_n(x)$ are recognized as Appél polynomials and satisfy the following recurrences.

Properties 2. *The properties below hold:*

$$(i)\ \frac{d}{dx} h_n(x) = n\, h_{n-1}(x), \quad \forall x \in \mathbb{R}, \tag{29}$$

$$(ii)\ h_{n+1}(x) = (x+1)\, h_n(x) + f_n(x), \, ,$$
$$f_n(x) := \sum_{s=1}^{n} \frac{n!}{(n-s)!} \frac{x^{n-s}}{(s+1)!} = \int_0^1 (x+y)^n dy - x^n, \quad \forall x \in \mathbb{R}. \tag{30}$$

Proof. The recurrence given in Equation (29) follows from the definition of the derivative itself since we treat h as an ordinary algebraic quantity. The proof of the identity (30) is slightly more elaborate; we note indeed that:

$$\begin{aligned} h_{n+1}(x) &= (x+\hat{h})(x+\hat{h})^n \varphi_{h_0} = (x+\hat{h}) \left(x^n + \sum_{s=1}^{n} \binom{n}{s} x^{n-s}\, \hat{h}^s \right) \varphi_{h_0} \\ &= x\, h_n(x) + 1 \cdot x^n + \sum_{s=1}^{n} \binom{n}{s} x^{n-s}\, \hat{h}^{s+1} \varphi_{h_0} \\ &= x\, h_n(x) + \left(x^n + \sum_{s=1}^{n} \binom{n}{s} x^{n-s}\, \hat{h}^s \right) \varphi_{h_0} + \sum_{s=1}^{n} \frac{n!\, x^{n-s}}{(n-s)!(s+1)!} \\ &= (x+1)\, h_n(x) + \sum_{s=1}^{n} \frac{n!\, x^{n-s}}{(n-s)!(s+1)!} \end{aligned}$$

and:

$$\begin{aligned} \sum_{s=1}^{n} \frac{n!}{(n-s)!} \frac{x^{n-s}}{(s+1)!} &= \sum_{s=1}^{n} \frac{n!}{s!(n-s)!} \frac{x^{n-s}}{s+1} y^{s+1} \bigg|_{y=1} \\ &= \sum_{s=1}^{n} \binom{n}{s} x^{n-s} \int_0^1 y^s dy = \int_0^1 \left(\sum_{s=0}^{n} \binom{n}{s} x^{n-s} y^s - x^n \right) dy \\ &= \int_0^1 (x+y)^n dy - x^n\,. \end{aligned}$$

□

Corollary 6. *The identity:*

$$h_n(-1) = (-1)^n \left(1 - \frac{1}{n}\right), \quad \forall n \in \mathbb{N}, \tag{31}$$

follows from the Equation (30) *after setting* $x = -1$.

The further relationship:

$$h_n = 1 + \sum_{s=1}^{n} \binom{n}{s} h_s(-1), \quad \forall n \in \mathbb{N}_0, \tag{32}$$

is a consequence of the fact that $\hat{h}^n = ((\hat{h} - 1) + 1)^n$.

The harmonic Hermite polynomials (touched on in [1,3,14]) can also be written as follows.

Definition 6.

$$\begin{gathered}\sum_{n=0}^{\infty} \frac{t^n}{n!}\, {}_hH_n(x) = e^{x\,t}\,{}_he(t^2), \quad \forall x, t \in \mathbb{R}, \\ {}_hH_n(x) := H_n(x, \hat{h})\varphi_{h_0} = e^{\hat{h}\partial_x^2} x^n \varphi_{h_0} = x^n + n! \sum_{r=1}^{\lfloor \frac{n}{2} \rfloor} \frac{x^{n-2\,r} h_r}{(n-2\,r)!\,r!}.\end{gathered} \tag{33}$$

Properties 3. *The recurrences identity of the umbral Hermite polynomials:*

$$\begin{aligned}&(i)\ \frac{d}{dx}\,{}_hH_n(x) = n\,{}_hH_{n-1}(x)\,, \quad \forall x \in \mathbb{R}, \\ &(ii)\ {}_hH_{n+1}(x) = \left(x + 2\,\hat{h}\,\frac{d}{dx}\right) {}_hH_n(x)\varphi_{h_0} = \left(x + 2\frac{d}{dx}\right) {}_hH_n(x) + 2\,\alpha_n'(x)\,, \\ &\qquad \alpha_n(x) = n! \sum_{s=1}^{\lfloor \frac{n}{2} \rfloor} \frac{x^{n-2\,s}}{(s+1)!\,(n-2s)!}\,, \\ &\qquad \alpha_n'(x) = \frac{d}{dx}\alpha_n(x) = n\,\alpha_{n-1}(x)\,,\end{aligned} \tag{34}$$

are a by-product of the previous identities and a consequence of the monomiality principle discussed in [15].

Corollary 7. *The umbral Hermite satisfies the second order non-homogeneous ODE:*

$$\left(x\frac{d}{dx} + 2\left(\frac{d}{dx}\right)^2\right) {}_hH_n(x) = n\,{}_hH_n(x) - 2\,\alpha_n''(x). \tag{35}$$

4. Truncated Exponential Numbers and Final Comments

Before closing the paper, we want to stress the possibility of extending the present procedure to the truncated exponential numbers, namely:

$$e_n := \sum_{r=0}^{n} \frac{1}{r!}\,, \quad \forall n \in \mathbb{N}. \tag{36}$$

The relevant integral representation is written [16]:

$$e_\alpha := \frac{1}{\Gamma(\alpha+1)} \int_0^\infty e^{-s}(1+s)^\alpha ds\,, \tag{37}$$

which holds for $\alpha \in \mathbb{R}$, as well. For example, we find:

Example 2.

$$e_{-\frac{1}{2}} = \frac{e}{\sqrt{\pi}}\Gamma\left(\frac{1}{2},1\right) \tag{38}$$

with $\Gamma\left(\frac{1}{2},1\right)$ *being the lower incomplete Gamma function.*

According to the previous discussion and to Equation (38), setting $\hat{e}^{\alpha} \leftrightarrow e_{\alpha}$, we also find that:

$$\begin{aligned} \int_{-\infty}^{+\infty} e^{-\hat{e}\,x^2}dx &= \sqrt{\pi}e_{-\frac{1}{2}}, \\ e^{-\hat{e}\,x^2} &= \sum_{r=0}^{\infty}(-1)^r \frac{e_r}{r!}x^{2r}\,. \end{aligned} \tag{39}$$

This last identity is a further proof that the implications offered by the topics treated in this paper are fairly interesting and deserve further and more detailed investigation, which will be more accurately treated elsewhere.

Author Contributions: G.D., B.G., S.L. and M.R.M. contributed to the realization of the article, as well as provided topics and obtained results.

References

1. Dattoli, G.; Srivastava, H.M. A Note on Harmonic Numbers, Umbral Calculus and Generating Functions. *Appl. Math. Lett.* **2008**, *21*, 686–693. [CrossRef]
2. Weisstein, E.W. *CRC Concise Encyclopedia of Mathematics*; Chapman and Hall/CRC: Boca Raton, FL, USA, 2003; p. 3115, ISBN 1-58488-347-2.
3. Coffey, M.W. Expressions for Harmonic Number Generating Functions. In *Contemporary Mathematics, 517, Gems in Experimental Mathematics*; Amdeberhan, T., Medina, L.A, Moll, V.H., Eds.; AMS Special Session, Experimental Mathematics: Washington, DC, USA, 2009.
4. Cvijović, D. The Dattoli-Srivastava Conjectures Concerning Generating Functions Involving the Harmonic Numbers. *Appl. Math. Comput.* **2010**, *215*, 4040–4043. [CrossRef]
5. Mezo, I. Exponential Generating Function of Hyper-Harmonic Numbers Indexed by Arithmetic Progressions. *Cent. Eur. J. Math.* **2013**, *11*, 931–939.
6. Conway, J.H.; Guy, R.K. *The Book of Numbers*; Springer: New York, NY, USA, 1996.
7. Rochowicz, J.A., Jr. Harmonic Numbers: Insights, Approximations and Applications. *Spreadsheets Educ. eJSiE* **2015**, *8*, 4.
8. Lagarias, J.C. Euler's Constant: Euler's Work and Modern Developments. *Bulletin (New Series) of the American Mathematical Society*; S 0273-0979(2013)01423-X; pp. 527–628. Available online: http://citeseerx.ist.psu.edu/viewdoc/download?doi=10.1.1.363.9527&rep=rep1&type=pdf (accessed on 19 July 2013).
9. Cartier, P. Mathemagics (A Tribute to L. Euler and R. Feynman). In *Noise, Oscillators and Algebraic Randomness. Lecture Notes in Physics*; Planat, M. Ed.; Springer: Berlin, Heidelberg, 2000; Volume 550.
10. Licciardi, S. Umbral Calculus, A Different Mathematical Language. Ph.D. Thesis, Department of Mathematics and Computer Sciences, XXIX Cycle, University of Catania, Catania, Italy, 2018.
11. Roman, S.M.; Rota, G.-C. The umbral calculus. *Adv. Math* **1978**, *27*, 95–188. [CrossRef]
12. Doetsch, G. *Handbuch der Laplace Transformation*; Birkhnauser: Basel, Switzerland, 1950.
13. Mező I.; Dil, A. Euler-Seidel method for certain combinatorial numbers and a new characterization of Fibonacci sequence, *Cent. Eur. J. Math.* **2009**, *7* , 310–321. [CrossRef]
14. Zhukovsky, K.; Dattoli, G. Umbral Methods, Combinatorial Identities And Harmonic Numbers. *Appl. Math.* **2011**, *1*, 46. [CrossRef]

15. Dattoli, G. Generalized Polynomials, Operational Identities and their Applications. *J. Comput. Appl. Math* **2000**, *118*, 111–123. [CrossRef]
16. Dattoli, G.; Ricci, P.E.; Marinelli, L. Generalized Truncated Exponential Polynomials and Applications. In *An International Journal of Mathematics*; Rendiconti dell'Istituto di Matematica dell'Universitá di Trieste: Trieste, Italy, 2002; Volume 34, pp. 9–18.

5

Functional Equation

Yang-Hi Lee [1] **and Gwang Hui Kim** [2,*]

1 Department of Mathematics Education, Gongju National University of Education, Gongju 32553, Korea; yanghi2@hanmail.net

2 Department of Mathematics, Kangnam University, Yongin 16979, Korea

* Correspondence: ghkim@kangnam.ac.kr

Abstract: We will prove the generalized Hyers–Ulam stability and the hyperstability of the additive functional equation $f(x_1+y_1, x_2+y_2, \ldots, x_n+y_n) = f(x_1, x_2, \ldots, x_n) + f(y_1, y_2, \ldots, y_n)$. By restricting the domain of a mapping f that satisfies the inequality condition used in the assumption part of the stability theorem, we partially generalize the results of the stability theorems of the additive function equations.

Keywords: additive (Cauchy) equation; additive mapping; Hyers–Ulam stability; generalized Hyers–Ulam stability; hyperstability

---

1. Introduction

In 1940, Ulam [1] gave the question concerning the stability of homomorphisms in a conference of the mathematics club of the University of Wisconsin as follows:

Let $(G, \cdot)$ be a group, and let $(G', \cdot, d)$ be a metric group with the metric d. Given $\delta > 0$, does there exist $\epsilon > 0$ such that if a mapping $h : G \to G'$ satisfies the inequality

$$d(h(xy), h(x)h(y)) \leq \delta$$

for all $x, y \in G$, then there is a homomorphism $H : G \to H$ with

$$d(h(x), H(x)) \leq \epsilon$$

for all $x \in G$?

Next year, the Ulam's conjecture was partially solved by Hyers [2] for the additive functional equation.

Theorem 1. *[2], Let X and Y be Banach spaces. Suppose that the mapping $f : X \to Y$ satisfies the inequality*

$$\|f(x+y) - f(x) - f(y)\| \leq \varepsilon, \quad \forall x, y \in X, \quad \varepsilon : constant.$$

Then, there exists a unique additive mapping

$$A(x+y) = A(x) + A(y),$$

such that $||f(x) - A(x)|| \leq \varepsilon$, where the limit $A(x) = \lim_{n\to\infty} 2^{-n} f(2^n x)$.

Thereafter, this phenomenon has been called the Hyers–Ulam stability.

Theorem 2. *Let X and Y be Banach spaces. Suppose that the mapping $f : X \to Y$ satisfies the inequality*

$$\|f(x+y) - f(x) - f(y)\| \leq \theta(\|x\|^p + \|y\|^p) \tag{1}$$

for all $x, y \in X\backslash\{0\}$, where θ and p are constants with $\theta > 0$ and $p \neq 1$. Then, there exists a unique additive mapping $T : X \to Y$ such that

$$\|f(x) - T(x)\| \leq \frac{\theta}{|1 - 2^{p-1}|}\|x\|^p \tag{2}$$

for all $x \in X\backslash\{0\}$.

Theorem 2 is due to Aoki [3] and Rassias [4] for $0 < p < 1$, Gajda [5] for $p > 1$, Hyers [2] for $p = 0$, and Rassias [6] for $p < 0$.

In 1994, Găvruta [7] generalized these results for additive mapping by replacing $\theta(\|x\|^p + \|y\|^p)$ in (1) by a general function $\varphi(x, y)$, which is called the 'generalized Hyers–Ulam stability' in this paper.

In 2001, the term hyperstability was used for the first time probably by G. Maksa and Z. Páles in [8]. However, in 1949, it seems to have created by D. G. Bourgin [9] that the first hyperstability result concerned the ring homomorphisms.

We say that a functional equation $\mathfrak{D}(f) = 0$ is hyperstable if any function f satisfying the equation $\mathfrak{D}(f) = 0$ approximately is a true solution of $\mathfrak{D}(f) = 0$, which is a phenomenon called hyperstability.

The hyperstability results for the additive (Cauchy) equation were investigated by Brzdęk [10,11].

In this paper, let V and W be vector spaces, X be a real normed space, and Y be a real Banach space. We denote the set of natural numbers by $\mathbb{N}$ and the set of real numbers by $\mathbb{R}$.

For a given mapping $f : V^n \to W$, where V^n denotes $V \times V \times \cdots \times V$, let us consider the additive functional equation

$$f(x_1 + y_1, x_2 + y_2, \ldots, x_n + y_n) = f(x_1, x_2, \ldots, x_n) + f(y_1, y_2, \ldots, y_n), \tag{3}$$

for all $x_i, y_i \in V$ $(i = 1, 2, \ldots, n)$.

Each solution of the additive functional Equation (3) is called an n-variable additive mapping. A typical example for the solutions of Equation (3) is the mapping $f : \mathbb{R}^n \to \mathbb{R}^l$ given by $f(x_1, x_2, \ldots, x_n) = (\sum_{i=1}^n a_{1i}x_i, \sum_{i=1}^n a_{2i}x_i, \ldots, \sum_{i=1}^n a_{li}x_i)$ with real constants a_{ij}.

In this paper, we will prove the generalized Hyers–Ulam stability of the additive functional Equation (3) in the spirit of Găvruta [7], and the hyperstability of the additive functional Equation (3).

2. Main Results

For a given mapping $f : V^n \to W$, we use the following abbreviation:

$$\begin{aligned} Df(x_1, y_1, x_2, y_2, \ldots, x_n, y_n) := \ & f(x_1 + y_1, x_2 + y_2, \ldots, x_n + y_n) \\ & - f(x_1, x_2, \ldots, x_n) - f(y_1, y_2, \ldots, y_n) \end{aligned}$$

for all $x_1, y_1, x_2, y_2, \ldots, x_n, y_n \in V$. We need the following lemma to prove main theorems.

Lemma 1. *If a mapping $f : V^n \to W$ satisfies* (3) *for all $x_1, y_1, x_2, y_2, \ldots, x_n, y_n \in V\backslash\{0\}$, then f satisfies* (3) *for all $x_1, y_1, x_2, y_2, \ldots, x_n, y_n \in V$.*

Proof. Let $x \in V\backslash\{0\}$ be a fixed element, and let $i \in \{1,2,\ldots,n\}$. For given $x_i, y_i \in V$, let $x_i^{(1)}$, $x_i^{(2)}$, $y_i^{(1)}, y_i^{(2)}$ be

$$\begin{aligned} &x_i^{(1)} = x, x_i^{(2)} = -x, y_i^{(1)} = x, y_i^{(2)} = -x && \text{if } x_i = 0 \text{ and } y_i = 0,\\ &x_i^{(1)} = y_i, x_i^{(2)} = -y_i, y_i^{(1)} = \frac{y_i}{2}, y_i^{(2)} = \frac{y_i}{2} && \text{if } x_i = 0 \text{ and } y_i \neq 0,\\ &x_i^{(1)} = \frac{x_i}{2}, x_i^{(2)} = \frac{x_i}{2}, y_i^{(1)} = x_i, y_i^{(2)} = -x_i && \text{if } x_i \neq 0 \text{ and } y_i = 0,\\ &x_i^{(1)} = \frac{x_i}{2}, x_i^{(2)} = \frac{x_i}{2}, y_i^{(1)} = (k+1)y_i, y_i^{(2)} = -ky_i && \text{if } x_i \neq 0 \text{ and } y_i \neq 0, \end{aligned}$$

where k is a fixed integer, such that $\frac{x_i}{2} + (k+1)y_i \neq 0$, $\frac{x_i}{2} - ky_i \neq 0$. Then, $x_i^{(1)}, x_i^{(2)}, y_i^{(1)}, y_i^{(2)}, x_i^{(1)} + y_i^{(1)}, x_i^{(2)} + y_i^{(2)} \in V\backslash\{0\}$ and $x_i^{(1)} + y_i^{(1)} + x_i^{(2)} + y_i^{(2)} = x_i + y_i$ for all $i = 1,2,\ldots,n$.

Hence, the equalities $Df(x_1^{(1)}, y_1^{(1)}, \ldots, x_n^{(1)}, y_n^{(1)}) = 0$, $Df(x_1^{(2)}, y_1^{(2)}, \ldots, x_n^{(2)}, y_n^{(2)}) = 0$, $Df(x_1^{(1)}, x_1^{(2)}, x_2^{(1)}, x_2^{(2)}, \ldots, x_n^{(1)}, x_n^{(2)}) = 0$, and $Df(y_1^{(1)}, y_1^{(2)}, y_2^{(1)}, y_2^{(2)}, \ldots, y_n^{(1)}, y_n^{(2)}) = 0$ hold for all $x_1, y_1, x_2, y_2, \ldots, x_n, y_n \in V$. Since the equality

$$\begin{aligned} &Df(x_1, y_1, x_2, y_2, \ldots, x_n, y_n)\\ &= Df(x_1^{(1)} + y_1^{(1)}, x_1^{(2)} + y_1^{(2)}, x_2^{(1)} + y_2^{(1)}, x_2^{(2)} + y_2^{(2)}, \ldots, x_n^{(1)} + y_n^{(1)}, x_n^{(2)} + y_n^{(2)})\\ &\quad + Df(x_1^{(1)}, y_1^{(1)}, x_2^{(1)}, y_2^{(1)}, \ldots, x_n^{(1)}, y_n^{(1)}) + Df(x_1^{(2)}, y_1^{(2)}, x_2^{(2)}, y_2^{(2)}, \ldots, x_n^{(2)}, y_n^{(2)})\\ &\quad - Df(x_1^{(1)}, x_1^{(2)}, x_2^{(1)}, x_2^{(2)}, \ldots, x_n^{(1)}, x_n^{(2)}) - Df(y_1^{(1)}, y_1^{(2)}, y_2^{(1)}, y_2^{(2)}, \ldots, y_n^{(1)}, y_n^{(2)}) \end{aligned}$$

holds for all $x_1, y_1, x_2, y_2, \ldots, x_n, y_n \in V$, we conclude that f satisfies $Df(x_1, y_1, \ldots, x_n, y_n) = 0$ for all $x_1, y_1, x_2, y_2, \ldots, x_n, y_n \in V$. □

Thereafter, let $i \in \{1,2,3,\ldots,n\}$. For a given element $(x_1, x_2, \ldots, x_n) \neq (0,0,\ldots,0)$, we can choose a fixed element $x' \neq 0$, such that $x' \in \{x_1, x_2, \ldots, x_n\}$. Moreover, let $x_i^{(1)}$, $x_i^{(2)} \in V\backslash\{0\}$ be the elements defined by

$$\begin{aligned} &x_i^{(1)} = x_i, x_i^{(2)} = x_i && \text{if } x_i \neq 0,\\ &x_i^{(1)} = x', x_i^{(2)} = -x' && \text{if } x_i = 0. \end{aligned} \tag{4}$$

By using Lemma 1, we can prove the following set of stability theorems.

Theorem 3. *Suppose that $f : V^n \to Y$ is a mapping for which there exists a function $\varphi : (V\backslash\{0\})^{2n} \to [0,\infty)$, such that*

$$\sum_{m=0}^{\infty} \frac{\varphi(2^m x_1, 2^m y_1, 2^m x_2, 2^m y_2, \ldots, 2^m x_n, 2^m y_n)}{2^m} < \infty \tag{5}$$

and

$$\|Df(x_1, y_1, x_2, y_2, \ldots, x_n, y_n)\| \leq \varphi(x_1, y_1, x_2, y_2, \ldots, x_n, y_n) \tag{6}$$

for all $x_1, y_1, x_2, y_2, \ldots, x_n, y_n \in V\backslash\{0\}$. Then, there exists a unique mapping $F : V^n \to Y$ that satisfies

$$DF(x_1, y_1, x_2, y_2, \ldots, x_n, y_n) = 0 \tag{7}$$

for all $x_1, y_1, x_2, y_2, \ldots, x_n, y_n \in V$ *and*

$$\|f(x_1, x_2, \ldots, x_n) - F(x_1, x_2, \ldots, x_n)\| \leq \sum_{m=0}^{\infty} \frac{\mu(2^m x_1, 2^m x_2, \ldots, 2^m x_n)}{2^{m+1}} \tag{8}$$

for all $(x_1, x_2, \ldots, x_n) \in V^n \setminus \{(0,0,\ldots,0)\}$*, where the function* $\mu : V^n \to \mathbb{R}$ *is defined by*

$$\begin{aligned}
&\mu(x_1, x_2, \ldots, x_n) \\
:=&\varphi\left(x_1^{(1)}, x_1^{(2)}, x_2^{(1)}, x_2^{(2)}, \ldots, x_n^{(1)}, x_n^{(2)}\right) + 2\varphi\left(\frac{x_1^{(1)}}{2}, \frac{x_1^{(2)}}{2}, \frac{x_2^{(1)}}{2}, \frac{x_2^{(2)}}{2}, \ldots, \frac{x_n^{(1)}}{2}, \frac{x_n^{(2)}}{2}\right) \\
&+ \varphi\left(\frac{x_1^{(1)}}{2}, \frac{x_1^{(1)}}{2}, \frac{x_2^{(1)}}{2}, \frac{x_2^{(1)}}{2}, \ldots, \frac{x_n^{(1)}}{2}, \frac{x_n^{(1)}}{2}\right) + \varphi\left(\frac{x_1^{(2)}}{2}, \frac{x_1^{(2)}}{2}, \frac{x_2^{(2)}}{2}, \frac{x_2^{(2)}}{2}, \ldots, \frac{x_n^{(2)}}{2}, \frac{x_n^{(2)}}{2}\right)
\end{aligned}$$

for all $(x_1, x_2, \ldots, x_n) \in V^n \setminus \{(0,0,\ldots,0)\}$.

Proof. From the inequality (6) and the equalities

$$\begin{aligned}
&f(2x_1, 2x_2, \ldots, 2x_n) - 2f(x_1, x_2, \ldots, x_n) \\
&\quad= f(2x_1, 2x_2, \ldots, 2x_n) - f\left(x_1^{(1)}, x_2^{(1)}, \ldots, x_n^{(1)}\right) - f\left(x_1^{(2)}, x_2^{(2)}, \ldots, x_n^{(2)}\right) \\
&\qquad - 2f(x_1, x_2, \ldots, x_n) + 2f\left(\frac{x_1^{(1)}}{2}, \frac{x_2^{(1)}}{2}, \ldots, \frac{x_n^{(1)}}{2}\right) + 2f\left(\frac{x_1^{(2)}}{2}, \frac{x_2^{(2)}}{2}, \ldots, \frac{x_n^{(2)}}{2}\right) \\
&\qquad + f\left(x_1^{(1)}, x_2^{(1)}, \ldots, x_n^{(1)}\right) - 2f\left(\frac{x_1^{(1)}}{2}, \frac{x_2^{(1)}}{2}, \ldots, \frac{x_n^{(1)}}{2}\right) \\
&\qquad + f\left(x_1^{(2)}, x_2^{(2)}, \ldots, x_n^{(2)}\right) - 2f\left(\frac{x_1^{(2)}}{2}, \frac{x_2^{(2)}}{2}, \ldots, \frac{x_n^{(2)}}{2}\right) \\
&\quad= Df\left(x_1^{(1)}, x_1^{(2)}, x_2^{(1)}, x_2^{(2)}, \ldots, x_n^{(1)}, x_n^{(2)}\right) - 2Df\left(\frac{x_1^{(1)}}{2}, \frac{x_1^{(2)}}{2}, \frac{x_2^{(1)}}{2}, \frac{x_2^{(2)}}{2}, \ldots, \frac{x_n^{(1)}}{2}, \frac{x_n^{(2)}}{2}\right) \\
&\qquad + Df\left(\frac{x_1^{(1)}}{2}, \frac{x_1^{(1)}}{2}, \frac{x_2^{(1)}}{2}, \frac{x_2^{(1)}}{2}, \ldots, \frac{x_n^{(1)}}{2}, \frac{x_n^{(1)}}{2}\right) \\
&\qquad + Df\left(\frac{x_1^{(2)}}{2}, \frac{x_1^{(2)}}{2}, \frac{x_2^{(2)}}{2}, \frac{x_2^{(2)}}{2}, \ldots, \frac{x_n^{(2)}}{2}, \frac{x_n^{(2)}}{2}\right)
\end{aligned} \tag{9}$$

for all $(x_1, x_2, \ldots, x_n) \in V^n \setminus \{(0,0,\ldots,0)\}$, we have

$$\begin{aligned}
&\left\| f(x_1, x_2, \ldots, x_n) - \frac{f(2x_1, 2x_2, \ldots, 2x_n)}{2} \right\| \\
&\quad\leq \left\| Df\left(x_1^{(1)}, x_1^{(2)}, x_2^{(1)}, x_2^{(2)}, \ldots, x_n^{(1)}, x_n^{(2)}\right) \right\| + 2\left\| Df\left(\frac{x_1^{(1)}}{2}, \frac{x_1^{(2)}}{2}, \frac{x_2^{(1)}}{2}, \frac{x_2^{(2)}}{2}, \ldots, \frac{x_n^{(1)}}{2}, \frac{x_n^{(2)}}{2}\right) \right\| \\
&\qquad + \left\| Df\left(\frac{x_1^{(1)}}{2}, \frac{x_1^{(1)}}{2}, \frac{x_2^{(1)}}{2}, \frac{x_2^{(1)}}{2}, \ldots, \frac{x_n^{(1)}}{2}, \frac{x_n^{(1)}}{2}\right) \right\| \\
&\qquad + \left\| Df\left(\frac{x_1^{(2)}}{2}, \frac{x_1^{(2)}}{2}, \frac{x_2^{(2)}}{2}, \frac{x_2^{(2)}}{2}, \ldots, \frac{x_n^{(2)}}{2}, \frac{x_n^{(2)}}{2}\right) \right\| \\
&\quad\leq \frac{1}{2}\mu(x_1, x_2, \ldots, x_n)
\end{aligned}$$

for all $(x_1, x_2, \ldots, x_n) \in V^n \backslash \{(0,0,\ldots,0)\}$. From the above inequality, we get the (following-4 palces) inequality

$$\begin{aligned}
&\left\| \frac{f\left(2^m x_1, \ldots, 2^m x_n\right)}{2^m} - \frac{f\left(2^{m+m'} x_1, \ldots, 2^{m+m'} x_n\right)}{2^{m+m'}} \right\| \\
&\quad \leq \sum_{k=m}^{m+m'-1} \left\| \frac{f\left(2^k x_1, \ldots, 2^k x_n\right)}{2^k} - \frac{f\left(2^{k+1} x_1, \ldots, 2^{k+1} x_n\right)}{2^{k+1}} \right\| \\
&\quad \leq \sum_{k=m}^{m+m'-1} \frac{\mu(2^k x_1, 2^k x_2, \ldots, 2^k x_n)}{2^{k+1}} \qquad (10)
\end{aligned}$$

for all $(x_1, x_2, \ldots, x_n) \in V^n \backslash \{(0,0,\ldots,0)\}$ and all positive integers m, m'. Thus, the sequence $\{\frac{f(2^n x_1,\ldots,2^n x_n)}{2^n}\}_{m \in \mathbb{N}}$ is a Cauchy sequence for all $(x_1, x_2, \ldots, x_n) \in V^n \backslash \{(0,0,\ldots,0)\}$. Since Y is a real Banach space and $\lim_{m\to\infty} \frac{f(2^m 0, 2^m 0, \ldots, 2^m 0)}{2^m} = 0$, we can define a mapping $F : V^n \to Y$ by

$$F\left(x_1, x_2, \ldots, x_n\right) = \lim_{m\to\infty} \frac{f\left(2^m x_1, 2^m x_2, \ldots, 2^m x_n\right)}{2^m}$$

for all $x_1, x_2, \ldots, x_n \in V$. By putting $m = 0$ and by letting $m' \to \infty$ in the inequalities (10), we can obtain the inequalities (8) for all $(x_1, x_2, \ldots, x_n) \in V^n \backslash \{(0,0,\ldots,0)\}$.

From the inequality (6), we can obtain

$$\left\| \frac{Df(2^m x_1, 2^m y_1, 2^m x_2, 2^m y_2, \ldots, 2^m x_n, 2^m y_n)}{2^m} \right\| \leq \frac{\varphi\left(2^m x_1, 2^m y_1, 2^m x_2, \ldots, 2^m x_n, 2^m y_n\right)}{2^m}$$

for all $x_1, y_1, x_2, y_2, \ldots, x_n, y_n \in V \backslash \{0\}$. Since the right-hand side in the above equality tends to zero as $m \to \infty$, and the equality

$$DF(x_1, y_1, x_2, y_2, \ldots, x_n, y_n) = \lim_{m\to\infty} \frac{Df(2^m x_1, 2^m y_1, 2^m x_2, 2^m y_2, \ldots, 2^m x_n, 2^m y_n)}{2^m}$$

holds, then F satisfies the equality (7) for all $x_1, y_1, \ldots, x_n, y_n \in V \backslash \{0\}$. By Lemma 1, F satisfies the equality (3) for all $x_1, y_1, x_2, y_2, \ldots, x_n, y_n \in V$. If $G : V^n \to Y$ is another n-variable additive mapping that satisfies (8), then we obtain $G(0,0,\ldots,0) = 0 = F(0,0\ldots,0)$ and

$$\begin{aligned}
&\left\| G\left(x_1, x_2, \ldots, x_n\right) - F\left(x_1, x_2, \ldots, x_n\right) \right\| \\
&\quad \leq \left\| \frac{G\left(2^k x_1, 2^k x_2, \ldots, 2^k x_n\right)}{2^k} - \frac{f\left(2^k x_1, 2^k x_2, \ldots, 2^k x_n\right)}{2^k} \right\| \\
&\qquad + \left\| \frac{f\left(2^k x_1, 2^k x_2, \ldots, 2^k x_n\right)}{2^k} - \frac{F\left(2^k x_1, 2^k x_2, \ldots, 2^k x_n\right)}{2^k} \right\| \\
&\quad \leq \sum_{m=k}^{\infty} \frac{\mu(2^m x_1, 2^m x_2, \ldots, 2^m x_n)}{2^m}
\end{aligned}$$

for all $(x_1, x_2, \ldots, x_n) \in V^n \backslash \{(0,0,\ldots,0)\}$ and all $k \in \mathbb{N}$. Since $\sum_{m=k}^{\infty} \frac{\mu(2^m x_1, 2^m x_2, \ldots, 2^m x_n)}{2^m} \to 0$ as $k \to \infty$, we have $G(x_1, x_2, \ldots, x_n) = F(x_1, x_2, \ldots, x_n)$ for all $x_1, x_2, \ldots, x_n \in V$. Hence, the mapping F is the unique n-variable additive mapping, as desired. □

The condition $x_1, y_1, x_2, y_2, \ldots, x_n, y_n \in V \backslash \{0\}$ used in the inequality (6) differs from the condition $(x_1, x_2, \ldots, x_n) \neq (0,0,\ldots,0)$ and $(y_1, y_2, \ldots, y_n) \neq (0,0,\ldots,0)$ handled by the other authors. If the function f satisfies the inequality (3.2) for all $(x_1, x_2, \ldots, x_n) \neq (0,0,\ldots,0)$ and $(y_1, y_2, \ldots, y_n) \neq$

$(0,0,\ldots,0)$, then the function f satisfies the inequality (3.2) for all $x_1, y_1, x_2, y_2, \ldots, x_n, y_n \in V\backslash\{0\}$. Therefore, the condition $x_1, y_1, x_2, y_2, \ldots, x_n, y_n \in V\backslash\{0\}$ used in the inequality (3.2) in this paper is a generalization of the conditions used in the inequality (3.2) in the well-known pre-results ([10,11]). This condition will apply until Corollary 1.

Theorem 4. *Suppose that $f : V^n \to Y$ is a mapping for which there exists a function $\varphi : (V\backslash\{0\})^{2n} \to [0,\infty)$ that satisfies*

$$\sum_{i=0}^{\infty} 2^i \varphi\left(\frac{x_1}{2^i}, \frac{y_1}{2^i}, \frac{x_2}{2^i}, \frac{y_2}{2^i}, \ldots, \frac{x_n}{2^i}, \frac{y_n}{2^i}\right) < \infty, \tag{11}$$

and (6) for all $x_1, y_1, x_2, y_2, \ldots, x_n, y_n \in V\backslash\{0\}$. Then, there exists a unique mapping $F : V^n \to Y$ that satisfies (7) for all $x_1, y_1, x_2, y_2, \ldots, x_n, y_n \in V$ and

$$\|f(x_1, x_2, \ldots, x_n) - F(x_1, x_2, \ldots, x_n)\| \leq \sum_{m=0}^{\infty} 2^m \mu\left(\frac{x_1}{2^{m+1}}, \frac{x_2}{2^{m+1}}, \ldots, \frac{x_n}{2^{m+1}}\right) \tag{12}$$

for all $(x_1, x_2, \ldots, x_n) \in V^n\backslash\{(0,0,\ldots,0)\}$, where the function $\mu : V^n \to \mathbb{R}$ is defined as Theorem 3.

Proof. By choosing a fixed element $x \in V\backslash\{0\}$, we can obtain

$$\begin{aligned}
\|f(0,0,\ldots,0)\| =& \left\| 2Df\left(\frac{x}{2^m}, \frac{-x}{2^m}, \ldots, \frac{x}{2^m}, \frac{-x}{2^m}\right) - Df\left(\frac{x}{2^{m-1}}, \frac{-x}{2^{m-1}}, \ldots, \frac{x}{2^{m-1}}, \frac{-x}{2^{m-1}}\right)\right. \\
& \left. - Df\left(\frac{x}{2^m}, \frac{x}{2^m}, \ldots, \frac{x}{2^m}, \frac{x}{2^m}\right) - Df\left(\frac{-x}{2^m}, \frac{-x}{2^m}, \ldots, \frac{-x}{2^m}, \frac{-x}{2^m}\right)\right\| \\
\leq& 2\varphi\left(\frac{x}{2^m}, \frac{-x}{2^m}, \ldots, \frac{x}{2^m}, \frac{-x}{2^m}\right) + \varphi\left(\frac{x}{2^{m-1}}, \frac{-x}{2^{m-1}}, \ldots, \frac{x}{2^{m-1}}, \frac{-x}{2^{m-1}}\right) \\
& + \varphi\left(\frac{x}{2^m}, \frac{x}{2^m}, \ldots, \frac{x}{2^m}, \frac{x}{2^m}\right) + \varphi\left(\frac{-x}{2^m}, \frac{-x}{2^m}, \ldots, \frac{-x}{2^m}, \frac{-x}{2^m}\right) \\
& \to 0 \quad \text{as} \quad m \to \infty,
\end{aligned}$$

so $f(0,0,\ldots,0) = 0$. Since the equality (9) holds for all $(x_1, x_2, \ldots, x_n) \in V\backslash\{(0,0,\ldots,0)\}$, the inequality (6) implies the inequality

$$\left\|f(x_1, x_2, \ldots, x_n) - 2f\left(\frac{x_1}{2}, \frac{x_2}{2}, \ldots, \frac{x_n}{2}\right)\right\| \leq \mu\left(\frac{x_1}{2}, \frac{x_2}{2}, \ldots, \frac{x_n}{2}\right)$$

for all $(x_1, x_2, \ldots, x_n) \in V^n\backslash\{(0,0,\ldots,0)\}$. From the above inequality, we can also obtain the inequality

$$\begin{aligned}
&\left\|2^m f\left(\frac{x_1}{2^m}, \frac{x_2}{2^m}, \ldots, \frac{x_n}{2^m}\right) - 2^{m+m'} f\left(\frac{x_1}{2^{m+m'}}, \frac{x_2}{2^{m+m'}}, \ldots, \frac{x_n}{2^{m+m'}}\right)\right\| \\
&\qquad \leq \sum_{k=m}^{m+m'-1} 2^k \mu\left(\frac{x_1}{2^{k+1}}, \frac{x_2}{2^{k+1}}, \ldots, \frac{x_n}{2^{k+1}}\right)
\end{aligned} \tag{13}$$

for all $(x_1, x_2, \ldots, x_n) \in V^n\backslash\{(0,0,\ldots,0)\}$ and all positive integers m, m'. Thus, the sequences $\{2^m f(\frac{x_1}{2^m}, \ldots, \frac{x_n}{2^m})\}_{m\in\mathbb{N}}$ is a Cauchy sequence for all $(x_1, \ldots, x_n) \in V^n\backslash\{(0,\ldots,0)\}$. Since $f(0,0,\ldots,0) = 0$ and Y is a real Banach space, we can define a mapping $F : V^n \to Y$ by

$$F(x_1, x_2, \ldots, x_n) = \lim_{m\to\infty} 2^m f\left(\frac{x_1}{2^m}, \frac{x_2}{2^m}, \ldots, \frac{x_n}{2^m}\right)$$

for all $x_1, x_2, \ldots, x_n \in V$. By putting $m = 0$ and by letting $m' \to \infty$ in the inequality (13), we can obtain the inequality (12) for all $(x_1, x_2, \ldots, x_n) \in V^n\backslash\{(0,0,\ldots,0)\}$.

From the inequality (6), we get

$$\left\| 2^m Df\left(\frac{x_1}{2^m}, \frac{y_1}{2^m}, \frac{x_2}{2^m}, \frac{y_2}{2^m}, \dots, \frac{x_n}{2^m}, \frac{y_n}{2^m}\right)\right\| \le 2^m \varphi\left(\frac{x_1}{2^m}, \frac{y_1}{2^m}, \frac{x_2}{2^m}, \frac{y_2}{2^m}, \dots, \frac{x_n}{2^m}, \frac{y_n}{2^m}\right)$$

for all $x_1, y_1, x_2, y_2, \dots, x_n, y_n \in V\backslash\{0\}$. Since the right-hand side in the above equality tends to zero as $m \to \infty$, then F satisfies the equality (7) for all $x_1, y_1, x_2, y_2, \dots, x_n, y_n \in V\backslash\{0\}$. By Lemma 1, F satisfies the equality (3) for all $x_1, y_1, x_2, y_2, \dots, x_n, y_n \in V$. If $G : V^n \to Y$ is another n-variable additive mapping satisfying (12), then we obtain $G(0,0,\dots,0) = 0 = F(0,0,\dots,0)$ and

$$\begin{aligned}
&\| G(x_1, x_2, \dots, x_n) - F(x_1, x_2, \dots, x_n) \| \\
&\le \left\| 2^k G\left(\frac{x_1}{2^k}, \frac{x_2}{2^k}, \dots, \frac{x_n}{2^k}\right) - 2^k f\left(\frac{x_1}{2^k}, \frac{x_2}{2^k}, \dots, \frac{x_n}{2^k}\right)\right\| \\
&\quad + \left\| 2^k f\left(\frac{x_1}{2^k}, \frac{x_2}{2^k}, \dots, \frac{x_n}{2^k}\right) - 2^k F\left(\frac{x_1}{2^k}, \frac{x_2}{2^k}, \dots, \frac{x_n}{2^k}\right)\right\| \\
&\le \sum_{m=k}^{\infty} 2^m \mu\left(\frac{x_1}{2^{m+1}}, \frac{x_2}{2^{m+1}}, \dots, \frac{x_n}{2^{m+1}}\right) \\
&\to 0 \quad \text{as} \quad k \to \infty
\end{aligned}$$

for all $(x_1, x_2, \dots, x_n) \in V^n\backslash\{(0,0,\dots,0)\}$. Hence, the mapping F is the unique n-variable additive mapping, as desired. □

The following corollary follows from Theorems 3 and 4.

Corollary 1. *Let $(X, |||\cdot|||)$ be a normed space, $\theta > 0$, and let p be a real number with $p \neq 1$. Suppose that $f : X^n \to Y$ is a mapping that satisfies*

$$\|Df(x_1, y_1, x_2, y_2, \dots, x_n, y_n)\| \le \theta(|||x_1|||^p + |||y_1|||^p + |||x_2|||^p + \dots + |||x_n|||^p + |||y_n|||^p) \tag{14}$$

for all $x_1, y_1, x_2, y_2, \dots, x_n, y_n \in X\backslash\{0\}$. Then, there exists a unique n-variable additive mapping $F : X^n \to Y$, such that

$$\|f(x_1, x_2, \dots, x_n) - F(x_1, x_2, \dots, x_n)\| \le \frac{4(2^p + 4)n\theta}{2^p|2 - 2^p|} \max_{x_i \neq 0}\{|||x_i|||^p \ : \ 1 \le i \le n\} \tag{15}$$

for all $(x_1, x_2, \dots, x_n) \in X^n\backslash\{(0,0,\dots,0)\}$.

Proof. Put $\varphi(x_1, y_1, x_2, y_2, \dots, x_n, y_n) := \theta(|||x_1|||^p + |||y_1|||^p + |||x_2|||^p + |||y_2|||^p + \dots + |||x_n|||^p + |||y_n|||^p)$ for all $x_1, y_1, x_2, y_2, \dots, x_n, y_n \in X\backslash\{0\}$, then $|||x_i^{(1)}|||, |||x_i^{(2)}||| \le \max_{x_i \neq 0}\{|||x_i|||^p \ : \ 1 \le i \le n\}$ for all i from (4). Hence, due to μ of Theorems 3 and 4, we obtain that

$$\begin{aligned}
&\mu(x_1, x_2, \dots, x_n) \\
&= \varphi\left(x_1^{(1)}, x_1^{(2)}, x_2^{(1)}, x_2^{(2)}, \dots, x_n^{(1)}, x_n^{(2)}\right) + 2\varphi\left(\frac{x_1^{(1)}}{2}, \frac{x_1^{(2)}}{2}, \frac{x_2^{(1)}}{2}, \frac{x_2^{(2)}}{2}, \dots, \frac{x_n^{(1)}}{2}, \frac{x_n^{(2)}}{2}\right) \\
&\quad + \varphi\left(\frac{x_1^{(1)}}{2}, \frac{x_1^{(1)}}{2}, \frac{x_2^{(1)}}{2}, \frac{x_2^{(1)}}{2}, \dots, \frac{x_n^{(1)}}{2}, \frac{x_n^{(1)}}{2}\right) + \varphi\left(\frac{x_1^{(2)}}{2}, \frac{x_1^{(2)}}{2}, \frac{x_2^{(2)}}{2}, \frac{x_2^{(2)}}{2}, \dots, \frac{x_n^{(2)}}{2}, \frac{x_n^{(2)}}{2}\right) \\
&\le (2n + \frac{8n}{2^p}) \max_{x_i \neq 0}\{|||x_i|||^p \ : \ 1 \le i \le n\}
\end{aligned}$$

for all $(x_1, x_2, \dots, x_n) \in X^n\backslash\{(0,0,\dots,0)\}$. Therefore, the inequality (15) can be obtained easily from (8) and (12) in Theorems 3 and 4. □

The following theorem for the hyperstability of n-variable additive functional equation follows from Corollary 1.

Theorem 5. *Let $(X, |||\cdot|||)$ be a normed space and p be a real number with $p < 0$. Suppose that $f : X^n \to Y$ is a mapping that satisfies (14) for all $x_1, y_1, x_2, y_2, \dots, x_n, y_n \in X\backslash\{0\}$. Then, f is an n-variable additive mapping itself.*

Proof. By Corollary 1, there exists a unique n-variable additive mapping $F : X^n \to Y$, such that (15) for all $x_1, x_2, \dots, x_n \in X^n\backslash\{(0,0,\dots,0)\}$ and $DF(x_1, y_1, x_2, y_2, \dots, x_n, y_n) = 0$ for all $x_1, y_1, x_2, y_2, \dots, x_n, y_n \in X$.

For a given $(x_1, x_2, \dots, x_n) \neq (0,0,\dots,0)$, let $x' \neq 0$ be a nonzero fixed element in $\{x_1, x_2, \dots, x_n\}$, and let

$$\begin{aligned} x_i^{(3)} &= (m+1)x_i, x_i^{(4)} = -mx_i && \text{when } x_i \neq 0, \\ x_i^{(3)} &= mx', x_i^{(4)} = -mx' && \text{when } x_i = 0. \end{aligned}$$

Then, we can easily show that $|||x_i^{(3)}|||, |||x_i^{(4)}||| \leq m^p \max_{x_i \neq 0}\{|||x_i|||^p : 1 \leq i \leq n\}$ for all i from (4). If $(x_1, x_2, \dots, x_n) \in X\backslash\{(0,0,\dots,0)\}$, then the equality $f(x_1, x_2, \dots, x_n) = F(x_1, x_2, \dots, x_n)$ follows from the inequalities

$$\begin{aligned} &\|f(x_1, x_2, \dots, x_n) - F(x_1, x_2, \dots, x_n)\| \\ &\quad = \Big\| Df\left(x_1^{(3)}, x_1^{(4)}, x_2^{(3)}, x_2^{(4)}, \dots, x_n^{(3)}, x_n^{(4)}\right) - DF\left(x_1^{(3)}, x_1^{(4)}, x_2^{(3)}, x_2^{(4)}, \dots, x_n^{(3)}, x_n^{(4)}\right) \\ &\qquad + f(x_1^{(3)}, x_2^{(3)}, \dots, (x_n^{(3)}) + f(x_1^{(4)}, x_2^{(4)}, \dots, x_n^{(4)}) \\ &\qquad - F(x_1^{(3)}, x_2^{(3)}, \dots, (x_n^{(3)}) - F(x_1^{(4)}, x_2^{(4)}, \dots, x_n^{(4)}) \Big\| \\ &\quad \leq m^p \cdot 2n\theta \max_{x_i \neq 0}\{|||x_i|||^p : 1 \leq i \leq n\} + \Big\| f(x_1^{(3)}, x_2^{(3)}, \dots, x_n^{(3)}) - F(x_1^{(3)}, x_2^{(3)}, \dots, x_n^{(3)}) \Big\| \\ &\qquad + \Big\| f(x_1^{(4)}, x_2^{(4)}, \dots, x_n^{(4)}) - F(x_1^{(4)}, x_2^{(4)}, \dots, x_n^{(4)}) \Big\| \\ &\quad \leq m^p \left(1 + \frac{4(2^p+4)}{2^p|2-2^p|}\right) 2n\theta \max_{x_i \neq 0}\{|||x_i|||^p : 1 \leq i \leq n\} \end{aligned}$$

as $m \to \infty$. For $(x_1, x_2, \dots, x_n) = (0,0,\dots,0)$, if we choose a fixed element of $x \in X\backslash\{0\}$, then the equality $f0,0,\dots,0) = F0,0,\dots,0)$ follows from the inequalities

$$\begin{aligned} &\|f(0,0,\dots,0) - F(0,0,\dots,0)\| \\ &\quad = \|Df(mx, -mx, mx, -mx, \dots, mx, -mx) - DF(mx, -mx, mx, \dots, mx, -mx) \\ &\qquad + f(mx, mx, \dots, mx) + f(-mx, -mx, \dots, -mx) \\ &\qquad - F(mx, mx, \dots, mx) - F(-mx, -mx, \dots, -mx)\| \\ &\quad \leq m^p \cdot 2n\theta\|x\|^p + \|f(mx, mx, \dots, mx) - F(mx, mx, \dots, mx)\| \\ &\qquad + \|f(-mx, -mx, \dots, -mx) - F(-mx, -mx, \dots, -mx)\| \\ &\quad \leq m^p \left(1 + \frac{4(2^p+4)}{2^p|2-2^p|}\right) 2n\theta|||x|||^p \end{aligned}$$

as $m \to \infty$. Therefore, f is an n-variable additive mapping itself. □

The following example follows from Theorem 5.

Example 1. *Let* $(\mathbb{R}, |\cdot|)$ *be a normed space with absolute value* $|\cdot|$, $(\mathbb{R}^l, \|\cdot\|)$ *be a Banach space with Euclid norm* $\|\cdot\|$, *and* $p < 0$ *be a real number. Suppose that* $f : \mathbb{R}^n \to \mathbb{R}^l$ *is a continuous mapping such that*

$$\|Df(x_1, y_1, x_2, y_2, \ldots, x_n, y_n)\| \leq \theta(|x_1|^p + |y_1|^p + |x_2|^p + |y_2|^p + \ldots + |x_n|^p + |y_n|^p)$$

for all $x_1, y_1, x_2, y_2, \ldots, x_n, y_n \in \mathbb{R}\backslash\{0\}$. *Then, the mapping* $f : \mathbb{R}^n \to \mathbb{R}^l$ *given by*

$$f(x_1, x_2, \ldots, x_n) = \left(\sum_{i=1}^{n} a_{1i}x_i, \sum_{i=1}^{n} a_{2i}x_i, \ldots, \sum_{i=1}^{n} a_{li}x_i \right), \tag{16}$$

where $a_{1i}, a_{2i}, \ldots, a_{li}$ *are real constants, indicates that*

$$\begin{aligned} f(1, 0, 0, \ldots, 0) &= (a_{11}, a_{21}, \ldots, a_{l1}), \\ f(0, 1, 0, \ldots, 0) &= (a_{12}, a_{22}, \ldots, a_{l2}), \\ \vdots \quad & \quad \vdots \\ f(0, \ldots, 0, 0, 1) &= (a_{1n}, a_{2n}, \ldots, a_{ln}). \end{aligned}$$

Proof. Since $f : \mathbb{R}^n \to \mathbb{R}^l$ is a continuous n-variable additive mapping by Theorem 5, then the function $f : \mathbb{R}^n \to \mathbb{R}^l$ is given by (16). □

In the following theorems, we replace the domain $(V\backslash\{0\})^{2n}$ of φ and Df in Theorems 3 and 4 with V^{2n}. Then, we can improve the result inequality (8).

Theorem 6. *Suppose that* $f : V^n \to Y$ *is a mapping for which there exists a function* $\varphi : V^{2n} \to [0, \infty)$ *satisfying* (5) *and* (6) *for all* $x_1, y_1, x_2, y_2, \ldots, x_n, y_n \in V$. *Then, there exists a unique mapping* $F : V^n \to Y$, *such that* (7) *for all* $x_1, y_1, x_2, y_2, \ldots, x_n, y_n \in V$ *and*

$$\|f(x_1, x_2, \ldots, x_n) - F(x_1, x_2, \ldots, x_n)\| \leq \sum_{m=0}^{\infty} \frac{\varphi(2^m x_1, 2^m x_1, 2^m x_2, \ldots, 2^m x_n, 2^m x_n)}{2^{m+1}} \tag{17}$$

for all $x_1, x_2, \ldots, x_n \in V$.

Proof. The equality

$$f(2x_1, 2x_2, \ldots, 2x_n) - 2f(x_1, x_2, \ldots, x_n) = Df(x_1, x_1, x_2, x_2, \ldots, x_n, x_n) \tag{18}$$

for all $x_1, x_2, \ldots, x_n \in V$ and the inequality (6) imply that the inequality

$$\left\| f(x_1, x_2, \ldots, x_n) - \frac{f(2x_1, 2x_2, \ldots, 2x_n)}{2} \right\| \leq \frac{1}{2}\varphi(x_1, x_1, x_2, x_2, \ldots, x_n, x_n)$$

for all $x_1, x_2, \ldots, x_n \in V$. From the above inequality, we can derive the inequalities

$$\begin{aligned} &\left\| \frac{f(2^m x_1, \ldots, 2^m x_n)}{2^m} - \frac{f\left(2^{m+m'} x_1, \ldots, 2^{m+m'} x_n\right)}{2^{m+m'}} \right\| \\ &\leq \sum_{k=m}^{m+m'-1} \frac{\varphi\left(2^k x_1, 2^k x_1, 2^k x_2, 2^k x_2, \ldots, 2^k x_n, 2^k x_n\right)}{2^{k+1}} \end{aligned} \tag{19}$$

for all $x_1, x_2, \ldots, x_n \in V$ and all positive integers m, m'. The remainder of the proof of this theorem developed after inequality (19) is omitted because it is similar to that of Theorem 3. □

Theorem 7. *Suppose that $f : V^n \to Y$ is a mapping for which there exists a function $\varphi : V^{2n} \to [0, \infty)$ satisfying (11) and (6) for all $x_1, y_1, x_2, y_2, \ldots, x_n, y_n \in V$. Then, there exists a unique mapping $F : V^n \to Y$ that satisfies (7) for all $x_1, y_1, x_2, y_2, \ldots, x_n, y_n \in V$ and*

$$\| f(x_1, \ldots, x_n) - F(x_1, \ldots, x_n) \| \le \sum_{m=0}^{\infty} 2^m \varphi\left(\frac{x_1}{2^{m+1}}, \frac{x_1}{2^{m+1}}, \frac{x_2}{2^{m+1}}, \ldots, \frac{x_n}{2^{m+1}}, \frac{x_n}{2^{m+1}}\right) \tag{20}$$

for all $x_1, x_2, \ldots, x_n \in V$.

Proof. The equality (18) for all $x_1, x_2, \ldots, x_n \in V$ and the inequality (6) imply that the inequality

$$\| f(x_1, x_2, \ldots, x_n) - 2f\left(\frac{x_1}{2}, \frac{x_2}{2}, \ldots, \frac{x_n}{2}\right) \| \le \varphi\left(\frac{x_1}{2}, \frac{x_1}{2}, \frac{x_2}{2}, \ldots, \frac{x_n}{2}, \frac{x_n}{2}\right)$$

for all $x_1, x_2, \ldots, x_n \in V$. From the above inequality, we can derive the inequality

$$\begin{aligned} &\left\| 2^m f\left(\frac{x_1}{2^m}, \frac{x_2}{2^m}, \ldots, \frac{x_n}{2^m}\right) - 2^{m+m'} f\left(\frac{x_1}{2^{m+m'}}, \frac{x_2}{2^{m+m'}}, \ldots, \frac{x_n}{2^{m+m'}}\right) \right\| \\ &\qquad \le \sum_{k=m}^{m+m'-1} 2^k \varphi\left(\frac{x_1}{2^{k+1}}, \frac{x_1}{2^{k+1}}, \frac{x_2}{2^{k+1}}, \ldots, \frac{x_n}{2^k}, \frac{x_n}{2^k}\right) \end{aligned} \tag{21}$$

for all $x_1, x_2, \ldots, x_n \in V$ and all positive integers m, m'. The remainder of the proof of this theorem developed after inequality (21) is omitted because it is similar to that of Theorem 4. □

The following corollary follows from Theorems 6 and 7.

Corollary 2. *Let $(X, ||| \cdot |||)$ be a normed space and p be a nonnegative real number with $p \neq 1$. Suppose that $f : X^n \to Y$ is a mapping satisfying (14) for all $x_1, y_1, x_2, y_2, \ldots, x_n, y_n \in X$. Then, there exists a unique n-variable additive mapping $F : X^n \to Y$ such that*

$$\| f(x_1, x_2, \ldots, x_n) - F(x_1, x_2, \ldots, x_n) \| \le \frac{2\theta}{|2 - 2^p|} (|||x_1|||^p + |||x_2|||^p + \ldots + |||x_n|||^p) \tag{22}$$

for all $x_1, x_2, \ldots, x_n \in X$.

Proof. By putting $\varphi(x_1, y_1, x_2, y_2, \ldots, x_n, y_n) := \theta(|||x_1|||^p + |||y_1|||^p + |||x_2|||^p + |||y_2|||^p + \cdots + |||x_n|||^p + |||y_n|||^p)$ for all $x_1, y_1, x_2, y_2, \ldots, x_n, y_n \in X$, then we easily obtain (22) from (17) and (20) of Theorems 6 and 7. □

3. Conclusions

We obtained two stability results.

Theorems 3 and 4 are the generalized Hyers–Ulam stability for the additive functional Equation (3) on V^n, which is a generalization for the stability of the Cauchy functional equation in papers of Aoki [3], Rassias [4], Gajda [5], Hyers [2], and Găvruta [7].

Theorems 6 and 7 are the hyperstablity of the additive functional Equation (3) on V^n, which is a generalization of the Brzdęk's results [10,11] for the Cauchy functional equation.

If the function f satisfies the inequality (6) for all $(x_1, x_2, \ldots, x_n) \neq (0, 0, \ldots, 0)$ and $(y_1, y_2, \ldots, y_n) \neq (0, 0, \ldots, 0)$, then the function f satisfies the inequality (6) for all $x_1, y_1, x_2, y_2, \ldots, x_n, y_n \in V \backslash \{0\}$. Therefore, the condition $x_1, y_1, x_2, y_2, \ldots, x_n, y_n \in V \backslash \{0\}$ used in the inequality (3.2) of this paper is a generalization of the conditions used in the inequality (6) in well-known pre-results ([10,11]).

Author Contributions: Conceptualization, Y.-H.L. and G.H.K.; Investigation, Y.-H.L. and G.H.K.

References

1. Ulam, S.M. *Problems in Modern Mathematics*; Wiley: New York, NY, USA, 1964.
2. Hyers, D.H. On the stability of the linear functional equation. *Proc. Natl. Acad. Sci. USA* **1941**, *27*, 222–224. [CrossRef] [PubMed]
3. Aoki, T. On the stability of the linear transformation in Banach algebras. *J. Math. Soc. Japan* **1950**, 2, 64–66. [CrossRef]
4. Rassias, T.M. On the stability of the linear mapping in Banach spaces. *Proc. Am. Math. Soc.* **1978**, *72*, 297–300. [CrossRef]
5. Gajda, Z. On stability of additive mappings. *Int. J. Math. Math. Sci.* **1991**, *14*, 431–434. [CrossRef]
6. Rassias, T.M.; Semrl, P. On the behavior of mappings which do not satisfy Hyers-Ulam stability. *Proc. Am. Math. Soc.* **1992**, *114*, 989–993. [CrossRef]
7. Găvruta, P. A generalization of the Hyers-Ulam-Rassias stability of approximately additive mappings. *J. Math. Anal. Appl.* **1994**, *184*, 431–436. [CrossRef]
8. Maksa, G.; Páles, Z. Hyperstability of a class of linear functional equations. *Acta Math. Acad. Paedagog. Nyházi* **2001**, *17*, 107–112.
9. Bourgin, D.G. Approximately isometric and multiplicative transformations on continuous function rings. *Duke Math. J.* **1949**, *16*, 385–397. [CrossRef]
10. Brzdęk, J. Remarks on hyperstability of the the Cauchy equation. *Aequationes Math.* **2013**, *86*, 255–267. [CrossRef]
11. Brzdęk, J. A hyperstability result for the Cauchy equation. *Bull. Aust. Math. Soc.* **2014**, *89*, 33–40. [CrossRef]

6

Pre-Metric Spaces Along with Different types of Triangle Inequalities

Hsien-Chung Wu

Department of Mathematics, National Kaohsiung Normal University, Kaohsiung 802, Taiwan; hcwu@nknucc.nknu.edu.tw

Abstract: The T_1-spaces induced by the pre-metric spaces along with many forms of triangle inequalities are investigated in this paper. The limits in pre-metric spaces are also studied to demonstrate the consistency of limit concept in the induced topologies.

Keywords: quasi-metric space; Hausdorff space; T_1-space; triangle inequalities

---

1. Introduction

Let X be a nonempty universal set, and let $d : X \times X \to \mathbb{R}_+$ be a nonnegative real-valued function defined on the product set $X \times X$. We say that (X, d) is a metric space if and only if the following conditions are satisfied:

- for any $x, y \in M$, $d(x, y) = 0$ implies $x = y$;
- (self-distance condition) for any $x \in M$, $d(x, x) = 0$;
- (symmetric condition) for any $x, y \in M$, $d(x, y) = d(y, x)$;
- (triangle inequality) for any $x, y, z \in M$, $d(x, z) \leq d(x, y) + d(y, z)$.

In the literature, different kinds of spaces are considered by weakening the above conditions. Wilson [1] says that (X, d) is a quasi-metric space when the symmetric condition is not satisfied; that is, the following conditions are satisfied:

- for any $x, y \in M$, $d(x, y) = 0$ if and only if $x = y$;
- for any $x, y, z \in M$, $d(x, z) \leq d(x, y) + d(y, z)$.

After that, many authors (referring to [2–15] and the references therein) also defined the quasi-metric space as follows:

- for any $x, y \in M$, $d(x, y) = 0 = d(y, x)$ if and only if $x = y$;
- for any $x, y, z \in M$, $d(x, z) \leq d(x, y) + d(y, z)$.

However, these two definitions are not equivalent. The reason is that $d(x, y) = 0$ does not necessarily imply $d(y, x) = 0$, since the symmetric condition is not satisfied. It is clear to see that, in the Wilson's sense, we also have $d(y, x) = 0$ if and only if $y = x$.

Wilson [16] also says that (X, d) is a semi-metric space when the triangle inequality is not satisfied; that is, the following conditions are satisfied:

- for any $x, y \in M$, $d(x, y) = 0$ if and only if $x = y$;
- for any $x, y \in M$, $d(x, y) = d(y, x)$.

On the other hand, Matthews [11] says that (X, d) is a partial metric space if and only if the following conditions are satisfied:

- for any $x,y \in M$, $x = y$ if and only if $d(x,x) = d(x,y) = d(y,y)$;
- for any $x,y \in M$, $d(x,x) \leq d(x,y)$;
- for any $x,y \in M$, $d(x,y) = d(y,x)$.
- for any $x,y,z \in M$, $d(x,z) \leq d(x,y) + d(y,z) - d(y,y)$.

The partial metric space does not assume the self-distance condition $d(x,x) = 0$.

In this paper, we shall consider a so-called pre-metric space in which we just assume that $d(x,y) = 0$ implies $x = y$ for any $x,y \in X$. In other words, the pre-metric space does not assume the self-distance condition and symmetric condition. Since the triangle inequality plays a very important role, without considering the symmetric condition, the triangle inequality can be considered in four forms, which was not discussed in the literature. Based on the four different kinds of triangle inequalities, we can induce the T_1-space space from the pre-metric space under some suitable conditions.

This paper is organized as follows. In Section 2 , we propose the so-called pre-metric space in which four forms of triangle inequalities are considered and studied. Many basic properties are also obtained for further investigation. In Section 3, we induce the T_1-space from a given pre-metric space under some suitable assumptions. In Section 4, the limits in pre-metric space are also studied. We present the consistency of limit concepts in the pre-metric space and the induced topologies.

2. Definitions and Properties

In this section, we shall introduce the concept of pre-metric space, and the four concepts of triangle inequalities. We also derive some interesting properties that will be used in the further study. Without considering the symmetric condition, we first introduce four types of triangle inequality as follows.

Definition 1. *Let X be a nonempty universal set, and let d be a mapping defined on $X \times X$ into $\mathbb{R}_+$.*

- *We say that d satisfies the* $\bowtie$-triangle inequality *if and only if the following inequality is satisfied:*

$$d(x,y) + d(y,z) \geq d(x,z) \text{ for all } x,y,z \in X.$$

- *We say that d satisfies the* $\triangleright$-triangle inequality *if and only if the following inequality is satisfied:*

$$d(x,y) + d(z,y) \geq d(x,z) \text{ for all } x,y,z \in X.$$

- *We say that d satisfies the* $\triangleleft$-triangle inequality *if and only if the following inequality is satisfied:*

$$d(y,x) + d(y,z) \geq d(x,z) \text{ for all } x,y,z \in X.$$

- *We say that d satisfies the* $\diamond$-triangle inequality *if and only if the following inequality is satisfied:*

$$d(y,x) + d(z,y) \geq d(x,z) \text{ for all } x,y,z \in X.$$

It is obvious that if d satisfies the symmetric condition, then the concepts of $\bowtie$-triangle inequality, $\triangleright$-triangle inequality, $\triangleleft$-triangle inequality and $\diamond$-triangle inequality are all equivalent.

Example 1. *We define a function $d : \mathbb{R}_+ \times \mathbb{R}_+ \to \mathbb{R}_+$ by $d(x,y) = \max\{x,y\}$. Then $d(x,x) = x$ for any $x \geq 0$, which also says that $d(x,x)$ is not always zero. It is not hard to check*

$$\max\{x,y\} + \max\{y,z\} \geq \max\{x,z\},$$

which also says that

$$d(x,y) + d(y,z) \geq d(x,z).$$

This shows that d satisfies the $\bowtie$-triangle inequality. Since d also satisfies the symmetric condition, it means that all the four forms of triangle inequalities are equivalent. However, since $d(x,x) > 0$ for $x > 0$, it says that $(\mathbb{R}_+, d)$ is still not a metric space.

Example 2. *We define a function $d : \mathbb{R}_+ \times \mathbb{R}_+ \to \mathbb{R}_+$ by*

$$d(x,y) = \begin{cases} x & \text{if } x \geq y \\ 2y - x & \text{if } x < y. \end{cases}$$

Then $d(x,x) = x$ for any $x \geq 0$, which also says that $d(x,x)$ is not always zero. For $x > y$, we see that $d(x,y) = x$ and $d(y,x) = 2x - y$, which says that $d(x,y) \neq d(y,x)$ in general; that is, the symmetric condition is not satisfied. It is not hard to check

$$d(x,y) + d(y,z) \geq d(x,z).$$

This shows that d also satisfies the $\bowtie$-triangle inequality.

Examples 1 and 2 say that $d(x,x) \neq 0$ for $x \neq 0$. Therefore, we propose the following definition.

Definition 2. *Let X be a nonempty universal set, and let d be a mapping defined on $X \times X$ into $\mathbb{R}_+$. We say that (X,d) is a* pre-metric space *if and only if $d(x,y) = 0$ implies $x = y$ for any $x, y \in X$.*

We see that (X,d) is a *quasi-metric space* if and only if (X,d) is a pre-metric space satisfying the $\bowtie$-triangle inequality and $d(x,x) = 0$ for all $x \in X$.

Example 3. *Examples 1 and 2 are pre-metric spaces, since it is not hard to check that $d(x,y) = 0$ implies $x = y = 0$ based on the nonnegativity.*

Remark 1. *Let (X,d) be a pre-metric space. Then $d(x,y) = 0$ implies $x = y$, which also implies $d(x,y) = 0 = d(x,x) = d(y,x)$ without needing the symmetric condition. We remark that this symmetric situation can only happen when $d(x,y) = 0$ or $d(y,x) = 0$. Therefore, if $d(x,y) > 0$ then we cannot have $d(x,y) = d(y,x)$ in general. On the other hand, we also see that $d(x,y) = 0$ or $d(y,x) = 0$ implies $d(x,x) = 0$. However, this situation does not say $d(x,x) = 0$ for all $x \in X$. We can just say that $d(x,x) = 0$ when $d(x,y) = 0$ for some $x, y \in X$. In other words, we can just say that $d(x,x) = 0$ for some $x \in X$. This situation can also be realized from Example 2.*

Proposition 1. *Let X be a nonempty universal set, and let d be a mapping defined on $X \times X$ into $\mathbb{R}_+$. Suppose that the following conditions are satisfied:*

- *$d(x,x) = 0$ for all $x \in X$;*
- *d satisfies the $\triangleright$-triangle inequality or the $\triangleleft$-triangle inequality or the $\diamond$-triangle inequality.*

Then d satisfies the symmetric condition.

Proof. Suppose that d satisfies the $\triangleright$-triangle inequality. Then, given any $x, y \in X$, we have

$$d(x,y) \leq d(x,x) + d(y,x) = d(y,x).$$

By interchanging the roles of x and y, we can also obtain $d(y,x) \leq d(x,y)$. This shows that $d(x,y) = d(y,x)$. The other cases of satisfying the $\triangleleft$-triangle inequality and the $\diamond$-triangle inequality can be similarly obtained. This completes the proof. □

Remark 2. *Suppose that $d(x,x) = 0$ for all $x \in X$, and that d satisfies the $\circ$-triangle inequality for some $\circ \in \{\triangleright, \triangleleft, \diamond\}$. Then, using Proposition 1, we see that all the four forms of triangle inequalities are equivalent.*

3. T_1-Space

We want to show that the pre-metric space along with the different kinds of triangle inequalities can induce the T_1-Space based on the concepts of open balls defined below.

Definition 3. *Let (X,d) be a pre-metric space. Given $r > 0$, the* open balls *centered at x are denoted and defined by*

$$B^{\triangleleft}(x;r) = \{y \in X : d(x,y) < r\}$$

and

$$B^{\triangleright}(x;r) = \{y \in X : d(y,x) < r\}.$$

Let $\mathcal{B}^{\triangleleft}$ denote the family of all open balls $B^{\triangleleft}(x;r)$, and let $\mathcal{B}^{\triangleright}$ denote the family of all open balls $B^{\triangleright}(x;r)$.

In the sequel, we also assume that the open balls $B^{\triangleleft}(x;r)$ and $B^{\triangleright}(x;r)$ are nonempty for each $x \in X$ and $r > 0$. In other words, given any $x \in X$ and $r > 0$, we assume that there exist y_1 and y_2 such that $d(x,y_1) < r$ and $d(x,y_2) < r$, respectively. It is also clear that if d satisfies the symmetric condition, then

$$B^{\triangleleft}(x;r) = B^{\triangleright}(x;r).$$

In this case, we simply write $B(x;r)$ to denote the open balls centered at x, and write $\mathcal{B}$ to denote the family of all open balls $B(x;r)$.

Proposition 2. *Let (X,d) be a pre-metric space.*

(i) *Given any $x \in X$, we have the following properties.*
   - *Suppose that $d(x,x) = 0$. Then $x \in B^{\triangleleft}(x;r) \in \mathcal{B}^{\triangleleft}$ and $x \in B^{\triangleright}(x;r) \in \mathcal{B}^{\triangleright}$ for all $r > 0$.*
   - *Suppose that $x \in B^{\triangleleft}(x;r)$ for all $r > 0$, or that $x \in B^{\triangleright}(x;r)$ for all $r > 0$. Then $d(x,x) = 0$.*

(ii) *If $x \neq y$, then there exist $r_1 > 0$ and $r_2 > 0$ such that $y \notin B^{\triangleleft}(x;r_1)$ and $y \notin B^{\triangleright}(x;r_2)$.*

(iii) *For each $x \in X$, we have the following properties.*
   - *Given any $B^{\triangleleft}(x;r) \in \mathcal{B}^{\triangleleft}$, there exists $n \in \mathbb{N}$ such that $B^{\triangleleft}(x;\frac{1}{n}) \subseteq B^{\triangleleft}(x;r)$.*
   - *Given any $B^{\triangleright}(x;r) \in \mathcal{B}^{\triangleright}$, there exists $n \in \mathbb{N}$ such that $B^{\triangleright}(x;\frac{1}{n}) \subseteq B^{\triangleright}(x;r)$.*

Proof. The first statement of part (i) is obvious. To prove the second statement of part (i), we take a sequence $\{r_n\}_{n=1}^{\infty}$ of positive numbers such that it is decreasing to zero. Then we have $d(x,x) < r_n$ for all n, which implies $d(x,x) = 0$ by taking $n \to \infty$. To prove part (ii), since $x \neq y$, it follows that $d(x,y) > 0$ and $d(y,x) > 0$ by the definition of pre-metric space. Using the denseness of $\mathbb{R}$, there exists $r_1 > 0$ such that $0 < r_1 < d(x,y)$, which also says that $y \notin B^{\triangleleft}(x;r_1)$. We also have $y \notin B^{\triangleright}(x;r_2)$ for some $r_2 > 0$ satisfying $0 < r_2 < d(y,x)$. Part (iii) follows from the existence of a positive integer n with $1/n < r$. This completes the proof. □

Proposition 3. *Let (X,d) be a pre-metric space. Then we have the following inclusions.*

(i) *Suppose that d satisfies the $\bowtie$-triangle inequality.*
   - *Given any $y \in B^{\triangleleft}(x;r)$, there exists $\bar{r} > 0$ such that $B^{\triangleleft}(y;\bar{r}) \subseteq B^{\triangleleft}(x;r)$.*
   - *Given any $y \in B^{\triangleright}(x;r)$, there exists $\bar{r} > 0$ such that $B^{\triangleright}(y;\bar{r}) \subseteq B^{\triangleright}(x;r)$.*

(ii) *Suppose that d satisfies the $\triangleright$-triangle inequality. Given any $y \in B^{\triangleleft}(x;r)$, there exists $\bar{r} > 0$ such that $B^{\triangleright}(y;\bar{r}) \subseteq B^{\triangleleft}(x;r)$ and $B^{\triangleright}(y;\bar{r}) \subseteq B^{\triangleright}(x;r)$.*

(iii) *Suppose that d satisfies the $\triangleleft$-triangle inequality. Given any $y \in B^{\triangleright}(x;r)$, there exists $\bar{r} > 0$ such that $B^{\triangleleft}(y;\bar{r}) \subseteq B^{\triangleright}(x;r)$ and $B^{\triangleleft}(y;\bar{r}) \subseteq B^{\triangleleft}(x;r)$.*

(iv) *Suppose that d satisfies the $\diamond$-triangle inequality.*

- *Given any $y \in B^{\lhd}(x;r)$, there exists $\bar{r} > 0$ such that $B^{\lhd}(y;\bar{r}) \subseteq B^{\rhd}(x;r)$.*
- *Given any $y \in B^{\rhd}(x;r)$, there exists $\bar{r} > 0$ such that $B^{\rhd}(y;\bar{r}) \subseteq B^{\lhd}(x;r)$.*

(v) Suppose that d satisfies the $\rhd$-triangle inequality and the $\lhd$-triangle inequality.

- *Given any $y \in B^{\lhd}(x;r)$, there exists $\bar{r} > 0$ such that $B^{\lhd}(y;\bar{r}) \subseteq B^{\lhd}(x;r)$.*
- *Given any $y \in B^{\rhd}(x;r)$, there exists $\bar{r} > 0$ such that $B^{\rhd}(y;\bar{r}) \subseteq B^{\rhd}(x;r)$.*

Proof. To prove part (i), for $y \in B^{\lhd}(x;r)$ and $z \in B^{\lhd}(y;\bar{r})$, let $\bar{r} \leq r - d(x,y)$. Using the $\bowtie$-triangle inequality, we have

$$d(x,z) \leq d(x,y) + d(y,z) < d(x,y) + \bar{r} \leq d(x,y) + r - d(x,y) = r,$$

which says that $z \in B^{\lhd}(x;r)$. Therefore we obtain the inclusion $B^{\lhd}(y;\bar{r}) \subseteq B^{\lhd}(x;r)$. For $y \in B^{\rhd}(x;r)$ and $z \in B^{\rhd}(y;\bar{r})$, let $\bar{r} \leq r - d(y,x)$. Then we can similarly obtain the inclusion $B^{\rhd}(y;\bar{r}) \subseteq B^{\rhd}(x;r)$.

To prove part (ii), for $y \in B^{\lhd}(x;r)$ and $z \in B^{\rhd}(y;\bar{r})$, let $\bar{r} \leq r - d(x,y)$. Using the $\rhd$-triangle inequality, we have

$$d(x,z) \leq d(x,y) + d(z,y) < d(x,y) + \bar{r} \leq d(x,y) + r - d(x,y) = r,$$

which says that $z \in B^{\lhd}(x;r)$. Therefore we obtain the inclusion $B^{\rhd}(y;\bar{r}) \subseteq B^{\lhd}(x;r)$. We can similarly obtain the inclusion $B^{\rhd}(y;\bar{r}) \subseteq B^{\rhd}(x;r)$.

Parts (iii) and (iv) can be similarly obtained. To prove the first statement of part (v), using part (ii), we can take $\bar{r}^* > 0$ such that $B^{\rhd}(y;\bar{r}^*) \subseteq B^{\lhd}(x;r)$. Using part (iii), we can also take $\bar{r} > 0$ such that $B^{\lhd}(y;\bar{r}) \subseteq B^{\rhd}(y;\bar{r}^*)$. This shows that $B^{\lhd}(y;\bar{r}) \subseteq B^{\lhd}(x;r)$. The second statement of part (v) can be similarly obtained. This completes the proof. □

Proposition 4. *Let (X,d) be a pre-metric space. Then we have the following inclusions.*

(i) Suppose that d satisfies the $\bowtie$-triangle inequality.

- *If $x \in B^{\lhd}(x_1,r_1) \cap B^{\lhd}(x_2,r_2)$, then there exists $r_3 > 0$ such that*

$$B^{\lhd}(x,r_3) \subseteq B^{\lhd}(x_1,r_1) \cap B^{\lhd}(x_2,r_2).$$

- *If $x \in B^{\rhd}(x_1,r_1) \cap B^{\rhd}(x_2,r_2)$, then there exists $r_3 > 0$ such that*

$$B^{\rhd}(x,r_3) \subseteq B^{\rhd}(x_1,r_1) \cap B^{\rhd}(x_2,r_2).$$

(ii) Suppose that d satisfies the $\rhd$-triangle inequality. If $x \in B^{\lhd}(x_1,r_1) \cap B^{\lhd}(x_2,r_2)$, then there exists $r_3 > 0$ such that

$$B^{\rhd}(x,r_3) \subseteq B^{\lhd}(x_1,r_1) \cap B^{\lhd}(x_2,r_2) \text{ and } B^{\rhd}(x,r_3) \subseteq B^{\rhd}(x_1,r_1) \cap B^{\rhd}(x_2,r_2).$$

(iii) Suppose that d satisfies the $\lhd$-triangle inequality. If $x \in B^{\rhd}(x_1,r_1) \cap B^{\rhd}(x_2,r_2)$, then there exists $r_3 > 0$ such that

$$B^{\lhd}(x,r_3) \subseteq B^{\rhd}(x_1,r_1) \cap B^{\rhd}(x_2,r_2) \text{ and } B^{\lhd}(x,r_3) \subseteq B^{\lhd}(x_1,r_1) \cap B^{\lhd}(x_2,r_2).$$

(iv) Suppose that d satisfies the $\diamond$-triangle inequality.

- *If $x \in B^{\lhd}(x_1,r_1) \cap B^{\lhd}(x_2,r_2)$, then there exists $r_3 > 0$ such that*

$$B^{\lhd}(x,r_3) \subseteq B^{\rhd}(x_1,r_1) \cap B^{\rhd}(x_2,r_2).$$

- *If $x \in B^{\rhd}(x_1,r_1) \cap B^{\rhd}(x_2,r_2)$, then there exists $r_3 > 0$ such that*

$$B^{\rhd}(x,r_3) \subseteq B^{\lhd}(x_1,r_1) \cap B^{\lhd}(x_2,r_2).$$

(v) *Suppose that d satisfies the* $\triangleright$*-triangle inequality and the* $\triangleleft$*-triangle inequality. We have the following inclusions.*

- *If* $x \in B^{\triangleleft}(x_1, r_1) \cap B^{\triangleleft}(x_2, r_2)$, *then there exists* $r_4 > 0$ *such that*
$$B^{\triangleleft}(x, r_4) \subseteq B^{\triangleleft}(x_1, r_1) \cap B^{\triangleleft}(x_2, r_2).$$
- *If* $x \in B^{\triangleright}(x_1, r_1) \cap B^{\triangleright}(x_2, r_2)$, *then there exists* $r_4 > 0$ *such that*
$$B^{\triangleright}(x, r_4) \subseteq B^{\triangleright}(x_1, r_1) \cap B^{\triangleright}(x_2, r_2).$$

Proof. To prove part (i), for $x \in B^{\triangleleft}(x_1, r_1)$ and $x \in B^{\triangleleft}(x_2, r_2)$, using part (i) of Proposition 3, there exist $\bar{r}_1$ and $\bar{r}_2$ such that
$$B^{\triangleleft}(x, \bar{r}_1) \subseteq B^{\triangleleft}(x_1, r_1) \text{ and } B^{\triangleleft}(x, \bar{r}_2) \subseteq B^{\triangleleft}(x_2, r_2).$$
We take $r_3 = \min\{\bar{r}_1, \bar{r}_2\}$. Then
$$B^{\triangleleft}(x, r_3) \subseteq B^{\triangleleft}(x, \bar{r}_1) \cap B^{\triangleleft}(x, \bar{r}_2) \subseteq B^{\triangleleft}(x_1, r_1) \cap B^{\triangleleft}(x_2, r_2).$$
Therefore we obtain the first inclusion. The second inclusion can be similarly obtained.

To prove part (ii), for $x \in B^{\triangleleft}(x_1, r_1)$ and $x \in B^{\triangleleft}(x_2, r_2)$, using part (ii) of Proposition 3, there exist $\bar{r}_1$ and $\bar{r}_2$ such that
$$B^{\triangleright}(x, \bar{r}_1) \subseteq B^{\triangleleft}(x_1, r_1) \text{ and } B^{\triangleright}(x, \bar{r}_1) \subseteq B^{\triangleright}(x_1, r_1).$$
and
$$B^{\triangleright}(x, \bar{r}_2) \subseteq B^{\triangleleft}(x_2, r_2) \text{ and } B^{\triangleright}(x, \bar{r}_2) \subseteq B^{\triangleright}(x_2, r_2).$$
Let $r_3 = \min\{\bar{r}_1, \bar{r}_2\}$. Then
$$B^{\triangleright}(x, r_3) \subseteq B^{\triangleright}(x, \bar{r}_1) \cap B^{\triangleright}(x, \bar{r}_2) \subseteq B^{\triangleleft}(x_1, r_1) \cap B^{\triangleleft}(x_2, r_2)$$
and
$$B^{\triangleright}(x, r_3) \subseteq B^{\triangleright}(x, \bar{r}_1) \cap B^{\triangleright}(x, \bar{r}_2) \subseteq B^{\triangleright}(x_1, r_1) \cap B^{\triangleright}(x_2, r_2).$$
Therefore we obtain the desired inclusions.

Parts (iii) and (iv) can be similarly obtained. To prove the first statement of part (v), using part (iii) of Proposition 3 and part (ii) of this proposition, we can find $r_4 > 0$ such that
$$B^{\triangleleft}(x, r_4) \subseteq B^{\triangleright}(x, r_3) \subseteq B^{\triangleleft}(x_1, r_1) \cap B^{\triangleleft}(x_2, r_2).$$
The second statement can be similarly obtained. This completes the proof. □

Proposition 5. *Let* (X, d) *be a pre-metric space. Suppose that* $x \neq y$. *Then we have the following properties.*

(i) *Suppose that d satisfies the* $\bowtie$*-triangle inequality or the* $\diamond$*-triangle inequality. Then* $B^{\triangleleft}(x; r) \cap B^{\triangleright}(y; r) = \emptyset$ *and* $B^{\triangleright}(x; r) \cap B^{\triangleleft}(y; r) = \emptyset$ *for some* $r > 0$.
(ii) *Suppose that d satisfies the* $\triangleright$*-triangle inequality. Then* $B^{\triangleleft}(x; r) \cap B^{\triangleleft}(y; r) = \emptyset$ *for some* $r > 0$.
(iii) *Suppose that d satisfies the* $\triangleleft$*-triangle inequality. Then* $B^{\triangleright}(x; r) \cap B^{\triangleright}(y; r) = \emptyset$ *for some* $r > 0$.

Proof. Since $x \neq y$, it says that $d(x, y) > 0$ and $d(y, x) > 0$. We consider the following cases.

- Suppose that d satisfies the $\triangleright$-triangle inequality. Let $r \leq d(x, y)/2$. We are going to prove $B^{\triangleleft}(x; r) \cap B^{\triangleleft}(y; r) = \emptyset$ by contradiction. Suppose that $z \in B^{\triangleleft}(x; r) \cap B^{\triangleleft}(y; r)$. Since d satisfies the $\triangleright$-triangle inequality, it follows that
$$d(x, y) \leq d(x, z) + d(y, z) < r + r = 2r \leq d(x, y),$$

which is a contradiction. Suppose that d satisfies the $\triangleleft$-triangle inequality. Then we can similarly obtain the desired result.

- Suppose that d satisfies the $\bowtie$-triangle inequality. Let $r \leq d(x,y)/2$. For $z \in B^{\triangleleft}(x;r) \cap B^{\triangleright}(y;r)$, it follows that

$$d(x,y) \leq d(x,z) + d(z,y) < r + r = 2r \leq d(x,y),$$

which is a contradiction. On the other hand, let $r \leq d(y,x)/2$, for $z \in B^{\triangleright}(x;r) \cap B^{\triangleleft}(y;r)$, it follows that

$$d(y,x) \leq d(y,z) + d(z,x) < r + r = 2r \leq d(y,x),$$

which is a contradiction. Suppose that d satisfies the $\diamond$-triangle inequality. Then we can similarly obtain the desired result.

This completes the proof. □

Theorem 1. *Let (X,d) be a pre-metric space. Define*

$$\tau^{\triangleleft} = \{O^{\triangleleft} \subseteq X : x \in O^{\triangleleft} \text{ if and only if there exist } r > 0 \text{ such that } x \in B^{\triangleleft}(x;r) \subseteq O^{\triangleleft}\}. \quad (1)$$

and

$$\tau^{\triangleright} = \{O^{\triangleright} \subseteq X : x \in O^{\triangleright} \text{ if and only if there exist } r > 0 \text{ such that } x \in B^{\triangleright}(x;r) \subseteq O^{\triangleright}\}. \quad (2)$$

Suppose that d satisfies the $\bowtie$-triangle inequality. Then we have the following results.

- *Assume additionally that $d(x,x) = 0$ for all $x \in X$, or that $x \in B^{\triangleleft}(x;r)$ for all $x \in X$ and $r > 0$. Then $(X,\tau^{\triangleleft})$ is a T_1-space such that $\mathcal{B}^{\triangleleft}$ is a base for the topology $\tau^{\triangleleft}$.*
- *Assume additionally that $d(x,x) = 0$ for all $x \in X$, or that $x \in B^{\triangleright}(x;r)$ for all $x \in X$ and $r > 0$. Then $(X,\tau^{\triangleright})$ is a T_1-space such that $\mathcal{B}^{\triangleright}$ is a base for the topology $\tau^{\triangleright}$.*

The T_1-spaces $(X,\tau^{\triangleleft})$ and $(X,\tau^{\triangleright})$ also satisfy the first axiom of countability. Moreover, $B^{\triangleleft}(x;r)$ is a $\tau^{\triangleleft}$-open set and $B^{\triangleright}(x;r)$ is a $\tau^{\triangleright}$-open set.

Proof. Using part (i) of Proposition 2 and part (i) of Proposition 4, we see that $\tau^{\triangleleft}$ is a topology such that $\mathcal{B}^{\triangleleft}$ is a base for $\tau^{\triangleleft}$. Part (ii) of Proposition 2 says that $(X,\tau^{\triangleleft})$ is a T_1-space. Part (iii) of Proposition 2 says that there exists a countable local base at each $x \in X$ for $\tau^{\triangleleft}$, which also says that $\tau^{\triangleleft}$ satisfies the first axiom of countability. Regarding $\tau^{\triangleright}$, we can similarly obtain the desired results. Finally, part (i) of Proposition 3 says that $B^{\triangleleft}(x;r)$ is a $\tau^{\triangleleft}$-open set and $B^{\triangleright}(x;r)$ is a $\tau^{\triangleright}$-open set. This completes the proof. □

We remark that, in Theorem 1, although we assume $d(x,x) = 0$ for all $x \in X$, (X,d) is not necessarily a metric space, since the symmetric condition is still not satisfied. The following example provides this observation.

Example 4. *We define a function $d : \mathbb{R}_+ \times \mathbb{R}_+ \to \mathbb{R}_+$ by*

$$d(x,y) = \begin{cases} x & \text{if } x > y \\ 0 & \text{if } x = y \\ 2y - x & \text{if } x < y. \end{cases}$$

Then $d(x,x) = 0$ for all $x \in X$. By referring to Example 2, we also see that the symmetric condition is not satisfied, and that d satisfies the $\bowtie$-triangle inequality. Using Theorem 1, we can induce two T_1-spaces $(\mathbb{R}_+,\tau^{\triangleleft})$ and $(\mathbb{R}_+,\tau^{\triangleright})$. Moreover, the spaces $(\mathbb{R}_+,\tau^{\triangleleft})$ and $(\mathbb{R}_+,\tau^{\triangleright})$ also satisfy the first axiom of countability.

4. Limits in Pre-Metric Space

Let (X, d) be a pre-metric space. Since the symmetric condition is not necessarily satisfied, the different concepts of limit are proposed below.

Definition 4. *Let (X, d) be a pre-metric space, and let $\{x_n\}_{n=1}^{\infty}$ be a sequence in X.*

- *We write $x_n \xrightarrow{d^{\triangleright}} x$ as $n \to \infty$ if and only if $d(x_n, x) \to 0$ as $n \to \infty$.*
- *We write $x_n \xrightarrow{d^{\triangleleft}} x$ as $n \to \infty$ if and only if $d(x, x_n) \to 0$ as $n \to \infty$.*
- *We write $x_n \xrightarrow{d} x$ as $n \to \infty$ if and only if*

$$\lim_{n\to\infty} d(x_n, x) = \lim_{n\to\infty} d(x, x_n) = 0.$$

The uniqueness of limits will be discussed below.

Proposition 6. *Let (X, d) be a pre-metric space, and let $\{x_n\}_{n=1}^{\infty}$ be a sequence in X.*

(i) Suppose that d satisfies the $\bowtie$-triangle inequality or $\diamond$-triangle inequality. If $x_n \xrightarrow{d^{\triangleleft}} x$ and $x_n \xrightarrow{d^{\triangleright}} y$, then $x = y$.

(ii) Suppose that d satisfies the $\triangleleft$-triangle inequality. If $x_n \xrightarrow{d^{\triangleright}} x$ and $x_n \xrightarrow{d^{\triangleright}} y$, then $x = y$. In other words, the $d^{\triangleright}$-limit is unique.

(iii) Suppose that d satisfies the $\triangleright$-triangle inequality. If $x_n \xrightarrow{d^{\triangleleft}} x$ and $x_n \xrightarrow{d^{\triangleleft}} y$, then $x = y$. In other words, the $d^{\triangleleft}$-limit is unique.

Proof. To prove part (i), we first assume that d satisfies the $\bowtie$-triangle inequality. Then

$$d(x, y) \leq d(x, x_n) + d(x_n, y) \to 0 + 0 = 0,$$

which says that $x = y$. Now suppose that d satisfies the $\diamond$-triangle inequality. Then

$$d(y, x) \leq d(x_n, y) + d(x, x_n) \to 0 + 0 = 0$$

which also says that $x = y$. The other cases can be similarly obtained. This completes the proof. □

Let (X, τ) be a topological space. The sequence $\{x_n\}_{n=1}^{\infty}$ in X converges to $x \in X$ with respect to the topology is denoted by $x_n \xrightarrow{\tau} x$ as $n \to \infty$.

Proposition 7. *Let (X, d) be a pre-metric space. Suppose that d satisfies the $\bowtie$-triangle inequality or the $\triangleright$-triangle inequality or the $\triangleleft$-triangle inequality. Assume that $d(x, x) = 0$ for all $x \in X$. Then the following statements hold true.*

(i) Let $\tau^{\triangleright}$ be the topology defined by (1) in Theorem 1, and let $\{x_n\}_{n=1}^{\infty}$ be a sequence in X. Then $x_n \xrightarrow{\tau^{\triangleright}} x$ as $n \to \infty$ if and only if $x_n \xrightarrow{d^{\triangleright}} x$ as $n \to \infty$.

(ii) Let $\tau^{\triangleleft}$ be the topology defined by (2) in Theorem 1, and let $\{x_n\}_{n=1}^{\infty}$ be a sequence in X. Then $x_n \xrightarrow{\tau^{\triangleleft}} x$ as $n \to \infty$ if and only if $x_n \xrightarrow{d^{\triangleleft}} x$ as $n \to \infty$.

Proof. Under the assumptions, Theorem 1 says that we can induce two topologies $\tau^{\triangleright}$ and $\tau^{\triangleleft}$. It suffices to prove part (i). Suppose that $x_n \xrightarrow{\tau^{\triangleright}} x$ as $n \to \infty$. Given any $\epsilon > 0$, there exits $n_\epsilon \in \mathbb{N}$ such that $x_n \in B^{\triangleright}(x; \epsilon)$ for all $n \geq n_\epsilon$, i.e., $d(x_n, x) < \epsilon$ for all $n \geq n_\epsilon$. This says that $d(x_n, x) \to 0$ as $n \to \infty$. Conversely, if $d(x_n, x) \to 0$ as $n \to \infty$, then, given any $\epsilon > 0$, there exists $n_\epsilon \in \mathbb{N}$ such that $d(x_n, x) < \epsilon$ for all $n \geq n_\epsilon$, which says that $x_n \in B^{\triangleright}(x; \epsilon)$ for all $n \geq n_\epsilon$. This shows that $x_n \xrightarrow{\tau^{\triangleright}} x$ as $n \to \infty$, and the proof is complete. □

Let (X, d) be a pre-metric space. We consider the following sets

$$\bar{B}^{\lhd}(x;r) = \{y \in X : d(x,y) \leq r\} \text{ and } \bar{B}^{\rhd}(x;r) = \{y \in X : d(y,x) \leq r\}.$$

If the symmetric condition is satisfied, then we simply write $\bar{B}(x;r)$. We are going to consider the closeness of $\bar{B}^{\lhd}(x;r)$ and $\bar{B}^{\rhd}(x;r)$. Let us recall that, given a topological space (X, τ), we say that a subset F of X is τ-closed if and only if $\tau\text{-cl}(F) = F$, where $\tau\text{-cl}(F)$ denotes the τ-closure of F.

Proposition 8. *Let (X, d) be a pre-metric space. Suppose that d satisfies $d(x,x) = 0$ for all $x \in X$ and the $\bowtie$-triangle inequality. We have the following results.*

- *$\bar{B}^{\lhd}(x;r)$ is $\tau^{\rhd}$-closed. In other words, we have $\tau^{\rhd}\text{-cl}(\bar{B}^{\lhd}(x;r)) = \bar{B}^{\lhd}(x;r)$.*
- *$\bar{B}^{\rhd}(x;r)$ is $\tau^{\lhd}$-closed. In other words, we have $\tau^{\lhd}\text{-cl}(\bar{B}^{\rhd}(x;r)) = \bar{B}^{\rhd}(x;r)$.*

Proof. Under the assumptions, Theorem 1 says that we can induce two topologies $\tau^{\rhd}$ and $\tau^{\lhd}$ satisfying the first axiom of countability. To prove the first statement, for $y \in \tau^{\rhd}\text{-cl}(\bar{B}^{\lhd}(x;r))$, since $(X, \tau^{\rhd})$ satisfies the first axiom of countability, there exists a sequence $\{y_n\}_{n=1}^{\infty}$ in $\bar{B}^{\lhd}(x;r)$ such that $y_n \xrightarrow{\tau^{\rhd}} y$ as $n \to \infty$. We also have $d(x, y_n) \leq r$ for all n. By part (i) of Proposition 7, we have $d(y_n, y) \to 0$ as $n \to \infty$ for all. The $\bowtie$-triangle inequality says that

$$d(x,y) \leq d(x,y_n) + d(y_n,y) \leq r + d(y_n,y) \to r \text{ as } n \to \infty,$$

which shows $y \in \bar{B}^{\lhd}(x;r)$. Therefore we obtain $\tau^{\rhd}\text{-cl}(\bar{B}^{\lhd}(x;r)) = \bar{B}^{\lhd}(x;r)$. The second statement can be similarly obtained. This completes the proof. □

Proposition 9. *Let (X, d) be a pre-metric space. Suppose that the following conditions are satisfied.*

- *d satisfies the $\rhd$-triangle inequality and the $\lhd$-triangle inequality simultaneously.*
- *$d(x,x) = 0$ for all $x \in X$.*

Then d satisfies the symmetric condition; that is, (X, d) is a metric space.

Proof. Using part (i) of Proposition 2 and part (v) of Proposition 4, we see that $\tau^{\lhd}$ defined by (1) in Theorem 1 is a topology such that $\mathcal{B}^{\lhd}$ is a base for $\tau^{\lhd}$. Part (iii) of Proposition 2 says that there exists a countable local base at each $x \in X$ for $\tau^{\lhd}$, which also says that $\tau^{\lhd}$ satisfies the first axiom of countability. For $y \in \tau^{\lhd}\text{-cl}(\bar{B}^{\lhd}(x;r))$, the first axiom of countability says that there exists a sequence $\{y_n\}_{n=1}^{\infty}$ in $\bar{B}^{\lhd}(x;r)$ such that $y_n \xrightarrow{\tau^{\lhd}} y$ as $n \to \infty$. We also have $d(x, y_n) \leq r$ for all n. By part (ii) of Proposition 7, we have $d(y, y_n) \to 0$ as $n \to \infty$. The $\rhd$-triangle inequality says that

$$d(x,y) \leq d(x,y_n) + d(y,y_n) \leq r + d(y,y_n) \to r \text{ as } n \to \infty,$$

which shows $y \in \bar{B}^{\lhd}(x;r)$, i.e., $\tau^{\lhd}\text{-cl}(\bar{B}^{\lhd}(x;r)) = \bar{B}^{\lhd}(x;r)$. On the other hand, we also have

$$d(y,x) \leq d(y,y_n) + d(x,y_n) \leq d(y,y_n) + r \to r \text{ as } n \to \infty,$$

which shows $y \in \bar{B}^{\rhd}(x;r)$. Therefore we obtain the inclusion $\tau^{\lhd}\text{-cl}(\bar{B}^{\lhd}(x;r)) \subseteq \bar{B}^{\rhd}(x;r)$, which also says that $\bar{B}^{\lhd}(x;r) \subseteq \bar{B}^{\rhd}(x;r)$. Now, given any $x, y \in X$ with $y \neq x$, we have $d(x,y) > 0$. Let $r = d(x,y)$. Then $y \in \bar{B}^{\lhd}(x;r)$. This also says that $y \in \bar{B}^{\rhd}(x;r)$, i.e.,

$$d(y,x) \leq r = d(x,y).$$

By interchanging the roles of x and y, we can similarly obtain $d(x,y) \leq d(y,x)$. This completes the proof. □

Remark 3. *Let (X, d) be a pre-metric space. Suppose that the conditions presented in Proposition 9 are satisfied. Then (X, d) turns into a metric space. It is well-known that the metric space (X, d) can induce a Hausdorff topological space. More precisely, using the notations in this paper, we see that $\tau^{\lhd} = \tau^{\rhd}$ that is simply written as τ. In other words, (X, τ) is a Hausdorff space such that $\mathcal{B}$ is a base for the topology τ, where $\mathcal{B} = \mathcal{B}^{\lhd} = \mathcal{B}^{\rhd}$. The Hausdorff space (X, τ) also satisfies the first axiom of countability. Moreover, $B(x; r)$ is a τ-open set and $\bar{B}(x; r)$ is a τ-closed set, where $B(x; r) = B^{\lhd}(x; r) = B^{\rhd}(x; r)$ and $\bar{B}(x; r) = \bar{B}^{\lhd}(x; r) = \bar{B}^{\rhd}(x; r)$.*

In a future study, we shall avoid to consider the conditions presented in Proposition 9. Otherwise, the study will become trivial, based on the results of conventional metric space.

References

1. Wilson, W.A. On Semi-metric spaces. *Am. J. Math.* **1931**, *53*, 361–373. [CrossRef]
2. Alegre, C.; Marín, J. Modified *w*-distances on quasi-metric spaces and a fixed point theorem on complete quasi-metric spaces. *Topol. Appl.* **2016**, *203*, 32–41. [CrossRef]
3. Ali-Akbari, M.; Pourmahdian, M. Completeness of hyperspaces of compact subsets of quasi-metric spaces. *Acta Math. Hung.* **2010**, *127*, 260–272. [CrossRef]
4. Cao, J.; Rodríguez-López, J. On hyperspace topologies via distance functionals in quasi-metric spaces. *Acta Math. Hung.* **2006**, *112*, 249–268. [CrossRef]
5. Cobzas, S. Completeness in quasi-metric spaces and Ekeland Variational Principle. *Topol. Appl.* **2011**, *158*, 1073–1084. [CrossRef]
6. Collins Agyingi, A.; Haihambo, P.; Künzi, H.-P. A. Endpoints in T_0-quasi-metric spaces. *Topol. Appl.* **2014**, *168*, 82–93. [CrossRef]
7. Doitchinov, D. On completeness in quasi-metric spaces. *Topol. Appl.* **1988**, *30*, 127–148. [CrossRef]
8. Künzi, H.-P.A. A construction of the B-completion of a T_0-quasi-metric space. *Topol. Appl.* **2014**, *170*, 25–39. [CrossRef]
9. Künzi, H.-P.A.; Kivuvu, C.M. The B-completion of a T_0-quasi-metric space. *Topol. Appl.* **2009**, *156*, 2070–2081. [CrossRef]
10. Künzi, H.-P.A.; Yildiz, F. Convexity structures in T_0-quasi-metric spaces. *Topol. Appl.* **2016**, *200*, 2–18. [CrossRef]
11. Matthews, S.G. Partial metric topology. *Ann. N. Y. Acad. Sci.* **1994**, *728*, 183–197. [CrossRef]
12. Romaguera, S.; Antonino, J.A. On convergence complete strong quasi-metrics. *Acta Math. Hung.* **1994**, *64*, 65–73. [CrossRef]
13. Triebel, H. A new approach to function spaces on quasi-metric spaces. *Rev. Mat. Complut.* **2005**, *18*, 7–48. [CrossRef]
14. Ume, J.S. A minimization theorem in quasi-metric spaces and its applications. *Int. J. Math. Math. Sci.* **2002**, *31*, 443–447. [CrossRef]
15. Vitolo, P. A representation theorem for quasi-metric spaces. *Topol. Appl.* **1995**, *65*, 101–104. [CrossRef]
16. Wilson, W.A. On Quasi-metric spaces. *Am. J. Math.* **1931**, *53*, 675–684. [CrossRef]

7

(p,q)-Hermite–Hadamard Inequalities for Double Integral and (p,q)-Differentiable Convex Functions

Julalak Prabseang [1], Kamsing Nonlaopon [1,*] and Jessada Tariboon [2]

[1] Department of Mathematics, Faculty of Science, Khon Kaen University, Khon Kaen 40002, Thailand; julalak.pra@kkumail.com
[2] Department of Mathematics, Faculty of Applied Science, King Mongkut's University of Technology North Bangkok, Bangkok 10800, Thailand; jessada.t@sci.kmutnb.ac.th
[*] Correspondence: nkamsi@kku.ac.th;

Abstract: The aim of this paper is to establish some new (p,q)-calculus of Hermite–Hadamard inequalities for the double integral and refinements of the Hermite–Hadamard inequality for (p,q)-differentiable convex functions.

Keywords: Hermite–Hadamard inequalities; (p,q)-derivative; (p,q)-integral; convex functions

---

1. Introduction

Quantum calculus is the study of calculus without limits and is sometimes called q-calculus. In q-calculus, we obtain the original mathematical formulas when q tends to one. The beginning of the study of q-calculus can be dated back to the era of Euler (1707–1783), who first launched the q-calculus in the tracks of Newton's work on infinite series. Then, in the early Twentieth Century, Jackson [1] defined an integral, which is known as the q-Jackson integral, and studied it in a systematic way. The subject of q-calculus has many applications in the field of mathematics and other areas such as number theory, special functions, combinatorics, basic hypergeometric functions, orthogonal polynomials, quantum theory, mechanics, and the theory of relativity and physics. In recent years, the topic of q-calculus has increasingly interested many researchers. For more details, see [2–9] and the references therein. Recently, Tunç and Göv [10–12] studied the concept of (p,q)-calculus over the intervals of $[a,b] \subset \mathbb{R}$. The (p,q)-derivative and (p,q)-integral were defined and some basic properties are given. Furthermore, they obtained some new result for the (p,q)-calculus of several important integral inequalities. Currently, the (p,q)-calculus is being investigated extensively by many researchers, and a variety of new results can be found in the literature [13–18] and the references cited therein.

Mathematical inequalities are important to the study of mathematics, as well as in other area of mathematics such as analysis, differential equations, geometry, etcetera.

In 1893, Hadamard [19] investigated one of the fundamental inequalities in analysis as:

$$f\left(\frac{a+b}{2}\right) \leq \frac{1}{b-a}\int_a^b f(x)dx \leq \frac{f(a)+f(b)}{2}, \tag{1}$$

which is now known as the Hermite–Hadamard inequality.

In 2014, Tariboon and Ntouyas [20] studied the extension to q-calculus on the finite interval of (1), which is called the q-Hermite–Hadamard inequality, and some important inequalities. Next, Alp et al. [21] approved the q-Hermite–Hadamard inequality and then obtained generalized q-Hermite–Hadamard inequalities.

In 2018, Mehmet Kunt et al. [22] proved the left-hand side of the (p,q)-Hermite–Hadamard's inequality of (1) through (p,q)-differentiable convex and quasi-convex functions, and then, they gave some new (p,q)-Hermite–Hadamard's inequalities.

In 2019, Prabseang et al. [23] established the q-calculus of Hermite–Hadamard inequalities for the double integral as:

$$f\left(\frac{a+b}{2}\right) \leq \frac{1}{(b-a)^2}\int_a^b \int_a^b f(tx+(1-t)y)dxdy \leq \frac{f(a)+f(b)}{2}, \tag{2}$$

which was given by Dragomir [24]. Moreover, they obtained refinements of the Hermite–Hadamard inequality for q-differentiable convex functions.

The aim of this paper is to present the (p,q)-calculus of Hermite–Hadamard inequalities for double integrals (2) and refinements of the Hermite–Hadamard inequality. These are obtained as special cases when $p = 1$ and $q \to 1$.

Before we proceed to our main theorem, the following definitions and some concepts require some clarifications.

2. Preliminaries

Throughout this paper, let $[a,b] \subseteq \mathbb{R}$ be an interval and $0 < q < p \leq 1$ be a constant. The following definitions for the (p,q)-derivative and (p,q)-integral were given in [10,11].

Definition 1. *Let $f : [a,b] \to \mathbb{R}$ be a continuous function, and let $x \in [a,b]$. Then, the (p,q)-derivative of f on $[a,b]$ at x is defined as:*

$$\begin{aligned} {}_aD_{p,q}f(x) &= \frac{f(px+(1-p)a)-f(qx+(1-q)a)}{(p-q)(x-a)}, \quad x \neq a \\ {}_aD_{p,q}f(a) &= \lim_{x\to a} {}_aD_{p,q}f(x). \end{aligned} \tag{3}$$

Obviously, a function f is (p,q)-differentiable on $[a,b]$ if ${}_aD_{p,q}f(x)$ exists for all $x \in [a,b]$. In Definition 1, if $a = 0$, then ${}_0D_{p,q}f = D_{p,q}f$, where $D_{p,q}f$ is defined by:

$$D_{p,q}f(x) = \frac{f(px)-f(qx)}{(p-q)x}, \quad x \neq 0. \tag{4}$$

Furthermore, if $p = 1$ in (4), then it reduces to D_qf, which is the q-derivative of the function f; see [5].

Example 1. *Define function $f : [a,b] \to \mathbb{R}$ by $f(x) = x^2+1$. Let $0 < q < p \leq 1$. Then, for $x \neq a$, we have:*

$$\begin{aligned} {}_aD_{p,q}(x^2+1) &= \frac{[(px+(1-p)a)^2+1]-[(qx+(1-q)a)^2+1]}{(p-q)(x-a)} \\ &= \frac{(p+q)x^2+2ax[1-(p+q)]+a^2[(p+q)-2]}{(x-a)} \\ &= \frac{x(p+q)(x-a)-a(p+q)(x-a)+2a(x-a)}{(x-a)} \\ &= (p+q)(x-a)+2a. \end{aligned} \tag{5}$$

Definition 2. *Let $f : [a,b] \to \mathbb{R}$ be a continuous function. Then, the (p,q)-integral on $[a,b]$ is defined by:*

$$\int_a^x f(t)\, {}_ad_{p,q}t = (p-q)(x-a)\sum_{n=0}^{\infty} \frac{q^n}{p^{n+1}} f\left(\frac{q^n}{p^{n+1}}x + \left(1-\frac{q^n}{p^{n+1}}\right)a\right), \tag{6}$$

for $x \in [a,b]$. If $a = 0$ and $p = 1$ in (6), then we have the classical q-integral [5].

Example 2. *Define function $f:[a,b]\to\mathbb{R}$ by $f(x)=2x$. Let $0<q<p\leq 1$. Then, we have:*

$$\begin{aligned}\int_a^b f(x)\ {}_ad_{p,q}x &= \int_a^b 2x_a d_{p,q}x \\ &= 2(p-q)(b-a)\sum_{n=0}^{\infty}\frac{q^n}{p^{n+1}}\left(\frac{q^n}{p^{n+1}}b+\left(1-\frac{q^n}{p^{n+1}}\right)a\right) \\ &= \frac{2(b-a)(b-a(1-p-q))}{p+q}.\end{aligned} \tag{7}$$

Theorem 1. *Let $f:[a,b]\to\mathbb{R}$ be a continuous function. Then, we have the following:*

(i) ${}_aD_{p,q}\int_a^x f(t)\ {}_ad_{p,q}t = f(x)$;
(ii) $\int_c^x {}_aD_{p,q}f(t)\ {}_ad_{p,q}t = f(x)-f(c)$ *for* $c\in(a,x)$.

Theorem 2. *Let $f,g:[a,b]\to\mathbb{R}$ be continuous functions and $\alpha\in\mathbb{R}$. Then, we have the following:*

(i) $\int_a^x [f(t)+g(t)]\ {}_ad_{p,q}t = \int_a^x f(t)\ {}_ad_{p,q}t + \int_a^x g(t)\ {}_ad_{p,q}t$;
(ii) $\int_a^x (\alpha f)(t)_a d_{p,q}t = \alpha\int_a^x f(t)_a d_{p,q}t$;
(iii) $\int_c^x f(pt+(1-p)a)_aD_{p,q}g(t)\ {}_ad_qt = (fg)|_c^x - \int_c^x g(qt+(1-q)a)_aD_{p,q}f(t)\ {}_ad_{p,q}t$ *for* $c\in(a,x)$.

For the proof properties of Theorems 1 and 2, we refer to [10,11].
The proofs of the following theorems were given in [22].

Theorem 3. *Let $f:[a,b]\to\mathbb{R}$ be a convex differentiable function on (a,b) and $0<q<p\leq 1$. Then, we have:*

$$f\left(\frac{qa+pb}{p+q}\right)\leq\frac{1}{p(b-a)}\int_a^{pb+(1-p)a} f(x)\ {}_ad_{p,q}x\leq\frac{qf(a)+pf(b)}{p+q}. \tag{8}$$

Theorem 4. *Let $f:[a,b]\to\mathbb{R}$ be a convex differentiable function on (a,b) and $0<q<p\leq 1$. Then, we have:*

$$\begin{aligned}f\left(\frac{pa+qb}{p+q}\right)+\frac{(p-q)(b-a)}{p+q}f'\left(\frac{pa+qb}{p+q}\right) &\leq \frac{1}{p(b-a)}\int_a^{pb+(1-p)a} f(x)\ {}_ad_{p,q}x \\ &\leq \frac{qf(a)+pf(b)}{p+q}.\end{aligned} \tag{9}$$

Theorem 5. *Let $f:[a,b]\to\mathbb{R}$ be a convex differentiable function on (a,b) and $0<q<p\leq 1$. Then, we have:*

$$\begin{aligned}f\left(\frac{a+b}{2}\right)+\frac{(p-q)(b-a)}{2(p+q)}f'\left(\frac{a+b}{2}\right) &\leq \frac{1}{p(b-a)}\int_a^{pb+(1-p)a} f(x)\ {}_ad_{p,q}x \\ &\leq \frac{qf(a)+pf(b)}{p+q}.\end{aligned} \tag{10}$$

Lemma 1. *Let $f:[a,b]\to\mathbb{R}$ be a convex continuous function on $[a,b]$ and $0<q<p\leq 1$. Then, we have:*

$$\begin{aligned}&f\left(\frac{1}{(pb-pa)^2}\int_a^{pb+(1-p)a}\int_a^{pb+(1-p)a}(tx+(1-t)y)\ {}_ad_{p,q}x_ad_{p,q}y\right) \\ &\leq \frac{1}{(pb-pa)^2}\int_a^{pb+(1-p)a}\int_a^{pb+(1-p)a} f(tx+(1-t)y)\ {}_ad_{p,q}x_ad_{p,q}y.\end{aligned} \tag{11}$$

Proof. The proof of this lemma can be obtained by Definition 2 and Jensen's inequality. □

3. Main Results

In this section, we present the (p,q)-Hermite–Hadamard inequality for double integrals and the refinement of Hermite–Hadamard inequalities on the interval $[a,b]$.

Theorem 6. *Let $f : [a,b] \to \mathbb{R}$ be a convex continuous function on $[a,b]$ and $0 < q < p \leq 1$. Then, we have:*

$$\begin{aligned} f\left(\frac{qa+pb}{p+q}\right) &\leq \frac{1}{(pb-pa)^2}\int_a^{pb+(1-p)a}\int_a^{pb+(1-p)a} f(tx+(1-t)y)\ {}_ad_{p,q}x\,{}_ad_{p,q}y \\ &\leq \frac{1}{p(b-a)}\int_a^{pb+(1-p)a} f(x)\ {}_ad_{p,q}x\,{}_ad_{p,q}y \\ &\leq \frac{qf(a)+pf(b)}{p+q}. \end{aligned} \tag{12}$$

Proof. Since f is convex on $[a,b]$, it follows that:

$$f(tx+(1-t)y) \leq tf(x)+(1-t)f(y) \tag{13}$$

for all $x, y \in [a,b]$ and $t \in [0,1]$. Taking the double (p,q)-integration on both sides for (13) on $[a, pb+(1-p)a] \times [a, pb+(1-p)a]$, we obtain:

$$\begin{aligned} &\int_a^{pb+(1-p)a}\int_a^{pb+(1-p)a} f(tx+(1-t)y)\ {}_ad_{p,q}x\,{}_ad_{p,q}y \\ &\quad\leq \int_a^{pb+(1-p)a}\int_a^{pb+(1-p)a} [tf(x)+(1-t)f(y)]\ {}_ad_{p,q}x\,{}_ad_{p,q}y \\ &\quad= (pb-pa)\int_a^{pb+(1-p)a} f(x)\,{}_ad_{p,q}x, \end{aligned} \tag{14}$$

which show the second part of (12) by using the right-hand side of the (p,q)-Hermite–Hadamard's inequality.

On the other hand, by Lemma 1, we have:

$$\begin{aligned} &f\left(\frac{1}{(pb-pa)^2}\int_a^{pb+(1-p)a}\int_a^{pb+(1-p)a} (tx+(1-t)y)\ {}_ad_{p,q}x\,{}_ad_{p,q}y\right) \\ &\leq \frac{1}{(pb-pa)^2}\int_a^{pb+(1-p)a}\int_a^{pb+(1-p)a} f(tx+(1-t)y)\ {}_ad_{p,q}x\,{}_ad_{p,q}y, \end{aligned}$$

and since:

$$\frac{1}{(pb-pa)^2}\int_a^{pb+(1-p)a}\int_a^{pb+(1-p)a} (tx+(1-t)y)\ {}_ad_{p,q}x\,{}_ad_{p,q}y = \frac{qa+pb}{p+q}.$$

This completes the proof. □

Remark 1. *If $p = 1$ and $q \to 1$, then (12) reduces to (2), that is,*

$$f\left(\frac{a+b}{2}\right) \leq \frac{1}{(b-a)^2}\int_a^b\int_a^b f(tx+(1-t)y)dxdy \leq \frac{f(a)+f(b)}{2}.$$

Corollary 1. *Let $f : [a,b] \to \mathbb{R}$ be a convex continuous function on $[a,b]$ and $0 < q < p \leq 1$. Then, we have:*

$$\begin{aligned} f\left(\frac{qa+pb}{p+q}\right) &\leq \frac{1}{(pb-pa)^2}\int_a^{pb+(1-p)a}\int_a^{pb+(1-p)a} f\left(\frac{x+y}{2}\right)\ {}_ad_{p,q}x\,{}_ad_{p,q}y \\ &\leq \frac{1}{p(b-a)}\int_a^{pb+(1-p)a} f(x)\ {}_ad_{p,q}x \\ &\leq \frac{qf(a)+pf(b)}{p+q}. \end{aligned} \tag{15}$$

Remark 2. *If $p = 1$ and $q \to 1$, then (15) reduces to:*

$$f\left(\frac{a+b}{2}\right) \leq \frac{1}{(b-a)^2}\int_a^b\int_a^b f\left(\frac{x+y}{2}\right)dxdy \leq \frac{1}{b-a}\int_a^b f(x)dx \leq \frac{f(a)+f(b)}{2},$$

which readily appeared in [25].

Theorem 7. *Let $f : [a,b] \to \mathbb{R}$ be a convex continuous function on $[a,b]$ and $0 < q < p \leq 1$. Then, we have:*

$$\begin{aligned} \frac{p}{(pb-pa)^2} & \int_a^{pb+(1-p)a}\int_a^{pb+(1-p)a} f\left(\frac{px+qy}{p+q}\right) {}_a d_{p,q}x {}_a d_{p,q}y \\ & \leq \frac{1}{(pb-pa)^2}\int_0^p\int_a^{pb+(1-p)a}\int_a^{pb+(1-p)a} f(tx+(1-t)y)\, {}_a d_{p,q}x {}_a d_{p,q}y {}_a d_{p,q}t \\ & \leq \frac{1}{pb-pa}\int_a^{pb+(1-p)a} f(x)\, {}_a d_{p,q}x. \end{aligned} \tag{16}$$

Proof. Let $g : [a,b] \to \mathbb{R}$ be given by:

$$g(t) = \frac{1}{(pb-pa)^2}\int_a^{pb+(1-p)a}\int_a^{pb+(1-p)a} f(tx+(1-t)y)\, {}_a d_{p,q}x {}_a d_{p,q}y.$$

For all $t_1, t_2 \in [0,1]$ and $\alpha, \beta \geq 0$ with $\alpha + \beta = 1$, we consider:

$$\begin{aligned} g(\alpha t_1 + \beta t_2) &= \frac{1}{(pb-pa)^2}\int_a^{pb+(1-p)a}\int_a^{pb+(1-p)a} f((\alpha t_1+\beta t_2)x + (1-(\alpha t_1+\beta t_2))y)\, {}_a d_{p,q}x {}_a d_{p,q}y \\ &\leq \frac{\alpha}{(pb-pa)^2}\int_a^{pb+(1-p)a}\int_a^{pb+(1-p)a} f(t_1x+(1-t_1)y)\, {}_a d_{p,q}x {}_a d_{p,q}y \\ &\quad + \frac{\beta}{(pb-pa)^2}\int_a^{pb+(1-p)a}\int_a^{pb+(1-p)a} f(t_2x+(1-t_2)y)\, {}_a d_{p,q}x {}_a d_{p,q}y \\ &= \alpha g(t_1) + \beta g(t_2), \end{aligned}$$

which show that g is convex on $[0,1]$. Using Theorem 3 for the convex function g, we have:

$$\begin{aligned} \frac{1}{(pb-pa)^2} & \int_a^{pb+(1-p)a}\int_a^{pb+(1-p)a} f\left(\frac{px+qy}{p+q}\right) {}_a d_q x {}_a d_q y \\ & = g\left(\frac{p}{p+q}\right) \leq \frac{1}{p}\int_0^p g(t)\, {}_a d_{p,q}t \\ & = \frac{1}{p(pb-pa)^2}\int_0^p\int_a^{pb+(1-p)a}\int_a^{pb+(1-p)a} f(tx+(1-t)y)\, {}_a d_{p,q}x {}_a d_{p,q}y {}_a d_{p,q}t \\ & \leq \frac{qg(0)+pg(1)}{p(p+q)} = \frac{1}{p(pb-pa)}\int_a^{pb+(1-p)a} f(x)\, {}_a d_{p,q}x. \end{aligned}$$

This completes the proof. □

Remark 3. *If $p = 1$ and $q \to 1$, then (16) reduces to:*

$$\begin{aligned} \frac{1}{(b-a)^2}\int_a^b\int_a^b f\left(\frac{x+y}{2}\right)dxdy &\leq \frac{1}{(b-a)^2}\int_0^1\int_a^b\int_a^b f(tx+(1-t)y)dxdydt \\ &\leq \frac{1}{b-a}\int_a^b f(x)dx, \end{aligned}$$

which readily appeared in [25].

Theorem 8. *Let $f : [a,b] \to \mathbb{R}$ be a (p,q)-differentiable convex continuous function and $0 < q < p \leq 1$, then the following inequalities:*

$$\begin{aligned} 0 &\leq \tfrac{p}{b-a} \int_a^{pb+(1-p)a} f(x) \ {}_a d_{p,q} x \\ &- \tfrac{1}{(b-a)^2} \int_a^{pb+(1-p)a} \int_a^{pb+(1-p)a} f(tx + (1-t)y) \ {}_a d_{p,q} x \, {}_a d_{p,q} y \\ &\leq t \left[\tfrac{p^2 f(a) + pq f(pb+(1-p)a)}{p+q} - \tfrac{p}{b-a} \int_a^{pb+(1-p)a} f(qx + (1-q)a) \ {}_a d_{p,q} x \right], \end{aligned} \tag{17}$$

are valid for all $t \in [0,1]$.

Proof. Since f is convex on J, it follows that:

$$f(tx + (1-t)y) \leq tf(x) + (1-t)f(y)$$

for all $x, y \in [a,b]$ and $t \in [0,1]$. Taking double (p,q)-integration on both sides of the above inequality on $[a, pb + (1-p)a] \times [a, pb + (1-p)a]$, we obtain:

$$\begin{aligned} &\int_a^{pb+(1-p)a} \int_a^{pb+(1-p)a} f(tx + (1-t)y) \ {}_a d_{p,q} x \, {}_a d_{p,q} y \\ &\qquad \leq \int_a^{pb+(1-p)a} \int_a^{pb+(1-p)a} [tf(x) + (1-t)f(y)] \ {}_a d_{p,q} x \, {}_a d_{p,q} y \\ &\qquad = p(b-a) \int_a^{pb+(1-p)a} f(x) {}_a d_{p,q} x. \end{aligned}$$

On the other hand, since f is (p,q)-differentiable convex on $[a,b]$ and $f' \geq {}_a D_{p,q} f$, we have:

$$f(tx + (1-t)y) - f(y) \geq t(x-y) \ {}_a D_{p,q} f(y)$$

for all $x, y \in [a,b]$ and $t \in [0,1]$. Taking the double (p,q)-integration on both sides of the above inequality on $[a, pb + (1-p)a] \times [a, pb + (1-p)a]$, we obtain:

$$\begin{aligned} &\int_a^{pb+(1-p)a} \int_a^{pb+(1-p)a} f(tx + (1-t)y) \ {}_a d_{p,q} x \, {}_a d_{p,q} y - (pb - pa) \int_a^{pb+(1-p)a} f(x) \ {}_a d_{p,q} x \\ &\qquad \geq t \int_a^{pb+(1-p)a} \int_a^{pb+(1-p)a} (x-y) {}_a D_{p,q} f(y) \ {}_a d_{p,q} x \, {}_a d_{p,q} y. \end{aligned} \tag{18}$$

Since,

$$\begin{aligned} &\int_a^{pb+(1-p)a} \int_a^{pb+(1-p)a} (x-y) {}_a D_{p,q} f(y) \ {}_a d_{p,q} x \, {}_a d_{p,q} y \\ &= (pb - pa) \int_a^{pb+(1-p)a} f(qx + (1-q)a) \ {}_a d_{p,q} x - (b-a)^2 \frac{[p^2 f(a) + pq f(pb + (1-p)a)]}{p+q}. \end{aligned}$$

Substituting the above inequality in (18), we have:

$$\begin{aligned} &(pb - pa) \int_a^{pb+(1-p)a} f(x) \ {}_a d_{p,q} x - \int_a^{pb+(1-p)a} \int_a^{pb+(1-p)a} f(tx + (1-t)y) \ {}_a d_{p,q} x \, {}_a d_{p,q} y \\ &\leq t \left[(b-a)^2 \frac{[p^2 f(a) + pq f(pb + (1-p)a)]}{p+q} - (pb - pa) \int_a^{pb+(1-p)a} f(qx + (1-q)a) \ {}_a d_{p,q} x \right] \end{aligned}$$

for all $t \in [0,1]$, which completes the proof. □

Remark 4. *If $p = 1$ and $q \to 1$, then (17) reduces to:*

$$0 \leq \frac{1}{b-a}\int_a^b f(x)dx - \frac{1}{(b-a)^2}\int_a^b\int_a^b f(tx+(1-t)y)dxdy$$
$$\leq t\left[\frac{f(a)+f(b)}{2} - \frac{1}{b-a}\int_a^b f(x)dx\right],$$

which readily appeared in [25,26].

Corollary 2. *Let $f : [a,b] \to \mathbb{R}$ be a (p,q)-differentiable convex continuous function and $0 < q < p \leq 1$. Then, we have:*

$$\begin{aligned} 0 &\leq \tfrac{p}{b-a}\int_a^{pb+(1-p)a} f(x)\ {}_ad_{p,q}x - \tfrac{1}{(b-a)^2}\int_a^{pb+(1-p)a}\int_a^{pb+(1-p)a} f\left(\tfrac{x+y}{2}\right)\ {}_ad_{p,q}x\,{}_ad_{p,q}y \\ &\leq \tfrac{1}{2}\left[\tfrac{p^2f(a)+pqf(pb+(1-p)a)}{p+q} - \tfrac{p}{b-a}\int_a^{pb+(1-p)a} f(qx+(1-q)a)\ {}_ad_{p,q}x\right]. \end{aligned} \tag{19}$$

Remark 5. *If $p = 1$ and $q \to 1$, then (19) reduces to:*

$$0 \leq \frac{1}{b-a}\int_a^b f(x)dx - \frac{1}{(b-a)^2}\int_a^b\int_a^b f\left(\frac{x+y}{2}\right)dxdy$$
$$\leq \frac{1}{2}\left[\frac{f(a)+f(b)}{2} - \frac{1}{b-a}\int_a^b f(x)dx\right],$$

which readily appeared in [25].

Theorem 9. *Let $f : [a,b] \to \mathbb{R}$ be a (p,q)-differentiable convex continuous function, which is defined at the point $\frac{qa+pb}{p+q} \in (a,b)$ and $0 < q < p \leq 1$. Then, the following inequalities:*

$$\begin{aligned} 0 &\leq \tfrac{1}{b-a}\int_a^{pb+(1-p)a} f(x)\ {}_ad_{p,q}x - \tfrac{1}{b-a}\int_a^{pb+(1-p)a} f\left(tx+(1-t)\tfrac{qa+pb}{p+q}\right)\ {}_ad_{p,q}x \\ &\leq (1-t)\left[\tfrac{pf(a)+qf(pb+(1-p)a)}{p+q} - \tfrac{1}{b-a}\int_a^{pb+(1-p)a} f(qx+(1-q)a)\ {}_ad_{p,q}x\right] \end{aligned} \tag{20}$$

are valid for all $t \in [0,1]$.

Proof. Since f is convex on $[a,b]$ and using Theorem 3, we have:

$$\frac{1}{p(b-a)}\int_a^{pb+(1-p)a} f\left(tx+(1-t)\frac{qa+pb}{p+q}\right)\ {}_ad_{p,q}x$$
$$\leq \frac{t}{p(b-a)}\int_a^{pb+(1-p)a} f(x)\ {}_ad_{p,q}x + (1-t)f\left(\frac{qa+pb}{p+q}\right)$$
$$\leq \frac{t}{p(b-a)}\int_a^{pb+(1-p)a} f(x)\ {}_ad_{p,q}x + \frac{1-t}{p(b-a)}\int_a^{pb+(1-p)a} f(x)\ {}_ad_{p,q}x$$

$$= \frac{1}{p(b-a)}\int_a^{pb+(1-p)a} f(x)\ {}_ad_{p,q}x$$

for all $t \in [0,1]$.

On the other hand, since f is the (p,q)-differentiable convex on $[a,b]$, we have:

$$f\left(tx+(1-t)\frac{qa+pb}{p+q}\right) - f(x) \geq (1-t)\left(\frac{qa+pb}{p+q} - x\right)\ {}_aD_{p,q}(x).$$

Taking the double (p,q)-integration on both sides of the above inequality on $[a,b]$, we obtain:

$$\begin{aligned} \frac{1}{p(b-a)} & \int_a^{pb+(1-p)a} f\left(tx+(1-t)\frac{qa+pb}{p+q}\right)\,{}_ad_{p,q}x - \frac{1}{p(b-a)}\int_a^{pb+(1-p)a} f(x)\,{}_ad_{p,q}x \\ & \geq \frac{(1-t)}{p(b-a)}\int_a^{pb+(1-p)a}\left(\frac{qa+pb}{p+q}-x\right)\,{}_aD_{p,q}f(x){}_ad_{p,q}x. \end{aligned} \tag{21}$$

Since,

$$\begin{aligned} \int_a^{pb+(1-p)a} & \left(\frac{qa+pb}{p+q}-x\right)\,{}_aD_{p,q}f(x){}_ad_{p,q}x \\ & = \int_a^{pb+(1-p)a} f(qx+(1-q)a){}_ad_{p,q}x-(b-a)\frac{pf(a)+qf(pb+(1-p)a)}{p+q}. \end{aligned} \tag{22}$$

This completes the proof. □

Corollary 3. *Let $f:[a,b]\to\mathbb{R}$ be a (p,q)-differentiable convex continuous function and $0<q<p\leq 1$. Then, we have:*

$$\begin{aligned} 0 & \leq \frac{1}{b-a}\int_a^{pb+(1-p)a} f(x)\,{}_ad_{p,q}x - \frac{2}{b-a}\int_{\frac{a(p+2q)+pb}{2(1+q)}}^{\frac{(p^2+pq)(b-a)+(p+2q)a+pb}{2(p+q)}} f(x)\,{}_ad_{p,q}x \\ & \leq \frac{1}{2}\left[\frac{pf(a)+qf(pb+(1-p)a)}{p+q}-\frac{1}{b-a}\int_a^{pb+(1-p)a} f(qx+(1-q)a)\,{}_ad_{p,q}x\right]. \end{aligned} \tag{23}$$

Theorem 10. *Let $f:[a,b]\to\mathbb{R}$ be a (p,q)-differentiable convex continuous function, which is defined at the point $\frac{pa+qb}{p+q}\in(a,b)$ and $0<q<p\leq 1$. Then, the following inequalities:*

$$\begin{aligned} (1-t) & \quad \frac{p(p-q)(b-a)}{p+q}f'\left(\frac{pa+qb}{p+q}\right) \\ & \leq \frac{1}{b-a}\int_a^{pb+(1-p)a} f(x)\,{}_ad_{p,q}x - \frac{1}{b-a}\int_a^{pb+(1-p)a} f\left(tx+(1-t)\frac{pa+qb}{p+q}\right)\,{}_ad_{p,q}x \\ & \leq (1-t)\left[\frac{qf(a)+pf(pb+(1-p)a)}{p+q}-\frac{1}{b-a}\int_a^{pb+(1-p)a} f(qx+(1-q)a)\,{}_ad_{p,q}x\right] \end{aligned} \tag{24}$$

are valid for all $t\in[0,1]$.

Proof. The proof of this theorem follows a similar procedure as Theorem 9 by using Theorem 4. □

Corollary 4. *Let $f:[a,b]\to\mathbb{R}$ be a (p,q)-differentiable convex continuous function and $0<q<p\leq 1$. Then, we have:*

$$\begin{aligned} & \frac{p(p-q)(b-a)}{2(p+q)}f'\left(\frac{pa+qb}{p+q}\right) \\ & \leq \frac{1}{b-a}\int_a^{pb+(1-p)a} f(x)\,{}_ad_{p,q}x - \frac{2}{b-a}\int_{\frac{2pa+q(a+b)}{2(p+q)}}^{\frac{(p^2+pq)(b-a)+(2p+q)a+qb}{2(1+q)}} f(x)\,{}_ad_{p,q}x \\ & \leq \frac{1}{2}\left[\frac{qf(a)+pf(pb+(1-p)a)}{p+q}-\frac{1}{b-a}\int_a^{pb+(1-p)a} f(qx+(1-q)a)\,{}_ad_{p,q}x\right]. \end{aligned} \tag{25}$$

Theorem 11. *Let $f:[a,b]\to\mathbb{R}$ be a (p,q)-differentiable convex continuous function, which is defined at the point $\frac{a+b}{2}\in(a,b)$ and $0<q<p\leq 1$. Then, the following inequalities:*

$$\begin{aligned} (1-t) & \quad \frac{p(p-q)(b-a)}{2(p+q)}f'\left(\frac{a+b}{2}\right) \\ & \leq \frac{1}{b-a}\int_a^{pb+(1-p)a} f(x)\,{}_ad_{p,q}x - \frac{1}{b-a}\int_a^{pb+(1-p)a} f\left(tx+(1-t)\frac{a+b}{2}\right)\,{}_ad_{p,q}x \\ & \leq (1-t)\left[\frac{f(a)+f(pb+(1-q)a)}{2}-\frac{1}{b-a}\int_a^{pb+(1-p)a} f(qx+(1-q)a)\,{}_ad_{p,q}x\right] \end{aligned} \tag{26}$$

are valid for all $t \in [0,1]$.

Proof. The proof of this theorem follows a similar procedure as Theorem 9 by using Theorem 5. □

Corollary 5. *Let* $f : [a,b] \to \mathbb{R}$ *be a* (p,q)*-differentiable convex continuous function and* $0 < q < p \leq 1$. *Then, we have:*

$$\begin{aligned} & \tfrac{p(p-q)(b-a)}{4(p+q)} f'\left(\tfrac{a+b}{2}\right) \\ & \leq \tfrac{1}{b-a} \int_a^{pb+(1-p)a} f(x)\ {}_a d_{p,q}x - \tfrac{2}{b-a} \int_{\frac{3a+b}{4}}^{\frac{2p(b-a)+3a+b}{4}} f(x)\ {}_a d_{p,q}x \\ & \leq \tfrac{1}{2}\left[\tfrac{qf(a)+pf(pb+(1-p)a)}{p+q} - \tfrac{1}{b-a}\int_a^{pb+(1-p)a} f(qx+(1-q)a)\ {}_a d_{p,q}x\right]. \end{aligned} \tag{27}$$

Remark 6. *If* $p = 1$ *and* $q \to 1$*, then (20), (24), and (26) reduce to:*

$$\begin{aligned} 0 & \leq \frac{1}{b-a}\int_a^b f(x)dx - \frac{1}{b-a}\int_a^b f\left(tx + (1-t)\frac{a+b}{2}\right)dx \\ & \leq (1-t)\left[\frac{f(a)+f(b)}{2} - \frac{1}{b-a}\int_a^b f(x)dx\right], \end{aligned}$$

which readily appeared in [25].

Remark 7. *If* $p = 1$ *and* $q \to 1$*, then (23), (25), and (27) reduce to:*

$$0 \leq \frac{1}{b-a}\int_a^b f(x)dx - \frac{2}{b-a}\int_{\frac{3a+b}{4}}^{\frac{a+3b}{4}} f(x)dx \leq \frac{1}{2}\left[\frac{f(a)+f(b)}{2} - \frac{1}{b-a}\int_a^b f(x)dx\right],$$

which readily appeared in [25].

4. Conclusions

In this paper, we have obtained some new results for the (p,q)-calculus of Hermite–Hadamard inequalities for the double integral and refinements of the Hermite–Hadamard inequality. Our work has improved the results of [23] and can be reduced to the classical inequality formulas in special cases when $p = 1$ and $q \to 1$. It is expected that this paper may stimulate further research in this field.

Author Contributions: The order of the author list reflects the contributions to the paper.

References

1. Jackson, F.H. On a q-definite integrals. *Quart. J. Pure Appl. Math.* **1910**, *41*, 193–203.
2. Bangerezako, G. Variational q-calculus. *J. Math. Anal. Appl.* **2004**, *289*, 650–665. [CrossRef]
3. Ernst, T. *The History of q-Calculus and a New Method*; Uppsala University: Uppsala, Sweden, 2000.
4. Jackson, F.H. q-Difference equations. *Am. J. Math.* **1910**, *32*, 305–314. [CrossRef]
5. Kac, V.; Cheung, P. *Quantum Calculus*; Springer: New York, NY, USA, 2002.
6. Gauchman, H. Integral inequalities in q-Calculus. *Comput. Math. Appl.* **2004**, *47*, 281–300. [CrossRef]
7. Noor, M.A.; Noor, K.I.; Awan, M.U. Some quantum estimates for Hermite–Hadamard inequalities. *Appl. Math. Comput.* **2015**, *251*, 675–679. [CrossRef]
8. Noor, M.A.; Noor, K.I.; Awan, M.U. Some quantum integral inequalities via preinvex functions. *Appl. Math. Comput.* **2015**, *269*, 242–251. [CrossRef]
9. Sudsutad, W.; Ntouyas, S.K.; Tariboon, J. Quantum integral inequalities for convex functions. *J. Math. Inequal.* **2015**, *9*, 781–793. [CrossRef]
10. Tunç, M.; Göv, E. (p,q)-Integral inequalities. *RGMIA Res. Rep. Coll.* **2016**, *19*, 97.
11. Tunç, M.; Göv, E. Some integral inequalities via (p,q)-calculus on finite intervals. *RGMIA Res. Rep. Coll.* **2016**, *19*, 95.

12. Tunç, M.; Göv, E. (p,q)-integral inequalities for convex functions. *RGMIA Res. Rep. Coll.* **2016**, *19*, 98.
13. Araci, S.; Duran, U.; Acikgoz, M.; Srivastava, H.M. A certain (p,q)-derivative operator and associated divided differences. *J. Inequal. Appl.* **2016**, *1*, 301. [CrossRef]
14. Duran, U.; Acikgoz, M.; Esi, A.; Araci, S. A note on the (p,q) Hermite polynomials. *Appl. Math. Inf. Sci.* **2018**, *12*, 227–231. [CrossRef]
15. Mursaleen, M.; Ansari, K.J.; Khan, A. Some Approximation Results by (p,q)-analogue of Bernstein-Stancu operators. *Appl. Math. Comput.* **2015**, *264*, 392–402. [Corrigendum: *Appl. Math. Comput.* **2015**, *269*, 744–746.] [CrossRef]
16. Sahai, V.; Yadav, S. Representations of two parameter quantum algebras and p,q-special functions. *J. Math. Anal. Appl.* **2007**, *335*, 268–279. [CrossRef]
17. Sadjang, P.N. On the fundamental theorem of (p,q)-calculus and some (p,q)-Taylor formulas. *Results Math.* **2018**, *73*, 39. [CrossRef]
18. Sadjang, P.N. On the (p,q)-Gamma and the (p,q)-Beta functions. *arXiv* **2015**, arXiv:1506.07394.
19. Hadamard, J. Etude sur les propriétés des fonctions entiéres et en particulier d'une fonction considérée par Riemann. *J. Math. Pures Appl.* **1893**, *9*, 171–216.
20. Tariboon, J.; Ntouyas, S.K. Quantum integral inequalities on finite intervals. *J. Inequal. App.* **2014**, *2014*, 121. [CrossRef]
21. Alp, N.; Sarıkaya, M.Z.; Kunt, M.; İşcan, İ. q-Hermite–Hadamard inequalities and quantum estimates for midpoint type inequalities via convex and quasi-convex functions. *J. King Saud Univ. Sci.* **2018**, *30*, 193–203. [CrossRef]
22. Kunt, M.; İşcan, İ.; Alp, N.; Sarikaya, M.Z. (p,q)-Hermite–Hadamard inequalities and (p,q)-estimates for midpoint typeinequalities via convex and quasi-convex functions. *Rev. Real Acad. Cienc. Exactas Fís. Nat. Ser. A Mat.* **2018**, *112*, 969–992. [CrossRef]
23. Prabseang, J.; Nonlaopon, K.; Tariboon, J. Quantum Hermite–Hadamard inequalities for double integral and q-differentiable convex functions. *J. Math. Inequal.* **2019**, *13*, 675–686.
24. Dragomir, S.S. Two refinements of Hadamard's inequalities. *Coll. Sci. Pap. Fac. Kragujevac* **1990**, *11*, 23–26.
25. Pachpatte, B.G. *Mathematical Inequalities*; North-Holland Library, Elsevier Science: Amsterdam, Holland, 2005; Volume 67.
26. Dragomir, S.S. Some integral inequalities for differentiable convex functions. *Contrib. Sec. Math. Technol. Sci.* **1992**, *13*, 13–17.

8

Subordination Properties for Multivalent Functions Associated with a Generalized Fractional Differintegral Operator

Hanaa M. Zayed [1,*], Mohamed Kamal Aouf [2] and Adela O. Mostafa [2]

1 Department of Mathematics, Faculty of Science, Menofia University, Shebin Elkom 32511, Egypt
2 Department of Mathematics, Faculty of Science, Mansoura University, Mansoura 35516, Egypt; mkaouf@mans.edu.eg (M.K.A.); adelaeg254@yahoo.com (A.O.M.)
* Correspondence: hanaa_zayed42@yahoo.com

Abstract: Using of the principle of subordination, we investigate some subordination and convolution properties for classes of multivalent functions under certain assumptions on the parameters involved, which are defined by a generalized fractional differintegral operator under certain assumptions on the parameters involved.

Keywords: differential subordination; p-valent functions; generalized fractional differintegral operator

---

1. Introduction and Definitions

Denote by $\mathcal{A}(p)$ the class of analytic and p-valent functions of the form:

$$f(z) = z^p + \sum_{n=1}^{\infty} a_{p+n} z^{p+n} \ (p \in \mathbb{N} = \{1, 2, ...\}; z \in \mathbb{U} = \{z \in \mathbb{C} : |z| < 1\}). \tag{1}$$

For functions f, g analytic in $\mathbb{U}$, f is subordinate to g, written $f(z) \prec g(z)$ if there exists a function w, analytic in $\mathbb{U}$ with $w(0) = 0$ and $|w(z)| < 1$, such that $f(z) = g(w(z))$, $z \in \mathbb{U}$. If g is univalent in $\mathbb{U}$, then (see [1,2]):

$$f(z) \prec g(z) \Leftrightarrow f(0) = g(0) \text{ and } f(\mathbb{U}) \subset g(\mathbb{U}).$$

If $\varphi(z)$ is analytic in $\mathbb{U}$ and satisfies:

$$H(\varphi(z), z\varphi'(z)) \prec h(z), \tag{2}$$

then φ is a solution of (2). The univalent function q is called dominant, if $\varphi(z) \prec q(z)$ for all φ. A dominant $\widetilde{q}$ is called the best dominant, if $\widetilde{q}(z) \prec q(z)$ for all dominants q.

Let ${}_2F_1(a, b; c; z)$ $(c \neq 0, -1, -2, \ldots)$ be the well-known (Gaussian) hypergeometric function defined by:

$${}_2F_1(a, b; c; z) := \sum_{n=0}^{\infty} \frac{(a)_n (b)_n}{(c)_n (1)_n} z^n, \ z \in \mathbb{U},$$

where:

$$(\lambda)_n := \frac{\Gamma(\lambda + n)}{\Gamma(\lambda)}.$$

We will recall some definitions that will be used in our paper.

Definition 1. *For $f(z) \in \mathcal{A}(p)$, the fractional integral and fractional derivative operators of order λ are defined by Owa [3] (see also [4]) as:*

$$D_z^{-\lambda} f(z) := \frac{1}{\Gamma(\lambda)} \int_0^z \frac{f(\zeta)}{(z-\zeta)^{1-\lambda}}\, d\zeta \quad (\lambda > 0),$$

$$D_z^{\lambda} f(z) := \frac{1}{\Gamma(1-\lambda)} \frac{d}{dz} \int_0^z \frac{f(\zeta)}{(z-\zeta)^{\lambda}}\, d\zeta \quad (0 \leq \lambda < 1),$$

where f is an analytic function in a simply-connected region of the complex z-plane containing the origin, and the multiplicity of $(z-\zeta)^{\lambda-1}$ $((z-\zeta)^{-\lambda})$ is removed by requiring $\log(z-\zeta)$ to be real when $z-\zeta > 0$.

Definition 2. *For $f(z) \in \mathcal{A}(p)$ and in terms of ${}_2F_1$, the generalized fractional integral and generalized fractional derivative operators defined by Srivastava et al. [5] (see also [6]) as:*

$$I_{0,z}^{\lambda,\mu,\eta} f(z) := \frac{z^{-\lambda-\mu}}{\Gamma(\lambda)} \int_0^z (z-\zeta)^{\lambda-1} f(\zeta)\, {}_2F_1\left(\mu+\lambda, -\eta; \lambda; 1-\frac{\zeta}{z}\right) d\zeta \;(\lambda > 0,\ \mu,\eta \in \mathbb{R}),$$

$$J_{0,z}^{\lambda,\mu,\eta} f(z) := \begin{cases} \frac{d}{dz}\left\{ \frac{z^{\lambda-\mu} \int_0^z (z-\zeta)^{-\lambda} f(\zeta) {}_2F_1\left(\mu-\lambda, 1-\eta; 1-\lambda; 1-\frac{\zeta}{z}\right) d\zeta}{\Gamma(1-\lambda)} \right\} & (0 \leq \lambda < 1), \\ \frac{d^n}{dz^n} J_{0,z}^{\lambda-n,\mu,\eta} f(z) \; (n \leq \lambda < n+1;\ n \in \mathbb{N}), & \end{cases}$$

where $f(z)$ is an analytic function in a simply-connected region of the complex $z-$plane containing the origin with the order $f(z) = O(|z|^{\varepsilon})$, $z \to 0$ when $\varepsilon > \max\{0, \mu-\eta\} - 1$, and the multiplicity of $(z-\zeta)^{\lambda-1}$ $((z-\zeta)^{-\lambda})$ is removed by requiring $\log(z-\zeta)$ to be real when $z-\zeta > 0$.

We note that:

$$I_{0,z}^{\lambda,-\lambda,\eta} f(z) = D_z^{-\lambda} f(z) \;(\lambda > 0) \text{ and } J_{0,z}^{\lambda,\lambda,\eta} f(z) = D_z^{\lambda} f(z) \;(0 \leq \lambda < 1),$$

where $D_z^{-\lambda} f(z)$ and $D_z^{\lambda} f(z)$ are the fractional integral and fractional derivative operators studied by Owa [3].

Goyal and Prajapat [7] (see also [8]) defined the operator:

$$S_{0,z}^{\lambda,\mu,\eta,p} f(z) = \begin{cases} \frac{\Gamma(p+1-\mu)\Gamma(p+1-\lambda+\eta)}{\Gamma(p+1)\Gamma(p+1-\mu+\eta)} z^{\mu} J_{0,z}^{\lambda,\mu,\eta} f(z) \; (0 \leq \lambda < \eta+p+1;\ z \in \mathbb{U}), \\ \frac{\Gamma(p+1-\mu)\Gamma(p+1-\lambda+\eta)}{\Gamma(p+1)\Gamma(p+1-\mu+\eta)} z^{\mu} I_{0,z}^{-\lambda,\mu,\eta} f(z) \; (-\infty < \lambda < 0;\ z \in \mathbb{U}). \end{cases}$$

For $f(z) \in \mathcal{A}(p)$, we have:

$$\begin{aligned} S_{0,z}^{\lambda,\mu,\eta,p} f(z) &= z^p {}_3F_2(1, 1+p, 1+p+\eta-\mu; 1+p-\mu, 1+p+\eta-\lambda; z) * f(z) \\ &= z^p + \sum_{n=1}^{\infty} \frac{(p+1)_n (p+1-\mu+\eta)_n}{(p+1-\mu)_n (p+1-\lambda+\eta)_n} a_{p+n} z^{p+n} \\ (p &\in \mathbb{N};\ \mu,\eta \in \mathbb{R};\ \mu < p+1;\ -\infty < \lambda < \eta+p+1), \end{aligned}$$

where "$*$" stands for convolution of two power series, and ${}_qF_s$ $(q \leq s+1;\ q,s \in \mathbb{N}_0 = \mathbb{N} \cup \{0\})$ is the well-known generalized hypergeometric function.

Let:

$$G^{\lambda}_{p,\eta,\mu}(z) = z^p + \sum_{n=1}^{\infty} \frac{(p+1)_n(p+1-\mu+\eta)_n}{(p+1-\mu)_n(p+1-\lambda+\eta)_n} z^{p+n}$$
$$(p \in \mathbb{N};\ \mu,\eta \in \mathbb{R};\ \mu < p+1;\ -\infty < \lambda < \eta + p + 1).$$

and:

$$G^{\lambda}_{p,\eta,\mu}(z) * \left[G^{\lambda}_{p,\eta,\mu}(z)\right]^{-1} = \frac{z^p}{(1-z)^{\delta+p}}\ (\delta > -p;\ z \in \mathbb{U}).$$

Tang et al. [9] (see also [10–15]) defined the operator $H^{\lambda,\delta}_{p,\eta,\mu} : \mathcal{A}(p) \to \mathcal{A}(p)$, where:

$$H^{\lambda,\delta}_{p,\eta,\mu} f(z) = z^p + \sum_{n=1}^{\infty} \frac{(\delta+p)_n(p+1-\mu)_n(p+1-\lambda+\eta)_n}{(1)_n(p+1)_n(p+1-\mu+\eta)_n} a_{p+n} z^{p+n}$$
$$(p \in \mathbb{N},\ \delta > -p,\ \mu,\eta \in \mathbb{R},\ \mu < p+1,\ -\infty < \lambda < \eta + p + 1).$$

It is easy to verify that:

$$z\left(H^{\lambda,\delta}_{p,\eta,\mu} f(z)\right)' = (\delta+p) H^{\lambda,\delta+1}_{p,\eta,\mu} f(z) - \delta H^{\lambda,\delta}_{p,\eta,\mu} f(z), \quad (3)$$

and:

$$z\left(H^{\lambda+1,\delta}_{p,\eta,\mu} f(z)\right)' = (p+\eta-\lambda) H^{\lambda,\delta}_{p,\eta,\mu} f(z) - (\eta-\lambda) H^{\lambda+1,\delta}_{p,\eta,\mu} f(z). \quad (4)$$

By using the operator $H^{\lambda,\delta}_{p,\eta,\mu}$, we introduce the following class.

Definition 3. *For A, B $(-1 \le B < A \le 1)$, $f \in \mathcal{A}(p)$ is in the class $\mathcal{T}^{\lambda,\delta}_{p,\eta,\mu}(A,B)$ if*

$$\frac{(H^{\lambda,\delta}_{p,\eta,\mu} f(z))'}{pz^{p-1}} \prec \frac{1+Az}{1+Bz}\ (z \in \mathbb{U};\ p \in \mathbb{N}),$$

which is equivalent to:

$$\left| \frac{\frac{(H^{\lambda,\delta}_{p,\eta,\mu} f(z))'}{pz^{p-1}} - 1}{B\frac{(H^{\lambda,\delta}_{p,\eta,\mu} f(z))'}{pz^{p-1}} - A} \right| < 1\ (z \in \mathbb{U}).$$

For convenience, we write $\mathcal{T}^{\lambda,\delta}_{p,\eta,\mu}\left(1 - \frac{2\xi}{p}, -1\right) = \mathcal{T}^{\lambda,\delta}_{p,\eta,\mu}(\xi)$ $(0 \le \xi < p)$, which satisfies the inequality:

$$\Re\left\{\frac{(H^{\lambda,\delta}_{p,\eta,\mu} f(z))'}{z^{p-1}}\right\} > \xi\ (0 \le \xi < p).$$

In this paper, we investigate some subordination and convolution properties for classes of multivalent functions, which are defined by a generalized fractional differintegral operator. The theory of subordination received great attention, particularly in many subclasses of univalent and multivalent functions (see, for example, [13,15–17]).

2. Preliminaries

To prove our main results, we shall need the following lemmas.

Lemma 1. *[18]. Let h be an analytic and convex (univalent) function in $\mathbb{U}$ with $h(0) = 1$. Additionally, let ϕ given by:*

$$\phi(z) = 1 + c_n z^n + c_{n+1} z^{n+1} + \ldots \quad (5)$$

be analytic in $\mathbb{U}$. *If:*

$$\phi(z) + \frac{z\phi'(z)}{\sigma} \prec h(z) \ (\Re(\sigma) \geqslant 0;\ \sigma \neq 0), \tag{6}$$

then:

$$\phi(z) \prec \psi(z) = \frac{\sigma}{n} z^{-\frac{\sigma}{n}} \int_0^z t^{\frac{\sigma}{n}-1} h(t) dt \prec h(z), \tag{7}$$

and ψ *is the best dominant of (6).*

Denote by $P(\varsigma)$ the class of functions Φ given by:

$$\Phi(z) = 1 + c_1 z + c_1 z^2 + ..., \tag{8}$$

which are analytic in $\mathbb{U}$ and satisfy the following inequality:

$$\Re\{\Phi(z)\} > \varsigma \ (0 \leq \varsigma < 1).$$

Using the well-known growth theorem for the Carathéodory functions (cf., e.g., [19]), we may easily deduce the following result:

Lemma 2. *[19]. If* $\Phi \in P(\varsigma)$. *Then*

$$\Re\{\Phi(\varsigma)\} \geqslant 2\varsigma - 1 + \frac{2(1-\varsigma)}{1+|z|} \ (0 \leq \varsigma < 1).$$

Lemma 3. *[20]. For* $0 \leq \varsigma_1, \varsigma_2 < 1$,

$$P(\varsigma_1) * P(\varsigma_2) \subset P(\varsigma_3) \ (\varsigma_3 = 1 - 2(1-\varsigma_1)(1-\varsigma_2)).$$

The result is the best possible.

Lemma 4. *[21]. Let* φ *be such that* $\varphi(0) = 1$ *and* $\varphi(z) \neq 0$ *and* $A, B \in \mathbb{C}$, *with* $A \neq B$, $|B| \leq 1, \nu \in \mathbb{C}^*$.

(i) If $\left|\frac{\nu(A-B)}{B} - 1\right| \leq 1$ or $\left|\frac{\nu(A-B)}{B} + 1\right| \leq 1, B \neq 0$ and $\varphi(z)$ satisfies:

$$1 + \frac{z\varphi'(z)}{\nu\varphi(z)} \prec \frac{1+Az}{1+Bz},$$

then:

$$\varphi(z) \prec (1+Bz)^{\nu\left(\frac{A-B}{B}\right)}$$

and this is the best dominant.

(ii) If $B = 0$ and $|\nu A| < \pi$ and if φ satisfies:

$$1 + \frac{z\varphi'(z)}{\nu\varphi(z)} \prec 1 + Az,$$

then:

$$\varphi(z) \prec e^{\nu A z},$$

and this is the best dominant.

Lemma 5. *[2]. Let $\Omega \subset \mathbb{C}$, $b \in \mathbb{C}$, $\Re(b) > 0$ and $\psi : \mathbb{C}^2 \times \mathbb{U} \to \mathbb{C}$ satisfy $\psi(ix, y; z) \notin \Omega$ for all $x, y \leq -\frac{|b-ix|^2}{2\Re(b)}$ and all $z \in \mathbb{U}$. If $p(z) = 1 + p_1 z + p_2 z^2 + \ldots$, is analytic in $\mathbb{U}$ and if:*

$$\psi\left(p(z), zp'(z); z\right) \in \Omega,$$

then $\Re\{p(z)\} > 0$ in $\mathbb{U}$.

Lemma 6. *[22]. Let $\psi(z)$ be analytic in $\mathbb{U}$ with $\psi(0) = 1$ and $\psi(z) \neq 0$ for all z. If there exist two points $z_1, z_2 \in \mathbb{U}$ such that:*

$$-\frac{\pi}{2}\rho_1 = \arg\{\psi(z_1)\} < \arg\{\psi(z)\} < \frac{\pi}{2}\rho_2 = \arg\{\psi(z_2)\}, \tag{9}$$

for some ρ_1 and ρ_2 $(\rho_1, \rho_2 > 0)$ and for all z $(|z| < |z_1| = |z_2|)$, then:

$$\frac{z_1\psi'(z_1)}{\psi(z_1)} = -i\left(\frac{\rho_1 + \rho_2}{2}\kappa\right) \text{ and } \frac{z_2\psi'(z_2)}{\psi(z_2)} = i\left(\frac{\rho_1 + \rho_2}{2}\kappa\right), \tag{10}$$

where:

$$\kappa \geq \frac{1 - |a|}{1 + |a|} \text{ and } a = i\tan\left(\frac{\rho_2 - \rho_1}{\rho_2 + \rho_1}\right). \tag{11}$$

3. Properties Involving $H_{p,\eta,\mu}^{\lambda,\delta}$

Unless otherwise mentioned, we assume throughout this paper that $p \in \mathbb{N}$, $\delta > -p$, $\mu, \eta \in \mathbb{R}$, $\mu < p + 1$, $-\infty < \lambda < \eta + p + 1$, $-1 \leq B < A \leq 1$, $\theta > 0$, and the powers are considered principal ones.

Theorem 1. *Let $f \in \mathcal{A}(p)$ satisfy:*

$$(1-\theta)\frac{\left(H_{p,\eta,\mu}^{\lambda,\delta} f(z)\right)'}{pz^{p-1}} + \theta\frac{\left(H_{p,\eta,\mu}^{\lambda,\delta+1} f(z)\right)'}{pz^{p-1}} \prec \frac{1 + Az}{1 + Bz}. \tag{12}$$

Then:

$$\Re\left(\frac{\left(H_{p,\eta,\mu}^{\lambda,\delta} f(z)\right)'}{pz^{p-1}}\right)^{\frac{1}{\tau}} > \left(\frac{\delta + p}{\theta}\int_0^1 u^{\frac{\delta+p}{\theta}-1}\left(\frac{1 - Au}{1 - Bu}\right)du\right)^{\frac{1}{\tau}}, \quad \tau \geq 1. \tag{13}$$

The estimate in (13) is sharp.

Proof. Let:

$$\phi(z) = \frac{\left(H_{p,\eta,\mu}^{\lambda,\delta} f(z)\right)'}{pz^{p-1}} \quad (z \in \mathbb{U}). \tag{14}$$

Then, ϕ is analytic in $\mathbb{U}$. After some computations, we get:

$$(1-\theta)\frac{\left(H_{p,\eta,\mu}^{\lambda,\delta} f(z)\right)'}{pz^{p-1}} + \theta\frac{\left(H_{p,\eta,\mu}^{\lambda,\delta+1} f(z)\right)'}{pz^{p-1}} = \phi(z) + \frac{\theta z\phi'(z)}{\delta + p} \prec \frac{1 + Az}{1 + Bz}.$$

Now, by using Lemma 1, we deduce that:

$$\frac{\left(H_{p,\eta,\mu}^{\lambda,\delta} f(z)\right)'}{pz^{p-1}} \prec \frac{\delta + p}{\theta} z^{-\frac{\delta+p}{\theta}} \int_0^z t^{\frac{\delta+p}{\theta}-1}\left(\frac{1 + At}{1 + Bt}\right)dt, \tag{15}$$

or, equivalently,

$$\frac{\left(H_{p,\eta,\mu}^{\lambda,\delta} f(z)\right)'}{pz^{p-1}} = \frac{\delta+p}{\theta} \int_0^1 u^{\frac{\delta+p}{\theta}-1} \left(\frac{1+Auw(z)}{1+Buw(z)}\right) du,$$

and so:

$$\Re\left(\frac{\left(H_{p,\eta,\mu}^{\lambda,\delta} f(z)\right)'}{pz^{p-1}}\right) > \left(\frac{\delta+p}{\theta} \int_0^1 u^{\frac{\delta+p}{\theta}-1} \left(\frac{1-Au}{1-Bu}\right) du\right). \tag{16}$$

Since:

$$\Re\left(\chi^{\frac{1}{\tau}}\right) \geq (\Re(\chi))^{\frac{1}{\tau}} \quad (\chi \in \mathbb{C},\ \Re\{\chi\} \geq 0,\ \tau \geq 1). \tag{17}$$

The inequality (13) now follows from (16) and (17). To prove that the result is sharp, let:

$$\frac{\left(H_{p,\eta,\mu}^{\lambda,\delta} f(z)\right)'}{pz^{p-1}} = \frac{\delta+p}{\theta} \int_0^1 u^{\frac{\delta+p}{\theta}-1} \left(\frac{1+Auz}{1+Buz}\right) du. \tag{18}$$

Now, for $f(z)$ defined by (18), we have:

$$(1-\theta)\frac{\left(H_{p,\eta,\mu}^{\lambda,\delta} f(z)\right)'}{pz^{p-1}} + \theta\frac{\left(H_{p,\eta,\mu}^{\lambda,\delta+1} f(z)\right)'}{pz^{p-1}} = \frac{1+Az}{1+Bz} \quad (z \in \mathbb{U}),$$

Letting $z \to -1$, we obtain:

$$\frac{\left(H_{p,\eta,\mu}^{\lambda,\delta} f(z)\right)'}{pz^{p-1}} \to \frac{\delta+p}{\theta} \int_0^1 u^{\frac{\delta+p}{\theta}-1} \left(\frac{1-Au}{1-Bu}\right) du,$$

which ends our proof. □

Putting $\theta = 1$ and using Lemma 1 for Equation (15) in Theorem 1, we obtain the following example.

Example 1. *Let the function $f(z) \in \mathcal{A}(p)$. Then, following containment property holds,*

$$\mathcal{T}_{p,\mu,\eta}^{\lambda,\delta+1}(A,B) \subset \mathcal{T}_{p,\mu,\eta}^{\lambda,\delta}(A,B).$$

Using (4) instead of (3) in Theorem 1, one can prove the following theorem.

Theorem 2. *Let $f \in \mathcal{A}(p)$ satisfy*

$$(1-\theta)\frac{\left(H_{p,\eta,\mu}^{\lambda+1,\delta} f(z)\right)'}{pz^{p-1}} + \theta\frac{\left(H_{p,\eta,\mu}^{\lambda,\delta} f(z)\right)'}{pz^{p-1}} \prec \frac{1+Az}{1+Bz}.$$

Then:

$$\Re\left(\frac{\left(H_{p,\eta,\mu}^{\lambda+1,\delta} f(z)\right)'}{pz^{p-1}}\right)^{\frac{1}{\tau}} > \left(\frac{p+\eta-\lambda}{\theta} \int_0^1 u^{\frac{p+\eta-\lambda}{\theta}-1} \left(\frac{1-Au}{1-Bu}\right) du\right)^{\frac{1}{\tau}},\ \tau \geq 1. \tag{19}$$

The result is sharp.

Putting $\theta = 1$ in Theorem 2, we obtain the following example.

Example 2. *Let the function $f(z) \in \mathcal{A}(p)$. Then, following inclusion property holds*

$$\mathcal{T}^{\lambda,\delta}_{p,\mu,\eta}(A,B) \subset \mathcal{T}^{\lambda+1,\delta}_{p,\mu,\eta}(A,B).$$

For a function $f \in \mathcal{A}(p)$, the generalized Bernardi–Libera–Livingston integeral operator $F_{p,\gamma}$ is defined by (see [23]):

$$\begin{aligned} F_{p,\gamma}f(z) &= \frac{\gamma+p}{z^p}\int_0^z t^{\gamma-1}f(t)dt \\ &= \left(z^p + \sum_{k=1}^{\infty}\frac{\gamma+p}{\gamma+p+k}z^{p+k}\right) * f(z)\ (\gamma > -p) \\ &= z^p\ {}_3F_2\left(1,1,\gamma+p;1,\gamma+p+1;z\right) * f(z). \end{aligned} \tag{20}$$

Lemma 7. *If $f \in \mathcal{A}(p)$, prove that:*
(i) $H^{\lambda,\delta}_{p,\eta,\mu}\left(F_{p,\gamma}f\right) = F_{p,\gamma}\left(H^{\lambda,\delta}_{p,\eta,\mu}f\right)$,
(ii)

$$z\left(H^{\lambda,\delta}_{p,\eta,\mu}F_{p,\gamma}f(z)\right)' = (p+\gamma)H^{\lambda,\delta}_{p,\eta,\mu}f(z) - \gamma H^{\lambda,\delta}_{p,\eta,\mu}F_{p,\gamma}f(z). \tag{21}$$

Proof. Since

$$\begin{aligned} H^{\lambda,\delta}_{p,\eta,\mu}\left(F_{p,\gamma}f\right) &= \left[z^p\ {}_3F_2\left(\delta+p,p+1-\mu,p+1-\lambda+\eta;p+1,p+1-\mu+\eta;z\right)\right] * \left(F_{p,\gamma}f\right) \\ &= \left[z^p\ {}_3F_2\left(\delta+p,p+1-\mu,p+1-\lambda+\eta;p+1,p+1-\mu+\eta;z\right)\right] * \\ &\quad \left[z^p\ {}_3F_2\left(1,1,\gamma+p;1,\gamma+p+1;z\right) * f(z)\right], \end{aligned}$$

and:

$$\begin{aligned} F_{p,\gamma}\left(H^{\lambda,\delta}_{p,\eta,\mu}f\right) &= z^p\ {}_3F_2\left(1,1,\gamma+p;1,\gamma+p+1;z\right) * \left(H^{\lambda,\delta}_{p,\eta,\mu}f\right) \\ &= z^p\ {}_3F_2\left(1,1,\gamma+p;1,\gamma+p+1;z\right) * \\ &\quad \left[z^p\ {}_3F_2\left(\delta+p,p+1-\mu,p+1-\lambda+\eta;p+1,p+1-\mu+\eta;z\right) * f(z)\right]. \end{aligned}$$

Now, the first part of this lemma follows. Furthermore,

$$z\left(F_{p,\gamma}f(z)\right)' = (p+\gamma)f(z) - \gamma F_{p,\gamma}f(z). \tag{22}$$

If we replace $f(z)$ by $H^{\lambda,\delta}_{p,\eta,\mu}f(z)$ and using the first part of this lemma, we get (21). □

Theorem 3. *Suppose that $p+\gamma > 0$, $f \in \mathcal{T}^{\lambda,\delta}_{p,\eta,\mu}(A,B)$ and $F_{p,\gamma}$ defined by (20). Then:*

$$\Re\left(\frac{\left(H^{\lambda,\delta}_{p,\eta,\mu}F_{p,\gamma}f(z)\right)'}{pz^{p-1}}\right)^{\frac{1}{\tau}} > \left((p+\gamma)\int_0^1 u^{p+\gamma-1}\left(\frac{1-Au}{1-Bu}\right)du\right)^{\frac{1}{\tau}},\ \tau \geq 1. \tag{23}$$

The result is sharp.

Proof. Let:

$$\phi(z) = \frac{\left(H^{\lambda,\delta}_{p,\eta,\mu}F_{p,\gamma}f(z)\right)'}{pz^{p-1}}\ (z \in \mathbb{U}). \tag{24}$$

Then, ϕ is analytic in $\mathbb{U}$. After some calculations, we have:

$$\frac{(H^{\lambda,\delta}_{p,\eta,\mu}f(z))'}{pz^{p-1}} = \phi(z) + \frac{z\phi'(z)}{p+\gamma} \prec \frac{1+Az}{1+Bz}.$$

Employing the same technique that was used in proving Theorem 1, the remaining part of the theorem can be proven. □

Theorem 4. *Let $-1 \le B_i < A_i \le 1$ $(i = 1,2)$. If each of the functions $f_i \in \mathcal{A}(p)$ satisfies:*

$$(1-\theta)\frac{H^{\lambda,\delta}_{p,\eta,\mu}f_i(z)}{z^p} + \theta\frac{H^{\lambda,\delta+1}_{p,\eta,\mu}f_i(z)}{z^p} \prec \frac{1+A_iz}{1+B_iz}\ (i=1,2), \tag{25}$$

then:

$$(1-\theta)\frac{H^{\lambda,\delta}_{p,\eta,\mu}F(z)}{z^p} + \theta\frac{H^{\lambda,\delta+1}_{p,\eta,\mu}F(z)}{z^p} \prec \frac{1+(1-2\varrho)z}{1-z}, \tag{26}$$

where:

$$F(z) = H^{\lambda,\delta}_{p,\eta,\mu}(f_1 * f_2)(z) \tag{27}$$

and:

$$\varrho = 1 - \frac{4(A_1-B_1)(A_2-B_2)}{(1-B_1)(1-B_2)}\left[1 - \frac{1}{2}\,{}_2F_1\left(1,1;\frac{\delta+p}{\theta}+1;\frac{1}{2}\right)\right]. \tag{28}$$

The result is possible when $B_1 = B_2 = -1$.

Proof. Suppose that $f_i \in \mathcal{A}(p)$ $(i=1,2)$ satisfy the condition (25). Setting:

$$p_i(z) = (1-\theta)\frac{H^{\lambda,\delta}_{p,\eta,\mu}f_i(z)}{z^p} + \theta\frac{H^{\lambda,\delta+1}_{p,\eta,\mu}f_i(z)}{z^p}\ (i=1,2), \tag{29}$$

we have:

$$p_i(z) \in P(\varsigma_i)\ \left(\varsigma_i = \frac{1-A_i}{1-B_i},\ i=1,2\right).$$

Thus, by making use of the identity (3) in (29), we get:

$$H^{\lambda,\delta}_{p,\eta,\mu}f_i(z) = \frac{\delta+p}{\theta}z^{p-\frac{\delta+p}{\theta}}\int_0^z t^{\frac{\delta+p}{\theta}-1}p_i(t)dt\ (i=1,2), \tag{30}$$

which, in view of F given by (27) and (30), yields:

$$H^{\lambda,\delta}_{p,\eta,\mu}F(z) = \frac{\delta+p}{\theta}z^{p-\frac{\delta+p}{\theta}}\int_0^z t^{\frac{\delta+p}{\theta}-1}F(t)dt, \tag{31}$$

where:

$$F(z) = (1-\theta)\frac{H^{\lambda,\delta}_{p,\eta,\mu}F(z)}{z^p} + \theta\frac{H^{\lambda,\delta+1}_{p,\eta,\mu}F(z)}{z^p} = \frac{\delta+p}{\theta}z^{-\frac{\delta+p}{\theta}}\int_0^z t^{\frac{\delta+p}{\theta}-1}(p_1 * p_2)(t)dt. \tag{32}$$

Since $p_i(z) \in P(\varsigma_i)$ $(i=1,2)$, it follows from Lemma 3 that:

$$(p_1 * p_2)(z) \in P(\varsigma_3)\ (\varsigma_3 = 1 - 2(1-\varsigma_1)(1-\varsigma_2)). \tag{33}$$

Now, by using (33) in (32) and then appealing to Lemma 2, we have:

$$\begin{aligned}\Re\{F(z)\} &= \frac{\delta+p}{\theta}\int_0^1 u^{\frac{\delta+p}{\theta}-1}\Re\{(p_1*p_2)(uz)\}\,du \\ &\geqslant \frac{\delta+p}{\theta}\int_0^1 u^{\frac{\delta+p}{\theta}-1}\left(2\varsigma_3-1+\frac{2(1-\varsigma_3)}{1+u\,|z|}\right)du \\ &> \frac{\delta+p}{\theta}\int_0^1 u^{\frac{\delta+p}{\theta}-1}\left(2\varsigma_3-1+\frac{2(1-\varsigma_3)}{1+u}\right)du \\ &= 1-\frac{4(A_1-B_1)(A_2-B_2)}{(1-B_1)(1-B_2)}\left[1-\tfrac{\delta+p}{\theta}\int_0^1 u^{\frac{\delta+p}{\theta}-1}(1+u)^{-1}du\right] \\ &= 1-\frac{4(A_1-B_1)(A_2-B_2)}{(1-B_1)(1-B_2)}\left[1-\frac{1}{2}\,{}_2F_1\left(1,1;\tfrac{\delta+p}{\theta}+1;\frac{1}{2}\right)\right]=\varrho.\end{aligned}$$

When $B_1 = B_2 = -1$, we consider the functions $f_i(z) \in \mathcal{A}(p)$ $(i = 1,2)$, which satisfy (25), are defined by:

$$H_{p,\eta,\mu}^{\lambda,\delta} f_i(z) = \tfrac{\delta+p}{\theta} z^{p-\frac{\delta+p}{\theta}} \int_0^z t^{\frac{\delta+p}{\theta}-1}\left(\frac{1+A_i t}{1-t}\right)dt \ (i=1,2).$$

Thus, it follows from (32) that:

$$F(z) = \tfrac{\delta+p}{\theta}\int_0^1 u^{\frac{\delta+p}{\theta}-1}\left[1-(1+A_1)(1+A_2)+\frac{(1+A_1)(1+A_2)}{(1-uz)}\right]du$$

$$=1-(1+A_1)(1+A_2)+(1+A_1)(1+A_2)(1-z)^{-1}\,{}_2F_1\left(1,1;\tfrac{\delta+p}{\theta}+1;\frac{z}{z-1}\right)$$

$$\to 1-(1+A_1)(1+A_2)+\frac{1}{2}(1+A_1)(1+A_2)\,{}_2F_1\left(1,1;\tfrac{\delta+p}{\theta}+1;\frac{1}{2}\right) \text{ as } z\to -1,$$

which evidently ends the proof. □

Theorem 5. *Let $v \in \mathbb{C}^*$, and let $A,\ B \in \mathbb{C}$ with $A \neq B$ and $|B| \leq 1$. Suppose that:*

$$\begin{array}{ll}\left|\frac{v(\delta+p)(A-B)}{B}-1\right|\leq 1 \text{ or } \left|\frac{v(\delta+p)(A-B)}{B}+1\right|\leq 1 & \text{if } B\neq 0,\\ |v(\delta+p)A|\leq\pi & \text{if } B=0.\end{array}$$

If $f \in \mathcal{A}(p)$ with $H_{p,\eta,\mu}^{\lambda,\delta} f(z) \neq 0$ for all $z \in \mathbb{U}^* = \mathbb{U}\backslash\{0\}$, then:

$$\frac{H_{p,\eta,\mu}^{\lambda,\delta+1} f(z)}{H_{p,\eta,\mu}^{\lambda,\delta} f(z)} \prec \frac{1+Az}{1+Bz},$$

implies:

$$\left(\frac{H_{p,\eta,\mu}^{\lambda,\delta} f(z)}{z^p}\right)^v \prec q(z),$$

where:

$$q(z) = \begin{cases}(1+Bz)^{v(\delta+p)(A-B)/B} & \text{if } B\neq 0,\\ e^{v(\delta+p)Az} & \text{if } B=0,\end{cases}$$

is the best dominant.

Proof. Putting:

$$\Delta(z) = \left(\frac{H_{p,\eta,\mu}^{\lambda,\delta} f(z)}{z^p}\right)^v \quad (z\in\mathbb{U}). \tag{34}$$

Then, Δ is analytic in $\mathbb{U}$, $\Delta(0) = 1$ and $\Delta(z) \neq 0$ for all $z \in \mathbb{U}$. Taking the logarithmic derivatives on both sides of (34) and using (3), we have:

$$1 + \frac{z\Delta'(z)}{v(\delta + p)\Delta(z)} = \frac{H_{p,\eta,\mu}^{\lambda,\delta+1} f(z)}{H_{p,\eta,\mu}^{\lambda,\delta} f(z)} \prec \frac{1 + Az}{1 + Bz}.$$

Now, the assertions of Theorem 5 follow by Lemma 4. □

Theorem 6. *Let* $0 \leq \alpha \leq 1$, $\zeta > 1$. *If* $f(z) \in \mathcal{A}(p)$ *satisfies:*

$$\Re\left((1-\alpha)\frac{\left(H_{p,\eta,\mu}^{\lambda,\delta+2} f(z)\right)'}{\left(H_{p,\eta,\mu}^{\lambda,\delta+1} f(z)\right)'} + \alpha\frac{\left(H_{p,\eta,\mu}^{\lambda,\delta+1} f(z)\right)'}{\left(H_{p,\eta,\mu}^{\lambda,\delta} f(z)\right)'}\right) < \zeta, \tag{35}$$

then:

$$\Re\left(\frac{H_{p,\eta,\mu}^{\lambda,\delta+1} f(z)}{H_{p,\eta,\mu}^{\lambda,\delta} f(z)}\right) < \beta,$$

where $\beta \in (1, \infty)$ *is the positive root of the equation:*

$$2(\delta + p + \alpha)\beta^2 - [2\zeta(\delta + p + 1) - (1-\alpha)]\beta - (1-\alpha) = 0. \tag{36}$$

Proof. Let:

$$\frac{H_{p,\eta,\mu}^{\lambda,\delta+1} f(z)}{H_{p,\eta,\mu}^{\lambda,\delta} f(z)} = \beta + (1-\beta)\varphi(z). \tag{37}$$

Then, φ is analytic in $\mathbb{U}$, $\varphi(0) = 1$ and $\varphi(z) \neq 0$ for all $z \in \mathbb{U}$. Taking the logarithmic derivatives on both sides of (37) and using the identity (3), we have:

$$(\delta + p + 1)\frac{\left(H_{p,\eta,\mu}^{\lambda,\delta+2} f(z)\right)'}{\left(H_{p,\eta,\mu}^{\lambda,\delta+1} f(z)\right)'} - (\delta + p)\frac{\left(H_{p,\eta,\mu}^{\lambda,\delta+1} f(z)\right)'}{\left(H_{p,\eta,\mu}^{\lambda,\delta} f(z)\right)'} = 1 + \frac{(1-\beta)z\varphi'(z)}{\beta + (1-\beta)\varphi(z)},$$

and so:

$$(1-\alpha)\frac{\left(H_{p,\eta,\mu}^{\lambda,\delta+2} f(z)\right)'}{\left(H_{p,\eta,\mu}^{\lambda,\delta+1} f(z)\right)'} + \alpha\frac{\left(H_{p,\eta,\mu}^{\lambda,\delta+1} f(z)\right)'}{\left(H_{p,\eta,\mu}^{\lambda,\delta} f(z)\right)'} = \alpha\beta + \frac{(1-\alpha)(\delta+p)\beta}{\delta+p+1}$$
$$+ \frac{(1-\beta)[\alpha + (1-\alpha)(\delta+p)]}{\delta+p+1}\varphi(z) + \frac{(1-\alpha)(1-\beta)}{[\beta + (1-\beta)\varphi(z)](\delta+p+1)}z\varphi'(z).$$

Let:

$$\begin{aligned}\Psi(r,s;z) &= \alpha\beta + \frac{(1-\alpha)(\delta+p)\beta}{\delta+p+1} + \frac{(1-\beta)[\alpha + (1-\alpha)(\delta+p)]}{\delta+p+1}r \\ &\quad + \frac{(1-\alpha)(1-\beta)}{[\beta + (1-\beta)\varphi(z)](\delta+p+1)}s,\end{aligned}$$

and:

$$\Omega = \{w \in \mathbb{C} : \Re(w) < \zeta\}.$$

Then, for $x, y \leq -\frac{1+x^2}{2}$, we have:

$$\begin{aligned}\Re\{\Psi(ix,y;z)\} &= \alpha\beta + \frac{(1-\alpha)(\delta+p)\beta}{\delta+p+1} + \frac{(1-\alpha)(1-\beta)\beta y}{\left[\beta^2+(1-\beta)^2x^2\right](\delta+p+1)} \\ &\geq \alpha\beta + \frac{(1-\alpha)(\delta+p)\beta}{\delta+p+1} - \frac{(1-\alpha)(1-\beta)}{2\beta(\delta+p+1)} = \zeta,\end{aligned}$$

where β is the positive root of Equation (36). Suppose that:

$$R(\beta) = 2(\delta+p+\alpha)\beta^2 - [2\zeta(\delta+p+1)-(1-\alpha)]\beta - (1-\alpha) = 0.$$

For $\beta = 0$, $R(0) = -(1-\alpha) \leq 0$ and for $\beta = 1$, $R(1) = 2(\delta+p)(1-\zeta)+2(\alpha-\zeta) \leq 0$. This proves that $\beta \in (1,\infty)$. Thus, for $z \in \mathbb{U}$, $\Psi(ix,y;z) \notin \Omega$, and so, we obtain the required result by an application of Lemma 5. □

Theorem 7. *Suppose that* $0 < \varepsilon_1, \varepsilon_2 \leq 1$. *If:*

$$-\frac{\pi}{2}\varepsilon_1 < \arg\left\{(1-\theta)\frac{\left(H_{p,\eta,\mu}^{\lambda,\delta}f(z)\right)'}{pz^{p-1}} + \theta\frac{\left(H_{p,\eta,\mu}^{\lambda,\delta+1}f(z)\right)'}{pz^{p-1}}\right\} < \frac{\pi}{2}\varepsilon_2, \tag{38}$$

then:

$$-\frac{\pi}{2}\xi_1 < \arg\left(\frac{\left(H_{p,\eta,\mu}^{\lambda,\delta}f(z)\right)'}{pz^{p-1}}\right) < \frac{\pi}{2}\xi_2, \tag{39}$$

where:

$$\varepsilon_1 = \xi_1 + \frac{2}{\pi}\arctan\left(\frac{(\xi_1+\xi_2)\theta}{2(\delta+p)}\frac{1-|a|}{1+|a|}\right), \ \varepsilon_2 = \xi_2 + \frac{2}{\pi}\arctan\left(\frac{(\xi_1+\xi_2)\theta}{2(\delta+p)}\frac{1-|a|}{1+|a|}\right). \tag{40}$$

Proof. Let:

$$\phi(z) = \frac{\left(H_{p,\eta,\mu}^{\lambda,\delta}f(z)\right)'}{pz^{p-1}} \quad (z \in \mathbb{U}).$$

Then, from Theorem 1, we have:

$$(1-\theta)\frac{\left(H_{p,\eta,\mu}^{\lambda,\delta}f(z)\right)'}{pz^{p-1}} + \theta\frac{\left(H_{p,\eta,\mu}^{\lambda,\delta+1}f(z)\right)'}{pz^{p-1}} = \phi(z) + \frac{\theta z\phi'(z)}{\delta+p}.$$

Let $U(z)$ be the function that maps $\mathbb{U}$ onto the domain:

$$\left\{w \in \mathbb{C} : -\frac{\pi}{2}\varepsilon_1 < \arg(w) < \frac{\pi}{2}\varepsilon_2\right\},$$

with $U(0) = 1$, then:

$$\phi(z) + \frac{\theta z\phi'(z)}{\delta+p} \prec U(z).$$

Assume that z_1, z_2 are two points in $\mathbb{U}$ such that the condition (9) is satisfied, then by Lemma 6, we obtain (10) under the constraint (11). Therefore,

$$\begin{aligned}\arg\left[(\delta+p)\,\phi(z_1)+\theta z_1\phi'(z_1)\right] &= \arg\phi(z_1)\left[(\delta+p)+\theta\frac{z_1\phi'(z_1)}{\phi(z_1)}\right] \\ &= \arg\phi(z_1)+\arg\left[(\delta+p)+\theta\frac{z_1\phi'(z_1)}{\phi(z_1)}\right] \\ &= -\frac{\pi}{2}\xi_1+\arg\left[(\delta+p)-i\theta\frac{(\xi_1+\xi_2)\,\kappa}{2}\right] \\ &= -\frac{\pi}{2}\xi_1-\arctan\left[\frac{(\xi_1+\xi_2)\,\theta\kappa}{2\,(\delta+p)}\right] \\ &\leq -\frac{\pi}{2}\xi_1-\arctan\left[\frac{(\xi_1+\xi_2)\,\theta}{2\,(\delta+p)}\frac{1-|a|}{1+|a|}\right],\end{aligned}$$

and:

$$\arg\left[(\delta+p)\,\phi(z_2)+\theta z_2\phi'(z_2)\right]\geq\frac{\pi}{2}\xi_2+\arctan\left[\frac{(\xi_1+\xi_2)\,\theta}{2\,(\delta+p)}\frac{1-|a|}{1+|a|}\right].$$

which contradicts the assumption (38). This evidently completes the proof of Theorem 7. □

Acknowledgments: The authors would like to thank all referees for their valuable comments which led to the improvement of this paper.

Author Contributions: All the authors read and approved the final manuscript as a consequence of the authors meetings.

References

1. Bulboacă, T. *Differential Subordinations and Superordinations, Recent Results*; House of Scientific Book Publ.: Cluj-Napoca, Romania, 2005.
2. Miller, S.S.; Mocanu, P.T. *Differential Subordination: Theory and Applications, Series on Monographs and Textbooks in Pure and Applied Mathematics*; Marcel Dekker Inc.: New York, NY, USA, 2000; Volume 225.
3. Owa, S. On the distortion theorems I. *Kyungpook Math. J.* **1978**, *18*, 53–59.
4. Owa, S.; Srivastava, H.M. Univalent and starlike generalized hypergeometric functions. *Can. J. Math.* **1987**, *39*, 1057–1077. [CrossRef]
5. Srivastava, H.M.; Saigo, M.; Owa, S. A class of distortion theorems involving certain operators of fractional calculus. *J. Math. Anal. Appl.* **1988**, *131*, 412–420. [CrossRef]
6. Prajapat, J.K.; Raina, R.K.; Srivastava, H.M. Some inclusion properties for certain subclasses of strongly starlike and strongly convex functions involving a family of fractional integral operators. *Integral Transform. Spec. Funct.* **2007**, *18*, 639–651. [CrossRef]
7. Goyal, S.P.; Prajapat, J.K. A new class of analytic p-valent functions with negative coefficients and fractional calculus operators. *Tamsui Oxf. J. Math. Sci.* **2004**, *20*, 175–186.
8. Prajapat, J.K.; Aouf, M.K. Majorization problem for certain class of p-valently analytic function defined by generalized fractional differintegral operator. *Comput. Math. Appl.* **2012**, *63*, 42–47. [CrossRef]
9. Tang, H.; Deng, G.; Li, S.; Aouf, M.K. Inclusion results for certain subclasses of spiral-like multivalent functions involving a generalized fractional differintegral operator. *Integral Transform. Spec. Funct.* **2013**, *24*, 873–883. [CrossRef]
10. Aouf, M.K.; Mostafa, A.O.; Zayed, H.M. Some characterizations of integral operators associated with certain classes of p-valent functions defined by the Srivastava-Saigo-Owa fractional differintegral operator. *Complex Anal. Oper. Theory* **2016**, *10*, 1267–1275. [CrossRef]
11. Aouf, M.K.; Mostafa, A.O.; Zayed, H.M. Subordination and superordination properties of p-valent functions defined by a generalized fractional differintegral operator. *Quaest. Math.* **2016**, *39*, 545–560. [CrossRef]
12. Aouf, M.K.; Mostafa, A.O.; Zayed, H.M. On certain subclasses of multivalent functions defined by a generalized fractional differintegral operator. *Afr. Mat.* **2017**, *28*, 99–107. [CrossRef]
13. Mostafa, A.O.; Aouf, M.K.; Zayed, H.M.; Bulboacă, T. Multivalent functions associated with Srivastava-Saigo -Owa fractional differintegral operator. *RACSAM* **2017**. [CrossRef]

14. Mostafa, A.O.; Aouf, M.K.; Zayed, H.M. Inclusion relations for subclasses of multivalent functions defined by Srivastava–Saigo–Owa fractional differintegral operator. *Afr. Mat.* **2018**, -018-0567-3. [CrossRef]
15. Mostafa, A.O.; Aouf, M.K.; Zayed, H.M. Subordinating results for p-valent functions associated with the Srivastava–Saigo–Owa fractional differintegral operator. *Afr. Mat.* **2018**. [CrossRef]
16. Mostafa, A.O.; Aouf, M.K. Some applications of differential subordination of p-valent functions associated with Cho-Kwon-Srivastava operator. *Acta Math. Sin. (Engl. Ser.)* **2009**, *25*, 1483–1496. [CrossRef]
17. Wang, Z.; Shi, L. Some properties of certain extended fractional differintegral operator. *RACSAM* **2017**, 1–11. [CrossRef]
18. Hallenbeck, D.Z.; Ruscheweyh, S. Subordination by convex functions. *Proc. Am. Math. Soc.* **1975**, *52*, 191–195. [CrossRef]
19. Pommerenke, C. *Univalent Functions*; Vandenhoeck & Ruprecht: Göttingen, Germany, 1975.
20. Stankiewicz, J.; Stankiewicz, Z. Some applications of Hadamard convolution in the theory of functions. *Ann. Univ. Mariae Curie-Sklodowska Sect. A* **1986**, *40*, 251–265.
21. Obradović, M.; Owa, S. On certain properties for some classes of starlike functions. *J. Math. Anal. Appl.* **1990**, *145*, 357–364. [CrossRef]
22. Takahashi, N.; Nunokawa, M. A certain connection between starlike and convex functions. *Appl. Math. Lett.* **2003**, *16*, 653–655. [CrossRef]
23. Choi, J.H.; Saigo, M.; Srivastava, H.M. Some inclusion properties of a certain family of integral operators. *J. Math. Anal. Appl.* **2002**, *276*, 432–445. [CrossRef]

9

Repeated Derivatives of Hyperbolic Trigonometric Functions and Associated Polynomials

Giuseppe Dattoli [1], Silvia Licciardi [1,*], Rosa Maria Pidatella [2] and Elio Sabia [1]

1. ENEA—Frascati Research Center, Via Enrico Fermi 45, 00044 Frascati (Rome), Italy; giuseppe.dattoli@enea.it (G.D.); elio.sabia@enea.it (E.S.)
2. Department of Mathematics and Computer Sciences, University of Catania, Viale A. Doria 6, 95125 Catania, Italy; rosa@dmi.unict.it

* Correspondence: silviakant@gmail.com;

Abstract: Elementary problems as the evaluation of repeated derivatives of ordinary transcendent functions can usefully be treated with the use of special polynomials and of a formalism borrowed from combinatorial analysis. Motivated by previous researches in this field, we review the results obtained by other authors and develop a complementary point of view for the repeated derivatives of $\sec(.)$, $\tan(.)$ and for their hyperbolic counterparts.

Keywords: operators theory 44A99, 47B99, 47A62; special functions 33C52, 33C65, 33C99, 33B10, 33B15; Stirling numbers and Touchard polynomials 11B73

---

1. Introduction

The problem of finding closed forms for the repeated derivatives of trigonometric functions as tangent and secant, even though being an apparently elementary issue, has been solved in relatively recent times in ref. [1]. Inspired by this work, a significant amount of research has been subsequently developed. In ref. [2] the proof of the results of [1] was reformulated in terms of a procedure exploiting the "Zeons" Algebra [3]. In [4,5] the authors addressed this study by employing a class of polynomials (the derivative polynomials (DP) introduced in refs. [6,7]) to reformulate the derivation and eventually get a set of fairly simple formulae, providing the successive derivatives of $\tan, \cot, \sec, \csc, \ldots$ along with those of their hyperbolic counterparts.

In this note we develop a point of view not dissimilar from that of ref. [4,5]. We provide straightforward results in terms of a single family of Legendre like polynomials and comment on the two forms of DP introduced in [3–5].

The repeated derivatives of composite functions $F(x) = f(g(x))$ is a well-established topic in calculus. The formulation of a procedure allowing the derivation of a formula comprising all the possible cases was established in the XIX century [8] and opened important avenue of research in combinatorics [9] and umbral calculus [10,11] as well.

The problem is particularly interesting, encompasses different topics in analysis, including special polynomials like those belonging to the Touchard family [12] and special numbers like the generalized Stirling forms [13,14] of crucial importance in combinatorial analysis. The repeated derivatives of the Gaussian function are those of a composite function in which $f(.)$ is an exponential and $g(.)$ a quadratic function. The relevant expression leads to the Hermite polynomials as auxiliary tool, to get a synthetic expression for any order of the derivative [15]. Within the same context, Bell polynomials emerge whenever one is interested in the derivatives of $F(x) = e^{g(x)}$ [9,16,17]. The generalization to the case of $f(.)$, provided by a generic infinitely differentiable function, and $g(.)$, a quadratic form, has been discussed in refs. [18,19] where the problem has been solved by the use of generalized nested forms of Hermite polynomials.

The hyperbolic secant is a composite function too in which $f(.) = (.)^{-1}$, $g(.) = \cosh(.)$. In refs. [2,4,5,20] interesting speculations have been presented on the study of the relevant repeated derivatives and of the associated auxiliary polynomials. In this paper we address the same problem, within a different context. Before entering the specific elements of the discussion we review the formalism we are going to exploit in this paper.

A pivotal role within the context of repeated derivatives is played by the Stirling number of second kind [13,14]. They will be introduced using the Touchard polynomials which are defined through the Rodriguez type formula [15,21] where, for $n \geq 0$, we set

$$T_n(x) = e^{-x}(x\hat{D}_x)^n e^x \tag{1}$$

where $\hat{D}_x$ is the derivative operator $\frac{d}{dx}$. They can be written in explicit form by the use of the following expression [13]

$$(x\hat{D}_x)^n = \sum_{r=0}^{n} S_2(n,r) x^r \hat{C}_x^r\,, \tag{2}$$

with $S_2(n,r)$ being the Stirling number of second kind (we introduce the notation $S_2(l,m)$ instead of the usually symbol $\left\{ {l \atop m} \right\}$; to indicate functions domains, we will use conventional symbols as $\mathbb{N}, \mathbb{Z}, \mathbb{R}$ for Natural, Integer or Real numbers respectively including the zero). Ref. [9,12] defined as

$$S_2(l,m) = \frac{1}{m!} \sum_{j=0}^{m} (-1)^{m-j} \binom{m}{j} j^l. \tag{3}$$

According to Equations (1) and (2), the Touchard polynomials are explicitly given by

$$T_n(x) = \sum_{r=0}^{n} S_2(n,r) x^r \tag{4}$$

and the numbers $S_2(n,k)$ are therefore the coefficients of the polynomials.

A fairly straightforward application of the previous computational tool is provided by the evaluation of a closed expression for higher order derivatives.

Theorem 1. *We have that,* $\forall m \in \mathbb{N}$,

$$\hat{D}_x^m f(e^x) = \sum_{r=0}^{m} S_2(m,r)\, e^{xr} f^{(r)}(e^x) \tag{5}$$

where $f^{(r)}(\xi)$ *denotes the r-order derivative .*

Proof. Let

$$I_m(x) = \hat{D}_x^m f(e^x), \tag{6}$$

by setting $e^x = \xi$, we find

$$\hat{D}_x^m f(e^x) = (\xi \hat{D}_\xi)^m f(\xi) = \sum_{r=0}^{m} S_2(m,r)\, \xi^r f^{(r)}(\xi) \tag{7}$$

and in conclusion we obtain

$$\hat{D}_x^m f(e^x) = \sum_{r=0}^{m} S_2(m,r)\, e^{xr} f^{(r)}(e^x)$$

in which it is understood that $f^{(r)}(e^x) = \hat{D}_\xi^r f(\xi)\mid_{\xi=e^x}$. □

We next define a family of auxiliary polynomials $P_n(x,y)$ and show how they can be very useful for the computation of the repeated derivatives of $\tan^{-1}(x)$ (we use the notation $\tan^{-1}(x)$ for $\arctan(x)$).

Theorem 2. *For any $x, y \in \mathbb{R},\ n \in \mathbb{N}$,*

$$P_n(x,y) = n! \sum_{r=0}^{\lfloor \frac{n}{2} \rfloor} \frac{x^{n-2r} y^r (n-r)!}{(n-2r)!r!}, \tag{8}$$

a family of two variable polynomials loosely ascribed to the Legendre [22,23] family, then

$$\hat{D}_x^n \left(\tan^{-1}(x)\right) = \frac{1}{1+x^2} P_{n-1}\left(-\frac{2x}{1+x^2}, -\frac{1}{1+x^2}\right), \qquad \forall n \in \mathbb{N}_0. \tag{9}$$

Proof. We start from the computation of the quantity

$$K_n(x) = \hat{D}_x^n \left(\frac{1}{1+x^2}\right), \quad \forall x \in \mathbb{R}, \forall n \in \mathbb{N} \tag{10}$$

which upon the use of the Laplace transform method can be written as

$$K_n(x) = \hat{D}_x^n \int_0^\infty e^{-s(1+x^2)} ds = \int_0^\infty e^{-s} \hat{D}_x^n e^{-sx^2} ds. \tag{11}$$

The n-th order derivative inside the integral sign can be explicetely worked in terms of two variable Hermite polynomials [15]

$$H_n(x,y) = n! \sum_{r=0}^{\lfloor \frac{n}{2} \rfloor} \frac{x^{n-2r} y^r}{(n-2r)!r!}, \tag{12}$$

namely the auxiliary polynomials for the repeated derivatives of Gaussian functions according to the identity

$$\hat{D}_x^n\, e^{-ax^2} = H_n(-2ax, -a)\, e^{-ax^2}. \tag{13}$$

Accordingly we find

$$\begin{aligned} K_n(x) &= \int_0^\infty e^{-s} H_n(-2xs, -s) e^{-sx^2} ds = \int_0^\infty e^{-\sigma} H_n\left(-2\frac{x\sigma}{1+x^2}, -\frac{\sigma}{1+x^2}\right) \frac{1}{1+x^2} d\sigma = \\ &= \frac{n!}{1+x^2} \sum_{r=0}^{\lfloor \frac{n}{2} \rfloor} \frac{\left(\frac{-2x}{1+x^2}\right)^{n-2r} \left(\frac{-1}{1+x^2}\right)^r}{(n-2r)!r!} \int_0^\infty e^{-\sigma} \sigma^{n-r} d\sigma = \frac{1}{1+x^2} P_n\left(-\frac{2x}{1+x^2}, -\frac{1}{1+x^2}\right). \end{aligned} \tag{14}$$

The proof of the identity (9) is readily achieved by noting that

$$\hat{D}_x \left(\tan^{-1}(x)\right) = \left(\frac{1}{1+x^2}\right), \tag{15}$$

which eventually provides

$$\hat{D}_x^n \left(\tan^{-1}(x)\right) = K_{n-1}(x) = \frac{1}{1+x^2} P_{n-1}\left(-\frac{2x}{1+x^2}, -\frac{1}{1+x^2}\right), \qquad \forall n \in \mathbb{N}_0.$$

□

Corollary 1. *If we introduce the further notation*

$$K_n^\nu(x) := \hat{D}_x^n \left(\frac{1}{(1+x^2)^\nu}\right), \tag{16}$$

an extension of the procedure we have just envisaged allows the further results

$$K_n^{\nu}(x) = \frac{1}{(1+x^2)^{\nu}} P_n^{\nu}\left(-\frac{2x}{1+x^2}, -\frac{1}{1+x^2}\right),$$
$$P_n^{\nu}(w,z) = \frac{n!}{\Gamma(\nu)} \sum_{r=0}^{\lfloor \frac{n}{2} \rfloor} \frac{w^{n-2r} z^r \, \Gamma(\nu+n-r)}{(n-2r)!r!}, \qquad \forall \nu \in \mathbb{R}. \tag{17}$$

Proof. $\forall \nu \in \mathbb{R}$

$$\begin{aligned}
K_n^{\nu}(x) &= \hat{D}_x^n \left(\frac{1}{(1+x^2)^{\nu}}\right) = \hat{D}_x^n \frac{1}{\Gamma(\nu)} \int_0^{\infty} s^{\nu-1} e^{-s(1+x^2)} ds = \frac{1}{\Gamma(\nu)} \int_0^{\infty} s^{\nu-1} e^{-s} \hat{D}_x^n e^{-sx^2} ds = \\
&= \frac{1}{\Gamma(\nu)} \int_0^{\infty} s^{\nu-1} e^{-s} H_n(-2xs, -s) e^{-sx^2} ds = \\
&= \frac{1}{\Gamma(\nu)} \int_0^{\infty} \left(\frac{\sigma}{1+x^2}\right)^{\nu-1} e^{-\sigma} H_n\left(-2\frac{x\sigma}{1+x^2}, -\frac{\sigma}{1+x^2}\right) \frac{1}{1+x^2} d\sigma = \\
&= \frac{n!}{\Gamma(\nu)(1+x^2)^{\nu}} \sum_{r=0}^{\lfloor \frac{n}{2} \rfloor} \frac{\left(\frac{-2x}{1+x^2}\right)^{n-2r} \left(\frac{-1}{1+x^2}\right)}{(n-2r)!r!} \int_0^{\infty} e^{-\sigma} \sigma^{\nu+n-r-1} d\sigma = \\
&= \frac{1}{(1+x^2)^{\nu}} P_n^{\nu}\left(-\frac{2x}{1+x^2}, -\frac{1}{1+x^2}\right).
\end{aligned}$$

□

Remark 1. *Corollary 1 can be exploited to write the repeated derivatives of* $\cos^{-1}(x)$ *in the form*

$$\hat{D}_x^n \cos^{-1}(x) = -\hat{D}_x^{n-1}\left(\frac{1}{\sqrt{1-x^2}}\right) = -\frac{1}{\sqrt{1-x^2}} P_{n-1}^{\frac{1}{2}}\left(\frac{2x}{1-x^2}, \frac{1}{1-x^2}\right), \qquad \forall n \in \mathbb{N}_0. \tag{18}$$

The previous examples have shown the interplay between special polynomials of the $P_n(x,y)$ and the derivatives of the inverse of trigonometric functions.

In the following we will see that the same polynomials are of central importance for the evaluation of the repetead derivatives of trigonometric functions. They play the role of auxiliary polynomials for the lorentzian type functions and inverse trigonometric functions as well. The relevant elements of contact with the *DP* will be discussed in the final section of the paper.

2. Higher Order Derivatives of Trigonometric Functions

The starting point of the discussion of this section is the derivation of a closed form for the quantity $\hat{D}_x^m\,(\mathrm{sech}(x)), \forall m \in \mathbb{N}_0$.

Proposition 1. $\forall m \in \mathbb{N}_0, \forall x \in [0, 2\pi]$,

$$\begin{aligned}
\hat{D}_x^m\,(\mathrm{sech}(x)) = \mathrm{sech}(x) \sum_{k=0}^{m} S_2(m,k)\, e^{(k-1)x} \Big(e^x\, P_k\Big(-\mathrm{sech}(x), -\frac{1}{2e^x}\mathrm{sech}(x)\Big) + \\
+ k\, P_{k-1}\Big(-\mathrm{sech}(x), -\frac{1}{2e^x}\mathrm{sech}(x)\Big)\Big).
\end{aligned} \tag{19}$$

Proof. Let

$$\mathrm{sech}(x) = \frac{2}{e^x + e^{-x}} \tag{20}$$

then, after setting $e^x = \xi$, we obtain $\forall m \in \mathbb{N}_0$

$$\hat{D}_x^m\left(\operatorname{sech}(x)\right)=\hat{D}_{\log\xi}^m\left(\frac{2\xi}{1+\xi^2}\right)=2\left(\xi\hat{D}_\xi\right)^m\left(\frac{\xi}{1+\xi^2}\right). \tag{21}$$

The straightforward applications of the proof of the Theorem 1 yields (we omit the argument of the P_n polynomials for simplicity)

$$2\left(\xi\hat{D}_\xi\right)^m\left(\frac{\xi}{1+\xi^2}\right)=\frac{2}{1+\xi^2}\sum_{k=0}^m S_2(m,k)\,\xi^k\left(\xi P_k(.,.)+kP_{k-1}(.,.)\right) \tag{22}$$

which, eventually, leads to

$$\hat{D}_x^m\left(\operatorname{sech}(x)\right)=\operatorname{sech}(x)\sum_{k=0}^m S_2(m,k)\,e^{(k-1)x}\left(e^x\,P_k\left(-\operatorname{sech}(x),-\frac{1}{2e^x}\operatorname{sech}(x)\right)+\right.$$
$$\left.+\,k\,P_{k-1}\left(-\operatorname{sech}(x),-\frac{1}{2e^x}\operatorname{sech}(x)\right)\right).$$

□

We can go even further and obtain a closed form for $\hat{D}_x^m\left(\operatorname{sech}(x)\right)^\nu,\forall m\in\mathbb{N},\forall\nu\in\mathbb{R}$.

Corollary 2. $\forall m\in\mathbb{N},\forall\nu\in\mathbb{R}$

$$\hat{D}_x^m\left(\operatorname{sech}^\nu(x)\right)=\operatorname{sech}^\nu(x)\sum_{r=0}^m S_2(m,r)\,e^{xr}\sum_{s=0}^r\binom{r}{s}\frac{n!}{(n-r+s)!}e^{-x(r-s)}\,P_s^\nu\left(-\operatorname{sech}(x),-\frac{1}{2e^x}\operatorname{sech}(x)\right) \tag{23}$$

where $P_m^k(.,.)$ are the polynomials in Equation (17).

By using the same techniques of the Proposition 1, Corollary 2 and the opportune identity $\sec(x)=\frac{2}{e^{ix}+e^{-ix}}$, we can move on and provide the higher order derivatives for other trigonometric circular functions.

Example 1. $\forall m\in\mathbb{N}_0,\forall x\in[0,2\pi]$

$$\hat{D}_x^m\left(\sec(x)\right)=i^m\sec(x)\sum_{r=0}^m S_2(m,r)\,e^{ix(r-1)}\left(e^{ix}P_r\left(-\sec(x),-\frac{1}{2e^{ix}}\sec(x)\right)+\right.$$
$$\left.+\,r\,P_{r-1}\left(-\sec(x),-\frac{1}{2e^{ix}}\sec(x)\right)\right). \tag{24}$$

Example 2. $\forall m\in\mathbb{N},\forall x\in[0,2\pi]:x\neq\frac{\pi}{2}+k\pi,\ k\in\mathbb{Z}$

$$\hat{D}_x^m\left(\tan(x)\right)=\sum_{s=0}^m\binom{m}{s}\sin\left(x+(m-s)\frac{\pi}{2}\right)\hat{D}_x^s(\sec(x)). \tag{25}$$

Example 3. $\forall m\in\mathbb{N},\forall x\in[0,2\pi]:x\neq k\pi,\ k\in\mathbb{Z}$

$$\hat{D}_x^m\left(\cot(x)\right)=\sum_{s=0}^m\binom{m}{s}\cos\left(x+(m-s)\frac{\pi}{2}\right)\hat{D}_x^s\left(\sec\left(x-\frac{\pi}{2}\right)\right). \tag{26}$$

Remark 2. *The use of the Leibniz rule in Equations (25) or (26) can be avoided by setting*

$$\tan(x)=i\,\frac{1-\xi^2}{1+\xi^2}=i\left(\frac{2}{1+\xi^2}-1\right),\quad \xi=e^{ix} \tag{27}$$

and eventually ending up with

$$\hat{D}_x^m\,(\tan(x)) = i^{m+1}\left(\sec(x)\sum_{r=0}^{m} S_2(m,r)\; e^{ix(r-1)} P_r\left(-\sec(x), -\frac{1}{2e^{ix}}\sec(x)\right) - \delta_{m,0}\right), \tag{28}$$

where $\delta_{m,0}$ is Kronecker's delta, indeed

$$\begin{aligned}
\hat{D}_x^m\,(\tan(x)) &= i^m\,(\xi\hat{D}_\xi)^m\left(i\left(\frac{2}{1+\xi^2}-1\right)\right) = 2\,i^{m+1}\,(\xi\hat{D}_\xi)^m\left(\frac{1}{1+\xi^2}-\frac{1}{2}\right) = \\
&= 2\,i^{m+1}\sum_{r=0}^{m} S_2(m,r)\;\xi^r\left(\frac{1}{1+\xi^2}-\frac{1}{2}\right)^{(r)} = \\
&= 2\,i^{m+1}\sum_{r=0}^{m} S_2(m,r)\;\xi^r\left(\left(\frac{1}{1+\xi^2}\right)^{(r)}+\left(-\frac{1}{2}\right)^{(r)}\right) = \\
&= 2\,i^{m+1}\left(\left(\sum_{r=0}^{m} S_2(m,r)\;\xi^r\frac{1}{1+\xi^2}P_r(.,.)\right)+\left(\sum_{r=0}^{m} S_2(m,r)\;\xi^r\left(-\frac{1}{2}\right)^{(r)}\right)\right) = \\
&= i^{m+1}\left(\sec(x)\sum_{r=0}^{m} S_2(m,r)\; e^{ix(r-1)} P_r\left(-\sec(x), -\frac{1}{2e^{ix}}\sec(x)\right) - \delta_{m,0}\right).
\end{aligned}$$

In the forthcoming section we will discuss the comparison with previous papers and present possible developments along the lines we have indicated.

3. Final Comments

In the previous two sections we have dealt with a general procedure useful to derive closed formulae for the repeated derivatives of circular and hyperbolic functions and of their inverse. We have underscored the importance of the polynomials $P_n(x,y)$ which play a role analogous to the *DP* introduced in refs. [6,7] and used in [4,5] for analogous occurrences.

The $P_n(x,y)$ are two variable polynomials defined in terms of the Laplace transform of the Hermite Kampé dé Fériét family [24]. They have been loosely defined Legendre-like and can be reduced to more familiar forms, by noting that

$$\begin{aligned}
P_n(x,y) &= y^{\frac{n}{2}} P_n\left(\frac{x}{\sqrt{-y}}\right), \\
P_n(z) &= n!\sum_{r=0}^{\lfloor \frac{n}{2}\rfloor}\frac{(-1)^r z^{n-2r}(n-r)!}{(n-2r)!r!}
\end{aligned} \tag{29}$$

with $P_n(z)$ satisfying the generating function

$$\sum_{n=0}^{\infty}\frac{t^n}{n!}P_n(z) = \frac{1}{1-tz+t^2}, \qquad \forall t,z\in\mathbb{R} : |\, t^2 - t\,z \,| < 1 \tag{30}$$

and

$$\frac{1}{n!}P_n(2x,-1) = \sum_{k=0}^{\lfloor \frac{n}{2}\rfloor}\frac{(-1)^k(n-k)!(2x)^{n-2k}}{(n-2k)!k!} = U_n(x). \tag{31}$$

The link with previous papers addressing the same problem treated here, in particular with that developed in refs. References [4,5] can be obtained by using the same steps suggested by Cvijović or Boyadzhiev. We assume that an identity of the type

$$\hat{D}_x^n(\tan(x)) = \Pi_n(\tan(x)) \tag{32}$$

with $\prod_n(.)$ being not yet specified polynomials, (in refs. [4,5] they are denoted by $P_n(x)$, we have used the capital Greek letter to avoid confusion with the $P_n(x, y)$ polynomials defined in this paper) does holds. The polynomials $\Pi_n(.)$ can be determined by the change of variable $\tan(x) = \xi$ which allows to transform Equation (32) into

$$\left[\left(1+\xi^2\right)\hat{D}_\xi\right]^n(\xi) = \Pi_n(\xi) \tag{33}$$

which is a kind of Rodrigues type relation [25] defining the polynomials $\Pi_n(\xi)$.

The relevant generating function can be obtained by multiplying both sides of Equation (33) by $\frac{t^n}{n!}$, by summing up over the index n and ending up with

$$e^{t\left[(1+\xi^2)\hat{D}_\xi\right]}\xi = \sum_{n=0}^{\infty}\frac{t^n}{n!}\Pi_n(\xi). \tag{34}$$

The use of the Lie derivative identity [26]

$$e^{t\left[(1+\xi^2)\hat{D}_\xi\right]}f(\xi) = f\left(\frac{\xi\cos(t)+\sin(t)}{\cos(t)-\xi\sin(t)}\right) \tag{35}$$

finally yields

$$\sum_{n=0}^{\infty}\frac{t^n}{n!}\Pi_n(\xi) = \frac{\xi+\tan(t)}{1-\xi\tan(t)} \tag{36}$$

which is the generating function given in [4,5].

The second family of DP can be defined by the use of the same procedure. According to refs. [4,5] they are implicitly defined by the condition

$$\hat{D}_x^n(\sec(x)) = \sec(x)Q_n(\tan(x)) \tag{37}$$

which, by the use of the same change of variable leading to Equation (33), yields

$$Q_n(\xi) = \frac{1}{\sqrt{1+\xi^2}}\left(\left(1+\xi^2\right)\hat{D}_\xi\right)^n\sqrt{1+\xi^2} \tag{38}$$

which can be straightforwardly exploited to derive the relevant properties. The use of the identity (35) eventually yields the associated generating functions

$$\sum_{n=0}^{\infty}\frac{t^n}{n!}Q_n(\xi) = \frac{\sec(t)}{1-\xi\tan(t)} \tag{39}$$

which is the same reported in refs. [4,5].

By recalling that the Hoppe formula writes

$$\begin{gathered}\hat{D}_t^m f(g(t)) = \sum_{k=0}^{m}\frac{1}{k!}f^{(k)}(\sigma)\mid_{\sigma=g(t)} A_{m,k}, \\ A_{m,k} = \sum_{j=0}^{k}\binom{k}{j}(-1)^{k-j}g(t)^{k-j}\hat{D}_t^m(g(t)^j), \quad \forall m,k\in\mathbb{N},\end{gathered} \tag{40}$$

for the case of $\sec(x)$, we find

$$\hat{D}_x^m\sec(x) = \sum_{k=0}^{m}\sec^{k+1}(x)\sum_{j=0}^{k}\binom{k}{j}(-1)^j\cos^{k-j}(x)\,\hat{D}_x^m(\cos^j(x)) \tag{41}$$

which appears impractical since it needs the repeated derivatives of an integer power of the cosine function.

Before closing the paper we consider worth to underscore the possible use of DP to evaluate the derivatives of the type

$$\lambda_{m,j}(x) = \hat{D}_x^m(\cos^j(x)). \tag{42}$$

Making the ansatz that

$$\hat{D}_x^m(\cos^j(x)) = (-1)^j \Lambda_{m,j}(\tan(x)) \tag{43}$$

and, by using the same procedure leading to Equation (33), we can identify the function

$$\Lambda_{m,j}(\xi) = \left[\left(1+\xi^2\right)\hat{D}_\xi\right]^m \left(\frac{1}{\left(\sqrt{1+\xi^2}\right)^j}\right) \tag{44}$$

which is specified by the generating function

$$\sum_{m=0}^{\infty} \frac{t^m}{m!} \Lambda_{m,j}(\xi) = \frac{[1-\xi\tan(t)]^j}{\left[(1+\xi^2)\left(1+\tan^2(t)\right)\right]^{\frac{j}{2}}}. \tag{45}$$

It is worth to consider the alternative assumption

$$\hat{D}_x^m(\cos^j(x)) = (-1)^m \Delta_{m,j}(\cos(x)) \tag{46}$$

which, after the change of variable $\xi = \cos(x)$, yields for the function $\Delta_{m,j}(.)$ the operational definition

$$\Delta_{m,j}(\xi) = \left(-\sqrt{1-\xi^2}\,\hat{D}_\xi\right)^m \xi^j \tag{47}$$

The use of the Lie derivative identity [26]

$$e^{-t\left(\sqrt{1-\xi^2}\,\hat{D}_\xi\right)} f(\xi) = f\left(\xi\cos(t) - \sqrt{1-\xi^2}\sin(t)\right) \tag{48}$$

yields, for $\Delta_{m,j}(\xi)$, the generating function

$$\sum_{m=0}^{\infty} \frac{t^m}{m!} \Delta_{m,j}(\xi) = \left(\xi - \sqrt{1-\xi^2}\,\tan(t)\right)^j \cos^j(t). \tag{49}$$

The two methods we have discussed in this paper, namely the procedure based on the DP of refs. [4,5] or on the support polynomials $P_n(x,y)$, are complementary. There are no prevailing reasons to prefer one or the other method. The formulae associated with the first are more synthetic but $\Pi_n(\xi)$, $Q_n(\xi)$ are not given explicitly and should be evaluated recursively (which becomes cumbersome for larger order of the derivative). On the other side, the use of the other procedure leads to less appealing formulae in terms of polynomials which are, however, explicitly given.

Author Contributions: Conceptualization: G.D.; methodology: G.D., S.L.; data curation: S.L.; validation: G.D., S.L., R.M.P.; formal analysis: G.D., S.L., R.M.P.; writing—original draft preparation: G.D., E.S.; writing—review and editing: S.L.

Acknowledgments: The work of S. Licciardi was supported by an Enea Research Center individual fellowship. R.M. Pidatella wants to thank the fund of University of Catania "Metodi gruppali e umbrali per modelli di diffusione e trasporto" included in "Piano della Ricerca 2016/2018 Linea di intervento 2" for partial support of this work.

References

1. Adamchik, V.S. On the Hurwitz function for rational arguments. *Appl. Math. Comput.* **2007**, *187*, 3–12. [CrossRef]
2. Neto, A.F. Higher Order Derivatives of Trigonometric Functions, Stirling Numbers of the Second Kind and Zeon Algebra. *J. Integer Seq.* **2014**, *17*, 14.9.3.
3. Feinsilver, P. Zeon algebra, Fock space, and Markov chains. *Commun. Stoch. Anal.* **2008**, *2*, 6. [CrossRef]
4. Cvijović, D. Derivative Polynomials and Closed-Form Higher Derivative Formulae. *Appl. Math. Comput.* **2009**, *215*, 3002–3006. [CrossRef]
5. Boyadzhiev, K.N. Derivative polynomials for tanh, tan, sech and sec in explicit form. *Fibonacci Q.* **2007**, *45*, 291–303.
6. Hoffman, M.E. Derivative polynomials for tangent and secant. *Am. Math. Mon.* **1995**, *102*, 23–30. [CrossRef]
7. Hoffman, M.E. Derivative polynomials, Euler polynomials and associated integer sequences. *Electron. J. Comb.* **1999**, *6*, R21.
8. Johnson, W.P. The Curious History of Faa di Bruno's Formula. *Am. Math. Mon.* **2002**, *109*, 217–234.
9. Comtet, L. *Advanced Combinatorics: The Art of Finite and Infinite Expansions*; Springer: Berlin/Heidelberg, Germany, 1974.
10. Roman, S. *The Umbral Calculus*; Dover Publications: New York, NY, USA, 2005.
11. Licciardi, S. Umbral Calculus, a Different Mathematical Language. Ph.D. Thesis, Department of Mathematics and Computer Sciences, XXIX Cycle, University of Catania, Catania, Italy, 2018.
12. Dattoli, G.; Germano, B.; Martinelli, M.R.; Ricci, P.E. Touchard like polynomials and generalized Stirling numbers. *Appl. Math. Comput.* **2012**, *218*, 6661–6665. [CrossRef]
13. Graham,R.L.; Knuth, D.E.; Patashnik, O. *Concrete Mathematics: A Foundation for Computer Science*, 2nd ed.; Addison-Wesley Longman Publishing Co., Inc.: Boston, MA, USA, 1994.
14. Mansour, T.; Schork, M. *Commutation Relations, Normal Ordering, and Stirling Numbers*; Chapman and Hall/CRC: Boca Raton, FL, USA, 2015; p. 504, ISBN 9781466579880.
15. Dattoli, G. Generalized polynomials, operational identities and their applications. *J. Comp. Appl. Math.* **2000**, *118*. 111–123. [CrossRef]
16. Riordan, J. *An Introduction to Combinatorial Analysis*; Wiley: New York, NY, USA, 1980.
17. Bell, E.T. Exponential Polynomials. *Ann. Math.* **1934**, *35*, 258–277. [CrossRef]
18. Dattoli, G.; Srivastava, H.M.; Sacchetti, D. Mixed and higher-order generating functions: An extension to new families. *Int. J. Math. Stat. Sci.* **1999**, *8*, 79–87.
19. Babusci, D.; Dattoli, G.; Gorska, K.; Penson, K. Repeated derivatives of composite functions and generalizations of the Leibniz rule. *Appl. Math. Comput.* **2014**, *241*, 193. [CrossRef]
20. Wintucky, E.G. *Formulas for nth Order Derivatives of Hyperbolic and Trigonometric Functions*; NASA Lewis Research Center: Cleveland, OH, USA, 1971.
21. Andrews, L.C. *Special Functions For Engeneers and Applied Mathematicians*; Mc Millan: New York, NY, USA, 1985.
22. Dattoli, G.; Germano, B.; Martinelli, M.R.; Ricci, P.E. A novel theory of Legendre polynomials. *Math. Comput. Modell.* **2011**, *54*, 80–87. [CrossRef]
23. Dattoli, G.; Germano, B.; Martinelli, M.R.; Ricci, P.E. Sheffer and Non-Sheffer Polynomial Families. *Int. J. Math. Math. Sci.* **2012**, *2012*, 323725. [CrossRef]
24. Appél, P.; Kampé de Fériét, J. *Fonctions Hypergeometriques and Hyperspheriques. Polynomes d'Hermite*; Gauthiers-Villars: Paris, France, 1926.
25. Rainville, E.D. *Special Functions*; Macmillan: New York, NY, USA, 1960.
26. Dattoli, G.; Ottaviani, P.L.; Torre, A.; Vazquez, L. Evolution operator equations-integration with algebraic and finite-difference methods-applications to physical problems in classical and quantum mechanics and quantum field theory. *La Rivista del Nuovo Cimento* **1997**, *20*, 3–133. [CrossRef]

Special Numbers and Polynomials Including their Generating Functions in Umbral Analysis Methods

Yilmaz Simsek

Department of Mathematics, Faculty of Science University of Akdeniz, TR-07070 Antalya, Turkey; ysimsek@akdeniz.edu.tr

Abstract: In this paper, by applying umbral calculus methods to generating functions for the combinatorial numbers and the Apostol type polynomials and numbers of order k, we derive some identities and relations including the combinatorial numbers, the Apostol-Bernoulli polynomials and numbers of order k and the Apostol-Euler polynomials and numbers of order k. Moreover, by using p-adic integral technique, we also derive some combinatorial sums including the Bernoulli numbers, the Euler numbers, the Apostol-Euler numbers and the numbers $y_1(n,k;\lambda)$. Finally, we make some remarks and observations regarding these identities and relations.

Keywords: Apostol-Bernoulli polynomials and numbers; Apostol-Euler polynomials and numbers; Sheffer sequences; Appell sequences; Fibonacci numbers; umbral algebra, p-adic integral

---

1. Introduction

In order to give the results presented in this paper, we use two techniques which are p-adic integral technique and the umbral calculus technique. In [1–5], we constructed generating functions for families of combinatorial numbers which are used in counting techniques and problems and also computing negative order of the first and the second kind Euler numbers and other combinatorial sums. In this paper, by applying umbral algebra and umbral analysis methods and their operators to generating functions of the combinatorial numbers and the Apostol type polynomials and numbers, we give many identities and relations including the Fibonacci numbers, the combinatorial numbers, the Apostol-Bernoulli polynomials and numbers of higher order and the Apostol-Euler polynomials and numbers of higher order.

Throughout this paper, we use the following notations, definitions and relations.

Here and in the following, let $\mathbb{C}$, $\mathbb{R}$, $\mathbb{Z}$, and $\mathbb{N}$ be the sets of complex numbers, real numbers, integers, and positive integers, respectively, and let $\mathbb{N}_0 := \mathbb{N}\cup\{0\}$. We assume $0^0 = 1$.

Moreover, throughout this paper, $\log z$ is tacitly assumed to denote the principal branch of the many-valued function $\log z$ with the imaginary part $(\log z)$ constrained by

$$-\pi < \operatorname{Im}(\log z) \leq \pi$$

(cf. [6–9]).

The Apostol-Bernoulli polynomials $\mathcal{B}_n^{(k)}(x,\lambda)$ of order k are defined by

$$F_{\mathcal{B}}(t,x;\lambda,k) = \left(\frac{t}{\lambda e^t - 1}\right)^k e^{tx} = \sum_{n=0}^{\infty} \mathcal{B}_n^{(k)}(x,\lambda)\frac{t^n}{n!}, \tag{1}$$

where λ is an arbitrary (real or complex) parameter and $x \in \mathbb{R}$, and $|t| < 2\pi$ when $\lambda = 1$ and $|t| < \frac{2\pi}{|\log \lambda|}$ when $\lambda \neq 1$. Moreover, $\mathcal{B}_n^{(k)}(\lambda) := \mathcal{B}_n^{(k)}(0,\lambda)$ denote the Apostol-Bernoulli numbers of order. $B_n^{(k)} := B_n^{(k)}(1)$ denote the Bernoulli numbers of order k and also $B_n := B_n^{(1)}$ denote the Bernoulli numbers (cf. see,for details, [6,8–14], and the references cited therein).

The Apostol-Euler polynomials $\mathcal{E}_n^{(k)}(x,\lambda)$ of order k are defined by

$$F_{\mathcal{E}}(t,x;\lambda,k) = \left(\frac{2}{\lambda e^t + 1}\right)^k e^{tx} = \sum_{n=0}^{\infty} \mathcal{E}_n^{(k)}(x,\lambda)\frac{t^n}{n!}, \tag{2}$$

where λ is an arbitrary (real or complex) parameter and $x \in \mathbb{R}$, and $|t| < \pi$ when $\lambda = 1$ and $|t| < \frac{\pi}{|\log \lambda|}$ when $\lambda \neq 1$. Moreover, $\mathcal{E}_n^{(k)}(\lambda) := \mathcal{E}_n^{(k)}(0,\lambda)$ denote the Apostol-Euler numbers of order k. $E_n^{(k)} := E_n^{(k)}(1)$ denote the Euler numbers of order k and also $E_n := E_n^{(1)}$ denote the Euler numbers (cf. see, for details, [6,8–15], and the references cited therein).

The λ-array polynomials $S_v^n(x;\lambda)$ are defined by

$$F_S(t,x,v;\lambda) = \frac{(\lambda e^t - 1)^v}{v!} e^{xt} = \sum_{n=0}^{\infty} S_v^n(x;\lambda)\frac{t^n}{n!} \tag{3}$$

where $v \in \mathbb{N}_0$ and $\lambda \in \mathbb{C}$ (cf. [16]). Furthermore,

$$S_2(n,v;\lambda) := S_v^n(0;\lambda)$$

where, as usual, $S_2(n,v;\lambda)$ denote the λ-Stirling numbers (cf. [8,12]). Substituting $\lambda = 1$ into (3), we have the array polynomials:

$$S_v^n(x) := S_v^n(x;1)$$

(cf. [16–18] and (Theorem 2 [19])).

In (cf. Equation (8) [1]), we defined the combinatorial numbers $y_1(n,k;\lambda)$ by means of the following generating function:

$$F_{y_1}(t,k;\lambda) = \frac{1}{k!}(\lambda e^t + 1)^k = \sum_{n=0}^{\infty} y_1(n,k;\lambda)\frac{t^n}{n!} \tag{4}$$

where $k \in \mathbb{N}_0$ and $\lambda \in \mathbb{C}$.

By using (4), we have

$$y_1(n,k;\lambda) = \frac{1}{k!}\sum_{j=0}^{k}\binom{k}{j} j^n \lambda^j \tag{5}$$

where $n \in \mathbb{N}_0$ (cf. Equation (9) [1]).

Relationships between the λ-array polynomials $S_v^n(x;\lambda)$ and the numbers $y_1(n,k;\lambda)$ and the Stirling numbers of the second kind $S_2(n,k)$ are given below, respectively:

$$S_k^n(0;\lambda) = S_2(n,v;\lambda) = (-1)^k y_1(n,k;-\lambda)$$

and

$$S_2(n,k) = (-1)^k y_1(n,k;-1) \tag{6}$$

(cf. [1,17,20–25]).

The Fibonacci numbers F_j are defined by the following generating function

$$\frac{t}{1-t-t^2} = \sum_{n=0}^{\infty} F_n t^n$$

(cf. (p. 229. [26])). We need the following well-known formulas for the Fibonacci numbers. Let $\lambda = \frac{1+\sqrt{5}}{2}$ and $\beta = \frac{1-\sqrt{5}}{2}$. Let $j \in \mathbb{N}$, we have

$$\lambda^j = \lambda F_j + F_{j-1} \tag{7}$$

and

$$\beta^j = \beta F_j + F_{j-1}$$

(cf. (p. 78, Lemma 5.1. [26])). Using the above identities, one easily derives the following Binet's formula:

$$F_j = \frac{\lambda^j - \beta^j}{\lambda - \beta}.$$

Substituting $-n$ with $n \in \mathbb{N}$, into the above equation, we easily have

$$F_{-n} = (-1)^{n+1} F_n$$

(cf. (p. 84 [26])).

1.1. *p-Adic Integrals*

In the last section, we will give some combinatorial sums with p-adic integrals technique. Hence, let us give definitions of these integrals and a few properties of them.

Let $f(x) \in C^1(\mathbb{Z}_p \to \mathbb{K})$, a set of continuous derivative functions, and $\mathbb{K}$ is a field with a complete valuation.

The Volkenborn integral (the bosonic p-adic integral) is defined by

$$\int_{\mathbb{Z}_p} f(x) d\mu_1(x) = \lim_{N\to\infty} \frac{1}{p^N} \sum_{x=0}^{p^N-1} f(x), \tag{8}$$

where $\mu_1(x) = \mu_1\left(x + p^N\mathbb{Z}_p\right)$ is the Haar distribution on $\mathbb{Z}_p$:

$$\mu_1\left(x + p^N\mathbb{Z}_p\right) = \frac{1}{p^N},$$

(cf. [27,28]). On the other hand, the p-adic fermionic integral is defined by

$$\int_{\mathbb{Z}_p} f(x)\, d\mu_{-1}(x) = \lim_{N\to\infty} \sum_{x=0}^{p^N-1} (-1)^x f(x) \tag{9}$$

where

$$\mu_{-1}(x) = \mu_{-1}\left(x + p^N\mathbb{Z}_p\right) = \frac{(-1)^x}{p^N}$$

(cf. [29]).

The Bernoulli numbers and the Euler numbers are related to the following p-adic integrals representations, respectively,

$$B_n = \int_{\mathbb{Z}_p} x^n d\mu_1(x), \tag{10}$$

(cf. [27,28]) and

$$E_n = \int_{\mathbb{Z}_p} x^n d\mu_{-1}(x), \tag{11}$$

(cf. [27]).

1.2. Umbral Algebra and Calculus

Throughout this section, we use the notations and definitions of the Roman's book (cf. [13]). Let $\mathbb{P} = \mathbb{C}[x]$ be the algebra of polynomials in the single variable x over the field of complex numbers. Let $\mathbb{P}^*$ be the vector space of all linear functionals on $\mathbb{P}$. Let $\langle L \mid p(x) \rangle$ be the action of a linear functional L on a polynomial $p(x)$. Let $\mathcal{F}$ denote the algebra of formal power series

$$f(t) = \sum_{k=0}^{\infty} a_k \frac{t^k}{k!},$$

(cf. [13]). Furthermore, for all $n \in \mathbb{N}_0$, one has

$$\langle f(t) \mid x^n \rangle = a_n \tag{12}$$

and also

$$\langle f(t)g(t) \mid p(x) \rangle = \langle f(t) \mid g(t)p(x) \rangle, \tag{13}$$

where $f(t)$, $g(t)$ are in $\mathcal{F}$ (cf. [13]).

For $p(x) \in \mathbb{P}$, as a *linear functional*, we have

$$\left\langle e^{yt} \mid p(x) \right\rangle = p(y). \tag{14}$$

and as a *linear operator*, we have

$$e^{yt} p(x) = p(x+y) \tag{15}$$

(cf. [13]). The Sheffer polynomials for pair $(g(t), f(t))$, where $g(t)$ must be invertible and $f(t)$ must be delta series. The Sheffer polynomials for pair $(g(t), t)$ is the Appell polynomials or Appell sequences for $g(t)$. The Appell polynomials are defined by means of the following generating function

$$\sum_{k=0}^{\infty} \frac{s_k(x)}{k!} t^k = \frac{1}{g(t)} e^{xt}, \tag{16}$$

(cf. [13]). Some properties of the Appell polynomials are given as follows.

$$s_n(x) = g(t)^{-1} x^n, \tag{17}$$

(p. 86, Theorem 2.5.5 [13]), derivative formula

$$t s_n(x) = n s_{n-1}(x) \tag{18}$$

(cf. p. 86, Theorem 2.5.6 [13]); and see also [6,30,31]).

We summarize the results presented in this paper as follows:

In Section 2, by applying the umbral algebra and umbral calculus methods to generating functions of the special numbers and polynomials, we derive some identities and relations including the numbers $y_1(n,k;\lambda)$, combinatorial sums, the Fibonacci numbers, Apostol-Bernoulli type numbers and polynomials, and the Apostol-Euler type numbers and polynomials. Finally, we give some remarks and observations.

In Section 3, by using the p-adic integrals, we give many combinatorial sums related to the Bernoulli numbers, the Euler numbers, the Apostol-Euler numbers and the numbers $y_1(n,k;\lambda)$.

2. Identities Including the Numbers $y_1(n,k;\lambda)$, Combinatorial Sums, and Apostol-Euler Type Numbers and Polynomials

In this section, by using the umbral algebra and umbral calculus methods, we derive many identities and relations containing the numbers $y_1(n,k;\lambda)$, combinatorial sums, the Fibonacci numbers, Apostol-Bernoulli type numbers and polynomials, and the Apostol-Euler type numbers and polynomials.

Theorem 1.

$$\frac{1}{k!}\sum_{j=0}^{k}\binom{k}{j}\lambda^j\sum_{v=0}^{m}\binom{m}{v}x^{m-v}j^v=\frac{1}{k!}\sum_{j=0}^{k}\binom{k}{j}\lambda^j(x+j)^m$$

or

$$\sum_{v=0}^{m}\frac{(m)_v}{v!}x^{m-v}y_1(v,k;\lambda)=\frac{1}{k!}\sum_{j=0}^{k}\binom{k}{j}\lambda^j(x+j)^m \tag{19}$$

Proof. By applying linear operators in (15) and (18) to (4), respectively, we obtain

$$\frac{1}{k!}\left(\lambda e^t+1\right)^k x^m=\frac{1}{k!}\sum_{j=0}^{k}\binom{k}{j}\lambda^j(x+j)^m \tag{20}$$

and

$$\begin{aligned}\frac{1}{k!}\left(\lambda e^t+1\right)^k x^m &= \sum_{n=0}^{\infty}y_1(n,k;\lambda)\frac{1}{n!}t^n x^m \\ &= \begin{cases} 0, & n>m \\ y_1(n,k;\lambda), & n=m \\ \sum_{v=0}^{m}\frac{(m)_v}{v!}x^{m-v}y_1(v,k;\lambda) & n<m. \end{cases}\end{aligned} \tag{21}$$

Combining (20) with (21), we get the desired results. □

Remark 1. *Substituting* $x=0$ *into* (19), *we arrive at* (5).

By applying the action of a linear operator $\left(\lambda e^t+1\right)^k$ to the Apostol-Euler polynomial $\mathcal{E}_n^{(a)}(x,\lambda)$, we obtain the following result.

Theorem 2.

$$\sum_{j=0}^{k}\binom{k}{j}\lambda^j\mathcal{E}_n^{(a)}(x+j,\lambda)=\sum_{j=0}^{k}\binom{k}{j}\sum_{v=0}^{j}(-1)^{j-v}\binom{j}{v}2^v\mathcal{E}_n^{(a-v)}(x,\lambda). \tag{22}$$

Proof. By applying the action of a linear operator $\left(\lambda e^t+1\right)^k$ to the Apostol-Euler polynomial $\mathcal{E}_n^{(a)}(x,\lambda)$, we obtain

$$\left(\lambda e^t+1\right)^k\mathcal{E}_n^{(a)}(x,\lambda)=\sum_{j=0}^{k}\binom{k}{j}\lambda^j e^{tj}\mathcal{E}_n^{(a)}(x,\lambda). \tag{23}$$

Applying linear operators in (15) to the above equation, we have

$$\left(\lambda e^t+1\right)^k\mathcal{E}_n^{(a)}(x,\lambda)=\sum_{j=0}^{k}\binom{k}{j}\lambda^j\mathcal{E}_n^{(a)}(x+j,\lambda). \tag{24}$$

Combining the following relation with (23)

$$\mathcal{E}_n^{(a)}(x,\lambda)=\left(\frac{2}{\lambda e^t+1}\right)^a x^n$$

(cf. p. 101 [13]), we have

$$\left(\lambda e^{t}+1\right)^{k}\mathcal{E}_{n}^{(a)}\left(x,\lambda\right)=\sum_{j=0}^{k}\binom{k}{j}\sum_{v=0}^{j}(-1)^{j-v}\binom{j}{v}2^{v}\left(\frac{2}{\lambda e^{t}+1}\right)^{a-v}x^{n}.$$

After some elementary calculation in the above equation, we have

$$\left(\lambda e^{t}+1\right)^{k}\mathcal{E}_{n}^{(a)}\left(x,\lambda\right)=\sum_{j=0}^{k}\binom{k}{j}\sum_{v=0}^{j}(-1)^{j-v}\binom{j}{v}2^{v}\mathcal{E}_{n}^{(a-v)}\left(x,\lambda\right). \tag{25}$$

Combining (24) and (25), we arrive at the desired result. □

Substituting $\lambda=k=1$ into (22), we arrive at the following corollary, which was proved by Roman (p. 103, Equation (4.2.11) [13]), see also (cf. [32]).

Corollary 1.

$$2E_{n}^{(a-1)}(x)=E_{n}^{(a)}(x+1)+E_{n}^{(a)}(x)$$

We assume that, $\lambda\neq 1$ and $a\in\mathbb{N}$, we have the following well-known relationships between the polynomials $\mathcal{B}_{n}^{(a)}\left(x,\lambda\right)$ and $\mathcal{E}_{n}^{(a)}\left(x,-\lambda\right)$:

$$\begin{aligned}\sum_{n=0}^{\infty}\mathcal{E}_{n}^{(a)}\left(x,-\lambda\right)\frac{t^{n}}{n!} &= \left(\frac{2}{-\lambda e^{t}+1}\right)^{a}e^{tx}\\ &= \left(-\frac{2}{t}\right)^{a}\sum_{n=0}^{\infty}\mathcal{B}_{n}^{(a)}\left(x,\lambda\right)\frac{t^{n}}{n!}.\end{aligned}$$

Therefore

$$(n)_{a}\mathcal{E}_{n-a}^{(a)}\left(x,-\lambda\right)=(-2)^{a}\mathcal{B}_{n}^{(a)}\left(x,\lambda\right)$$

or

$$\mathcal{E}_{n}^{(a)}\left(x,\lambda\right)=\frac{(-2)^{a}}{(n+a)_{a}}\mathcal{B}_{n+a}^{(a)}\left(x,-\lambda\right).$$

Substituting the above relation into (22), we get the following result.

Theorem 3.

$$\begin{aligned}&\sum_{j=0}^{k}\binom{k}{j}\lambda^{j}\frac{(-2)^{a}}{(n+a)_{a}}\mathcal{B}_{n+a}^{(a)}\left(x+j,-\lambda\right)\\ =&\sum_{j=0}^{k}\binom{k}{j}\sum_{v=0}^{j}(-1)^{j+a}\binom{j}{v}\frac{2^{a}}{(n+a-v)_{a-v}}\mathcal{B}_{n+a-v}^{(a-v)}\left(x,-\lambda\right).\end{aligned} \tag{26}$$

Setting $k=1$ in (26), we get the following corollary.

Corollary 2.

$$\mathcal{B}_{n+a}^{(a)}\left(x,-\lambda\right)+\lambda\mathcal{B}_{n+a}^{(a)}\left(x+1,-\lambda\right)=-\left(n+a\right)\mathcal{B}_{n+a-1}^{(a-1)}\left(x,-\lambda\right). \tag{27}$$

Remark 2. *Another proof of the Equation* (27) *is given by Dere et al.* [6] *and see also (cf. [32]).*

Remark 3. *Substituting $n+a=m$ and $\lambda=-1$ into* (27)*, we get*

$$B_{m}^{(a)}(x+1)=B_{m}^{(a)}(x)+mB_{m-1}^{(a-1)}(x)$$

(cf. p. 95, Equation (4.2.6) [13]).

The following theorem was proved in (cf. [1]). Here, we give a proof different from that in (cf. [1]).

Theorem 4. *Let n and k be nonnegative integers. Then we have*

$$y_1(n,k;\lambda) = \frac{2^k}{k!}\mathcal{E}_n^{(-k)}(0,\lambda).$$

Proof. Using (12), we obtain

$$y_1(n,k;\lambda) = \frac{1}{k!}\left\langle \left(\lambda e^t+1\right)^k \mid x^n \right\rangle. \tag{28}$$

From the above equation, we have

$$\begin{aligned} y_1(n,k;\lambda) &= \frac{2^k}{k!}\left\langle \left(\frac{\lambda e^t+1}{2}\right)^k \mid x^n \right\rangle \\ &= \frac{2^k}{k!}\left\langle t^0 \mid \left(\frac{\lambda e^t+1}{2}\right)^{-k} x^n \right\rangle \\ &= \frac{2^k}{k!}\mathcal{E}_n^{(-k)}(0,\lambda). \end{aligned}$$

Therefore, we arrive at the desired result. □

Theorem 5.

$$\frac{2^k}{k!}\sum_{j=0}^{n}\binom{n}{j}\mathcal{E}_j^{(a)}(\lambda)\,\mathcal{E}_{n-j}^{(-k)}(\lambda) = \sum_{j=0}^{n}\binom{n}{j}\mathcal{E}_j^{(a)}(\lambda)\,y_1(n-j,k;\lambda)$$

Proof. We set

$$\begin{aligned} \frac{1}{k!}\left\langle \left(\lambda e^t+1\right)^k \mid \mathcal{E}_n^{(a)}(\lambda)\right\rangle &= \frac{1}{k!}\left\langle \left(\lambda e^t+1\right)^k \mid \left(\sum_{j=0}^{n}\binom{n}{j}\mathcal{E}_j^{(a)}(\lambda)\right)x^{n-j}\right\rangle \\ &= \frac{1}{k!}\sum_{j=0}^{n}\binom{n}{j}\mathcal{E}_j^{(a)}(\lambda)\left\langle \left(\lambda e^t+1\right)^k \mid x^{n-j}\right\rangle \end{aligned}$$

Combining the above equation with (28), we get

$$\frac{1}{k!}\left\langle \left(\lambda e^t+1\right)^k \mid \mathcal{E}_n^{(a)}(\lambda)\right\rangle = \frac{1}{k!}\sum_{j=0}^{n}\binom{n}{j}\mathcal{E}_j^{(a)}(\lambda)\,y_1(n-j,k;\lambda). \tag{29}$$

On the other hand

$$\begin{aligned} &\frac{1}{k!}\sum_{j=0}^{n}\binom{n}{j}\mathcal{E}_j^{(a)}(\lambda)\left\langle \left(\lambda e^t+1\right)^k \mid x^{n-j}\right\rangle \\ &= \frac{2^k}{k!}\sum_{j=0}^{n}\binom{n}{j}\mathcal{E}_j^{(a)}(\lambda)\left\langle t^0 \mid \left(\frac{\lambda e^t+1}{2}\right)^{-k} x^{n-j}\right\rangle. \end{aligned} \tag{30}$$

Therefore, combining (29) with (30), we arrive at the desired result. □

Theorem 6.

$$y_1(m,k;\lambda) = \frac{1}{2k!}\sum_{j=0}^{k}\binom{k}{j}\frac{\lambda^j}{2}\left(E_m(j+1)+E_m(j)\right).$$

Proof. We set the following functional equation

$$F_{y1}(t,k;\lambda) = \frac{1}{2k!}\sum_{j=0}^{k}\binom{k}{j}\lambda^j\left(F_{\mathcal{E}}(t,j+1;1,k)+F_{\mathcal{E}}(t,j;1,1)\right).$$

By combining the above equation with (4) and (2), we get

$$\sum_{n=0}^{\infty} y_1(n,k;\lambda)\frac{t^n}{n!} = \sum_{n=0}^{\infty}\left(\frac{1}{2k!}\sum_{j=0}^{k}\binom{k}{j}\lambda^j\left(E_n(j+1)+E_n(j)\right)\right)\frac{t^n}{n!}.$$

Comparing the coefficients of $\frac{t^n}{n!}$ on both sides of the above equation yields the desired result. □

Theorem 7. *Let $m \in \mathbb{N}$. Then we have*

$$S_2(m-1,k;\lambda) = \frac{1}{mk!}\sum_{j=0}^{k}(-1)^{k-j}\binom{k}{j}\lambda^j\left(\mathcal{B}_m(j+1,\lambda)-\mathcal{B}_m(j,\lambda)\right).$$

Proof. We also set the following functional equation

$$F_S(t,0,k;\lambda) = \frac{1}{tk!}\sum_{j=0}^{k}(-1)^{k-j}\binom{k}{j}\lambda^j\left(F_{\mathcal{B}}(t,j+1;\lambda,1)-F_{\mathcal{B}}(t,j;\lambda,1)\right).$$

By combining the above equation with (1) and (3), we get

$$\begin{aligned}&\sum_{m=0}^{\infty} S_2(m,k;\lambda)\frac{t^m}{m!}\\ =\;&\sum_{m=0}^{\infty}\left(\frac{1}{tk!}\sum_{j=0}^{k}(-1)^{k-j}\binom{k}{j}\lambda^j\left(\mathcal{B}_m(j+1,\lambda)-\mathcal{B}_m(j,\lambda)\right)\right)\frac{t^m}{m!}.\end{aligned}$$

Comparing the coefficients of $\frac{t^m}{m!}$ on both sides of the above equation yields the desired result. □

Theorem 8. *Let $2\lambda = 1+\sqrt{5}$. Then we have*

$$\sum_{j=0}^{n}\binom{n}{j}\mathcal{E}_j^{(a)}(\lambda)\,y_1(n-j,k;\lambda) = \frac{1}{k!}\sum_{j=0}^{k}\binom{k}{j}\left(\lambda F_j+F_{j-1}\right)\mathcal{E}_n^{(a)}(j,\lambda).$$

Proof. We define the following functional equation:

$$F_{y1}(t,k;\lambda)\,F_{\mathcal{E}}(t,0;\lambda,a) = \frac{1}{k!}\sum_{j=0}^{k}\binom{k}{j}\lambda^j\left(F_{\mathcal{E}}(t,j;\lambda,a)\right).$$

By combining the above equation with (4), (2), and (7), we get

$$\begin{aligned}&\sum_{n=0}^{\infty}\sum_{j=0}^{n}\binom{n}{j}\mathcal{E}_j^{(a)}(\lambda)\,y_1(n-j,k;\lambda)\frac{t^n}{n!}\\ =\;&\sum_{n=0}^{\infty}\frac{1}{k!}\sum_{j=0}^{k}\binom{k}{j}\left(\lambda F_j+F_{j-1}\right)\mathcal{E}_n^{(a)}(j,\lambda)\frac{t^n}{n!},\end{aligned}$$

where $\lambda = \frac{1+\sqrt{5}}{2}$. Comparing the coefficients of $\frac{t^n}{n!}$ on both sides of the above equation yields the desired result. □

3. Combinatorial Sums via p-Adic Integral

In this section, by using the p-adic integrals, we derive some combinatorial sums containing the Bernoulli numbers, the Euler numbers, the Apostol-Euler numbers and the numbers $y_1(n,k;\lambda)$.

Theorem 9.

$$B_n = \frac{k!}{2^k}\sum_{j=0}^{n}\binom{n}{j} y_1(n-j,k;\lambda)\sum_{v=0}^{j}\binom{j}{v}\mathcal{E}_v^{(k)}(\lambda)\,B_{j-v}.$$

Proof. Combining (2) with (4), we set the following functional equation:

$$F_{y1}(t,k;\lambda)\,F_{\mathcal{E}}(t,x;\lambda,k) = \frac{2^k}{k!}e^{tx}.$$

By using the above equation, we get

$$\sum_{n=0}^{\infty}\mathcal{E}_n^{(k)}(x,\lambda)\frac{t^n}{n!}\sum_{n=0}^{\infty} y_1(n,k;\lambda)\frac{t^n}{n!} = \frac{2^k}{k!}\sum_{n=0}^{\infty}x^n\frac{t^n}{n!}.$$

Therefore

$$\sum_{n=0}^{\infty}\sum_{j=0}^{n}\binom{n}{j} y_1(n-j,k;\lambda)\sum_{v=0}^{j}\binom{j}{v}x^{j-v}\mathcal{E}_v^{(k)}(\lambda)\frac{t^n}{n!} = \frac{2^k}{k!}\sum_{n=0}^{\infty}x^n\frac{t^n}{n!}.$$

Comparing the coefficients of $\frac{t^n}{n!}$ on both sides of the above equation yields the following relation:

$$\sum_{j=0}^{n}\binom{n}{j} y_1(n-j,k;\lambda)\sum_{v=0}^{j}\binom{j}{v}x^{j-v}\mathcal{E}_v^{(k)}(\lambda) = \frac{2^k}{k!}x^n. \tag{31}$$

By applying the Volkenborn integral to (31), we get

$$\sum_{j=0}^{n}\binom{n}{j} y_1(n-j,k;\lambda)\sum_{v=0}^{j}\binom{j}{v}\mathcal{E}_v^{(k)}(\lambda)\int_{\mathbb{Z}_p}x^{j-v}d\mu_1(x) = \frac{2^k}{k!}\int_{\mathbb{Z}_p}x^n d\mu_1(x).$$

Combining the above equation with (10), we arrive at the desired result. □

Remark 4. *Replacing x by k and λ by λ^2, the Equation (31) is reduced to the following relation:*

$$\sum_{j=0}^{n}\binom{n}{j} y_1\left(n-j,k;\lambda^2\right)\mathcal{E}_j^{(k)}\left(k,\lambda^2\right) = \frac{2^k}{k!}k^n. \tag{32}$$

Since

$$\lambda^k 2^n\mathcal{E}_n^{(k)}(k,\lambda^2) = \sum_{m=0}^{n}\binom{n}{m}k^m E_{n-m}^{*(k)}(\lambda)$$

where $E_n^{*(k)}(\lambda)$ denote the Apostol-type Euler numbers of the second kind of order k (cf. [25,33]), the Equation (32) yields

$$\sum_{j=0}^{n}\binom{n}{j}\frac{y_1(n-j,k;\lambda^2)}{2^j}\sum_{m=0}^{j}\binom{j}{m}k^m E_{j-m}^{*(k)}(\lambda) = \frac{(2\lambda)^k}{k!}k^n.$$

By combining the above equation with the following identity

$$\sum_{m=0}^{n}\binom{n}{m}2^{m}y_1\left(m,k;\lambda^2\right)E_{n-m}^{*(k)}(\lambda)=\frac{(2\lambda)^k}{k!}k^n$$

(cf. [33]), we get the following combinatorial sums

$$\sum_{j=0}^{n}\binom{n}{j}\frac{y_1\left(n-j,k;\lambda^2\right)}{2^j}\sum_{m=0}^{j}\binom{j}{m}k^m E_{j-m}^{*(k)}(\lambda)=\sum_{m=0}^{n}\binom{n}{m}2^{m}y_1\left(m,k;\lambda^2\right)E_{n-m}^{*(k)}(\lambda).$$

Theorem 10.

$$E_n=\frac{k!}{2^k}\sum_{j=0}^{n}\binom{n}{j}y_1(n-j,k;\lambda)\sum_{v=0}^{j}\binom{j}{v}\mathcal{E}_v^{(k)}(\lambda)E_{j-v}.$$

Proof. By applying the fermionic p-adic integral to (31), we have

$$\sum_{j=0}^{n}\binom{n}{j}y_1(n-j,k;\lambda)\sum_{v=0}^{j}\binom{j}{v}\mathcal{E}_v^{(k)}(\lambda)\int_{\mathbb{Z}_p}x^{j-v}d\mu_{-1}(x)=\frac{2^k}{k!}\int_{\mathbb{Z}_p}x^n d\mu_{-1}(x).$$

Combining the above equation with (11), we arrive at the desired result. □

Theorem 11.

$$\sum_{j=0}^{n}\binom{n}{j}y_1(n-j,k;\lambda)\sum_{v=0}^{j}\binom{j}{v}\frac{\mathcal{E}_v^{(k)}(\lambda)}{j+1-v}=\frac{2^k}{(n+1)k!}.$$

Proof. Integrate Equation (31) with respect to x from 0 to 1, we obtain

$$\sum_{j=0}^{n}\binom{n}{j}y_1(n-j,k;\lambda)\sum_{v=0}^{j}\binom{j}{v}\mathcal{E}_v^{(k)}(\lambda)\int_0^1 x^{j-v}dx=\frac{2^k}{k!}\int_0^1 x^n dx.$$

After some calculations, we get the desired result. □

Theorem 12.

$$\sum_{j=0}^{n}\binom{n}{j}y_1(j,k;\lambda)\left(B_{n-j}-E_{n-j}\right)=\frac{1}{k!}\sum_{j=0}^{k}\binom{k}{j}\lambda^j\left(B_n(j)-E_n(j)\right).$$

Proof. Setting

$$e^{tx}F_{y1}(t,k;\lambda)=\frac{1}{k!}\sum_{j=0}^{k}\binom{k}{j}\lambda^j e^{(x+j)t}.$$

Combining (4), we have

$$\sum_{n=0}^{\infty}\sum_{j=0}^{n}\binom{n}{j}x^{n-j}y_1(j,k;\lambda)\frac{t^n}{n!}=\sum_{n=0}^{\infty}\frac{1}{k!}\sum_{j=0}^{k}\binom{k}{j}\lambda^j(x+j)^n\frac{t^n}{n!}.$$

Comparing the coefficients of $\frac{t^n}{n!}$ on both sides of the above equation yields the following relation:

$$\sum_{j=0}^{n}\binom{n}{j}x^{n-j}y_1(j,k;\lambda)=\frac{1}{k!}\sum_{j=0}^{k}\binom{k}{j}\lambda^j(x+j)^n. \tag{33}$$

By applying the bosonic p-adic integral to (33) and combining with (10), we have

$$\sum_{j=0}^{n}\binom{n}{j}B_{n-j}y_1(j,k;\lambda)=\frac{1}{k!}\sum_{j=0}^{k}\binom{k}{j}\lambda^j B_n(j). \tag{34}$$

By applying the fermionic p-adic integral to (33) and combining with (11), we obtain

$$\sum_{j=0}^{n}\binom{n}{j}E_{n-j}y_1(j,k;\lambda)=\frac{1}{k!}\sum_{j=0}^{k}\binom{k}{j}\lambda^j E_n(j). \tag{35}$$

Subtracting both sides of Equations (34) and (35), after some elementary calculations, we arrive at the desired result. □

Theorem 13.

$$\sum_{j=0}^{n}\binom{n}{j}\frac{y_1(j,k;\lambda)}{n+1-j}=\frac{1}{k!}\sum_{j=0}^{k}\binom{k}{j}\lambda^j\frac{(j+1)^{n+1}-j^{n+1}}{n+1}.$$

Proof. Integrate Equation (33) with respect to x from 0 to 1, we obtain

$$\sum_{j=0}^{n}\binom{n}{j}y_1(j,k;\lambda)\int_0^1 x^{n-j}dx=\frac{1}{k!}\sum_{j=0}^{k}\binom{k}{j}\lambda^j\int_0^1(x+j)^n dx.$$

After some calculations, we get the desired result. □

Acknowledgments: The present investigation was supported by the Scientific Research Project Administration of Akdeniz University. The author would like to thank to all referees for their valuable comments.

References

1. Simsek, Y. New families of special numbers for computing negative order Euler numbers and related numbers and polynomials. *arXiv* **2018**, *12*, arxiv:1604.05601.
2. Simsek, Y. Analysis of the Bernstein basis functions: an approach to combinatorial sums involving binomial coefficients and Catalan numbers. *Math. Method Appl. Sci.* **2015**, *38*, 3007–3021.
3. Simsek, Y. Identities and relations related to combinatorial numbers and polynomials. *Proc. Jangjeon Math. Soc.* **2017**, *20*, 127–135.
4. Simsek, Y. Apostol type Daehee numbers and polynomials. *Adv. Stud. Contemp. Math.* **2016**, *26*, 555–566.
5. Simsek, Y. Construction of some new families of Apostol-type numbers and polynomials via Dirichlet character and p-adic q-integrals. *Turk. J. Math.* **2018**, *42*, 557–577, doi:10.3906/mat-1703-114.
6. Dere, R.; Simsek, Y.; Srivastava, H.M. A unified presentation of three families of generalized Apostol type polynomials based upon the theory of the umbral calculus and the umbral algebra. *J. Number Theory* **2013**, *133*, 3245–3263.
7. Srivastava, H.M.; Kurt, B.; Simsek, Y. Some families of Genocchi type polynomials and their interpolation functions. *Integral Transforms Spec. Funct.* **2012**, *24*, 919–938.
8. Srivastava, H.M. Some generalizations and basic (or q-) extensions of the Bernoulli, Euler and Genocchi polynomials. *Appl. Math. Inform. Sci.* **2011**, *5*, 390–444.
9. Srivastava, H.M.; Choi, J. *Zeta and q-Zeta Functions and Associated Series and Integrals*; Elsevier Science Publishers: Amsterdam, The Netherlands; London, UK; New York, NY, USA, 2012.
10. Luo, Q.-M.; Srivastava, H.M. Some generalizations of the Apostol-Bernoulli and Apostol-Euler polynomials. *J. Math. Anal. Appl.* **2005**, *308*, 290–302.
11. Luo, Q.-M.; Srivastava, H.M. Some relationships between the Apostol-Bernoulli and Apostol-Euler polynomials. *Comput. Math. Appl.* **2006**, *51*, 631–642.

12. Luo, Q.-M.; Srivastava, H.M. Some generalizations of the Apostol-Genocchi polynomials and the Stirling numbers of the second kind. *Appl. Math. Comput.* **2011**, *217*, 5702–5728.
13. Roman, S. *The Umbral Calculus*; Dover Publication Inc.: New York, NY, USA, 2005.
14. Srivastava, H.M.; Kim, T.; Simsek, Y. q-Bernoulli numbers and polynomials associated with multiple q-zeta functions and basic L-series. *Russ. J. Math. Phys.* **2005**, *12*, 241–268.
15. Khan, N.U.; Usman, T.; Choi, J. A new generalization of Apostol type Laguerre-Genocchi polynomials. *C. R. Acad. Sci. Paris Ser. I* **2017**, doi:10.1016/j.crma.2017.04.010
16. Simsek, Y. Generating functions for generalized Stirling type numbers, array type polynomials, Eulerian type polynomials and their applications. *Fixed Point Theory Appl.* **2013**, *2013*, 1–28.
17. Bayad, A.; Simsek, Y.; Srivastava, H.M. Some array type polynomials associated with special numbers and polynomials. *Appl. Math. Comput.* **2014**, 244, 149–157.
18. Cakic, N.P.; Milovanovic, G.V. On generalized Stirling numbers and polynomials. *Math. Balk.* **2004**, *18*, 241–248.
19. Simsek, Y. *Interpolation Function of Generalized q-Bernstein Type Polynomials and Their Application*; Curves and Surfaces 2011, LNCS 6920; Boissonnat, J.-D.; Chenin, P.; Cohen, A.; Gout, C.; Lyche, T.; Mazure, M.-L.; Schumaker, L.L., Eds.; Springer: Berlin/Heidelberg, Germany, 2012; pp. 647–662.
20. Bona, M. *Introduction to Enumerative Combinatorics*; The McGraw-Hill Companies Inc.: New York, NY, USA, 2007.
21. Riordan, J. *Introduction to Combinatorial Analysis*; Princeton University Press: Princeton, NJ, USA, 1958.
22. Simsek, Y. On parametrization of the q-Bernstein Basis functions and Their Applications. *J. Inequal. Spec. Funct.* **2017**, *8*, 158–169.
23. Spivey, M.Z. Combinatorial Sums and Finite Differences. *Discrete Math.* **2007**, *307*, 3130–3146.
24. Yuluklu, E.; Simsek, Y.; Komatsu, T. Identities Related to Special Polynomials and Combinatorial Numbers. *Filomat* **2017**, *31*, 4833–4844.
25. Simsek, Y. Computation methods for combinatorial sums and Euler type numbers related to new families of numbers. *Math. Method Appl. Sci.* **2017**, *40*, 2347–2361.
26. Koshy, T. *Fibonacci and Lucas Numbers with Applications; A Wiley-Interscience Publication*; John Wiley & Sons, Inc.: New York, NY, USA; Chichester, UK; Weinheim, Germany; Brisbane, Australial; Singapore; Toronto, ON, Canada, 2001.
27. Kim, T. q-Volkenborn integration. *Russ. J. Math. Phys.* **2002**, *19*, 288–299.
28. Schikhof, W.H. *Ultrametric Calculus: An Introduction to p-adic Analysis*; Cambridge Studies in Advanced Mathematics 4; Cambridge University Press: Cambridge, UK, 1984.
29. Kim, T. q-Euler numbers and polynomials associated with p-adic q-integral and basic q-zeta function. *Trend Math. Inf. Cent. Math. Sci.* **2006**, *9*, 7–12.
30. Dattoli, G.; Migliorati, M.; Srivastava, H.M. Sheffer polynomials, monomiality principle, algebraic methods and the theory of classical polynomials. *Math. Comput. Model.* **2007**, 45, 1033–1041.
31. Dere, R.; Simsek, Y. Genocchi polynomials associated with the Umbral algebra. *Appl. Math. Comput.* **2011**, *218*, 756–761.
32. Komatsu, T.; Simsek, Y. Identities related to the Stirling numbers and modified Apostol-type numbers on Umbral Calculus. *Miskolc Math. Notes* **2017**, *18*, 905–916, doi:10.18514/MMN.2017.1458.
33. Kucukoglu, I.; Simsek, Y. Identities and derivative formulas for the combinatorial and Apostol-Euler type numbers by their generating functions. preprint.

11

Slice Holomorphic Functions in Several Variables with Bounded L-Index in Direction

Andriy Bandura [1,*] and Oleh Skaskiv [2]

[1] Department of Advanced Mathematics, Ivano-Frankivsk National Technical University of Oil and Gas, 76019 Ivano-Frankivsk, Ukraine

[2] Department of Mechanics and Mathematics, Ivan Franko National University of Lviv, 79000 Lviv, Ukraine

* Correspondence: andriykopanytsia@gmail.com

Abstract: In this paper, for a given direction $\mathbf{b} \in \mathbb{C}^n \setminus \{\mathbf{0}\}$ we investigate slice entire functions of several complex variables, i.e., we consider functions which are entire on a complex line $\{z^0 + t\mathbf{b} : t \in \mathbb{C}\}$ for any $z^0 \in \mathbb{C}^n$. Unlike to quaternionic analysis, we fix the direction $\mathbf{b}$. The usage of the term slice entire function is wider than in quaternionic analysis. It does not imply joint holomorphy. For example, it allows consideration of functions which are holomorphic in variable z_1 and continuous in variable z_2. For this class of functions there is introduced a concept of boundedness of L-index in the direction $\mathbf{b}$ where $\mathbf{L} : \mathbb{C}^n \to \mathbb{R}_+$ is a positive continuous function. We present necessary and sufficient conditions of boundedness of L-index in the direction. In this paper, there are considered local behavior of directional derivatives and maximum modulus on a circle for functions from this class. Also, we show that every slice holomorphic and joint continuous function has bounded L-index in direction in any bounded domain and for any continuous function $L : \mathbb{C}^n \to \mathbb{R}_+$.

Keywords: bounded index; bounded L-index in direction; slice function; entire function; bounded l-index

---

1. Introduction

In recent years, analytic functions of several variables with bounded index have been intensively investigated. The main objects of investigations are such function classes: entire functions of several variables [1–3], functions analytic in a polydisc [4], in a ball [5] or in the Cartesian product of the complex plane and the unit disc [6].

For entire functions and analytic function in a ball there were proposed two approaches to introduce a concept of index boundedness in a multidimensional complex space. They generate so-called functions of bounded L-index in a direction, and functions of bounded $\mathbf{L}$-index in joint variables.

Let us introduce some notations and definitions.

Let $\mathbb{R}_+ = (0, +\infty)$, $\mathbb{R}_+^* = [0, +\infty)$, $\mathbf{0} = (0, \dots, 0)$, $\mathbf{b} = (b_1, \dots, b_n) \in \mathbb{C}^n \setminus \{\mathbf{0}\}$ be a given direction, $L : \mathbb{C}^n \to \mathbb{R}_+$ be a continuous function, $F : \mathbb{C}^n \to \mathbb{C}$ an entire function. The slice functions on a line $\{z^0 + t\mathbf{b} \colon t \in \mathbb{C}\}$ for fixed $z^0 \in \mathbb{C}^n$ we will denote as $g_{z^0}(t) = F(z^0 + t\mathbf{b})$ and $l_{z^0}(t) = L(z^0 + t\mathbf{b})$.

Definition 1 ([7]). *An entire function $F : \mathbb{C}^n \to \mathbb{C}$ is called a function of bounded L-index in a direction* $\mathbf{b}$, *if there exists $m_0 \in \mathbb{Z}_+$ such that for every $m \in \mathbb{Z}_+$ and for all $z \in \mathbb{C}^n$ one has*

$$\frac{|\partial_{\mathbf{b}}^m F(z)|}{m! L^m(z)} \le \max_{0 \le k \le m_0} \frac{|\partial_{\mathbf{b}}^k F(z)|}{k! L^k(z)}, \tag{1}$$

where $\partial_{\mathbf{b}}^0 F(z) = F(z), \partial_{\mathbf{b}} F(z) = \sum_{j=1}^{n} \frac{\partial F(z)}{\partial z_j} b_j, \partial_{\mathbf{b}}^k F(z) = \partial_{\mathbf{b}}\left(\partial_{\mathbf{b}}^{k-1} F(z)\right), k \geq 2.$

The least such integer number m_0, obeying (1), is called the L-index in the direction $\mathbf{b}$ of the function F and is denoted by $N_{\mathbf{b}}(F, L)$. If such m_0 does not exist, then we put $N_{\mathbf{b}}(F, L) = \infty$, and the function F is said to be of unbounded L-index in the direction $\mathbf{b}$ in this case. If $L(z) \equiv 1$, then the function F is said to be of bounded index in the direction $\mathbf{b}$ and $N_{\mathbf{b}}(F) = N_{\mathbf{b}}(F, 1)$ is called the index in the direction $\mathbf{b}$. Let $l : \mathbb{C} \to \mathbb{R}_+$ be a continuous function. For $n = 1$, $\mathbf{b} = 1$, $L(z) \equiv l(z)$, $z \in \mathbb{C}$ inequality (1) defines a function of bounded l-index with the l-index $N(F, l) \equiv N_1(F, l)$ [8,9], and if in addition $l(z) \equiv 1$, then we obtain a definition of index boundedness with index $N(F) \equiv N_1(F, 1)$ [10,11]. It is also worth to mention paper [12], which introduces the concept of generalized index. It is quite close to the bounded l-index. Let $N_{\mathbf{b}}(F, L, z^0)$ stands for the L-index in the direction $\mathbf{b}$ of the function F at the point z^0, i.e., it is the least integer m_0, for which inequality (1) is satisfied at this point $z = z^0$. By analogy, the notation $N(f, l, z^0)$ is defined if $n = 1$, i.e., in the case of functions of one variable.

The concept of L-index boundedness in direction requires to consider a slice $\{z^0 + t\mathbf{b} : t \in \mathbb{C}\}$. We fixed $z^0 \in \mathbb{C}^n$ and used considerations from one-dimensional case. Then we construct uniform estimates above all z^0. This is a nutshell of the method.

In view of this, Prof S. Yu. Favorov (2015) posed the following **problem** in a conversation with one of the authors.

Problem 1 ([13]). *Let* $\mathbf{b} \in \mathbb{C}^n \setminus \{\mathbf{0}\}$ *be a given direction,* $L : \mathbb{C}^n \to \mathbb{R}_+$ *be a continuous function. Is it possible to replace the condition "F is holomorphic in* $\mathbb{C}^n$*" by the condition "F is holomorphic on all slices* $z^0 + t\mathbf{b}$*" and to deduce all known properties of entire functions of bounded L-index in direction for this function class?*

There is a negative answer to Favorov's question [13]. This relaxation of restrictions by the function F does not allow the proving of some theorems. Here by $\overline{D}$ we denote a closure of domain D. There was proved the following proposition.

Proposition 1 ([13], Theorem 5). *For every direction* $\mathbf{b} \in \mathbb{C}^n \setminus \{\mathbf{0}\}$ *there exists a function* $F(z)$ *and a bounded domain* $D \subset \mathbb{C}^n$ *with following properties:*

(1) F is holomorphic function of bounded index on every slice $\{z^0 + t\mathbf{b} : t \in \mathbb{C}\}$ *for each fixed* $z^0 \in \mathbb{C}^n$*;*
(2) F is not entire function in $\mathbb{C}^n$*;*
(3) F does not satisfy (1) *in* $\overline{D}$*, i.e., for any* $p \in \mathbb{Z}_+$ *there exists* $m \in \mathbb{Z}_+$ *and* $z_p \in \overline{D}$

$$\frac{|\partial_{\mathbf{b}}^m F(z_p)|}{m!} > \max\left\{\frac{|\partial_{\mathbf{b}}^k F(z_p)|}{k!} : 0 \leq k \leq p\right\}.$$

Let D be a bounded domain in $\mathbb{C}^n$. If inequality (1) holds for all $z \in D$ instead $\mathbb{C}^n$, then F is called *function of bounded L-index in the direction* $\mathbf{b}$ *in the domain D*. The least such integer m_0 is called the *L-index in the direction* $\mathbf{b} \in \mathbb{C}^n \setminus \{\mathbf{0}\}$ *in the domain D* and is denoted by $N_{\mathbf{b}}(F, L, D) = m_0$.

Proposition 2 ([13], Theorem 2). *Let D be a bounded domain in* $\mathbb{C}^n$*,* $\mathbf{b} \in \mathbb{C}^n \setminus \{\mathbf{0}\}$ *be arbitrary direction. If* $L\colon \mathbb{C}^n \to \mathbb{R}_+$ *is continuous function and* $F(z)$ *is an entire function such that* $(\forall z^0 \in \overline{D})\colon\ F(z^0 + t\mathbf{b}) \not\equiv 0$*, then* $N_{\mathbf{b}}(F, L, D) < \infty$.

Hence, if we replace holomorphy in $\mathbb{C}^n$ by holomorphy on the slices $\{z^0 + t\mathbf{b} : t \in \mathbb{C}\}$, then conclusion of Proposition 2 is not valid. Thus, Proposition 1 shows impossibility to replace joint holomorphy by slice holomorphy without additional hypothesis. The proof of Proposition 2 uses continuity in joint variables (see [13], Equation (6)). It leads to the following question (see [14], where it is also formulated. There was considered a case $L(z) \equiv 1$).

Problem 2. *What are additional conditions providing validity of Proposition 2 for slice holomorphic functions?*

A main goal of this investigation is to deduce an analog of Proposition 2 for slice holomorphic functions. Please note that the positivity and continuity of the function L are weak restrictions to deduce constructive results. Thus, we assume additional restrictions by the function L.

Let us denote

$$\lambda_{\mathbf{b}}(\eta) = \sup_{z\in\mathbb{C}^n} \sup_{t_1,t_2\in\mathbb{C}} \left\{ \frac{L(z+t_1\mathbf{b})}{L(z+t_2\mathbf{b})} : |t_1-t_2| \le \frac{\eta}{\min\{L(z+t_1\mathbf{b}), L(z+t_2\mathbf{b})\}} \right\}.$$

By $Q^n_{\mathbf{b}}$ we denote a class of positive continuous function $L : \mathbb{C}^n \to \mathbb{R}_+$, satisfying the condition

$$(\forall \eta \ge 0): \ \lambda_{\mathbf{b}}(\eta) < +\infty, \tag{2}$$

Moreover, it is sufficient to require validity of (2) for one value $\eta > 0$.

For a positive continuous function $l(t)$, $t \in \mathbb{C}$, and $\eta > 0$ we define $\lambda(\eta) \equiv \lambda_1^{\mathbf{b}}(\eta)$ in the cases when $\mathbf{b} = 1$, $n = 1$, $L \equiv l$. As in [15], let $Q \equiv Q_1^1$ be a class of positive continuous functions $l(t)$, $t \in \mathbb{C}$, obeying the condition $0 < \lambda(\eta) < +\infty$ for all $\eta > 0$.

Besides, we denote by $\langle a, c\rangle = \sum\limits_{j=1}^{n} a_j\overline{c_j}$ the scalar product in $\mathbb{C}^n$, where $a, c \in \mathbb{C}^n$.

Let $\widetilde{\mathcal{H}}^n_{\mathbf{b}}$ be a class of functions which are holomorphic on every slices $\{z^0 + t\mathbf{b} : t \in \mathbb{C}\}$ for each $z^0 \in \mathbb{C}^n$ and let $\mathcal{H}^n_{\mathbf{b}}$ be a class of functions from $\widetilde{\mathcal{H}}^n_{\mathbf{b}}$ which are joint continuous. The notation $\partial_{\mathbf{b}}F(z)$ stands for the derivative of the function $g_z(t)$ at the point 0, i.e., for every $p \in \mathbb{N}$ $\partial^p_{\mathbf{b}}F(z) = g_z^{(p)}(0)$, where $g_z(t) = F(z + t\mathbf{b})$ is entire function of complex variable $t \in \mathbb{C}$ for given $z \in \mathbb{C}^n$. In this research, we will often call this derivative as directional derivative because if F is entire function in $\mathbb{C}^n$ then the derivatives of the function $g_z(t)$ matches with directional derivatives of the function F.

Please note that if $F \in \mathcal{H}^n_{\mathbf{b}}$ then for every $p \in \mathbb{N}$ $\partial_{\mathbf{b}}F \in \mathcal{H}^n_{\mathbf{b}}$. It can be proved by using of Cauchy's formula.

Together the hypothesis on joint continuity and the hypothesis on holomorphy in one direction do not imply holomorphy in whole n-dimensional complex space. We give some examples to demonstrate it. For $n = 2$ let $f : \mathbb{C} \to \mathbb{C}$ be an entire function, $g : \mathbb{C} \to \mathbb{C}$ be a continuous function. Then $f(z_1)g(z_2)$, $f(z_1) \pm g(z_2)$, $f(z_1 \cdot g(z_2))$ are functions which are holomorphic in the direction $(1, 0)$ and are joint continuous in $\mathbb{C}^2$. Moreover, if we have performed an affine transformation

$$\begin{cases} z_1 = b_2z_1' + b_1z_2', \\ z_2 = b_2z_1' - b_1z_2' \end{cases}$$

then the appropriate new functions are also holomorphic in the direction (b_1, b_2) and are joint continuous in $\mathbb{C}^2$, where $b_1 \ne 0$, $b_2 \ne 0$.

A function $F \in \widetilde{\mathcal{H}}^n_{\mathbf{b}}$ is said to be of *bounded L-index in the direction* $\mathbf{b}$, if there exists $m_0 \in \mathbb{Z}_+$ such that for all $m \in \mathbb{Z}_+$ and each $z \in \mathbb{C}^n$ inequality (1) is true. All notations, introduced above for entire functions of bounded L-index in direction, keep for functions from $\widetilde{\mathcal{H}}^n_{\mathbf{b}}$.

2. Sufficient Sets

Now we prove several assertions that establish a connection between functions of bounded L-index in direction and functions of bounded l-index of one variable. The similar results for entire functions of several variables were obtained in [7,16]. The next proofs use ideas from the mentioned papers. The proofs of Propositions 3, 4 and Theorems 1, 2 literally repeat arguments from proofs of corresponding propositions for entire functions [7,16]. Therefore, we omit these proofs.

Proposition 3. *If a function* $F \in \widetilde{\mathcal{H}}^n_{\mathbf{b}}$ *has bounded L-index in the direction* $\mathbf{b}$ *then for every* $z^0 \in \mathbb{C}^n$ *the entire function* $g_{z^0}(t)$ *is of bounded* l_{z^0}*-index and* $N(g_{z^0}, l_{z^0}) \le N_{\mathbf{b}}(F, L)$.

Proposition 4. *If a function* $F \in \widetilde{\mathcal{H}}^n_{\mathbf{b}}$ *has bounded L-index in the direction* $\mathbf{b}$ *then*

$$N_{\mathbf{b}}(F, L) = \max\left\{N(g_{z^0}, l_{z^0}) : z^0 \in \mathbb{C}^n\right\}.$$

Theorem 1. *A function* $F \in \widetilde{\mathcal{H}}^n_{\mathbf{b}}$ *has bounded L-index in the direction* $\mathbf{b}$ *if and only if there exists a number* $M > 0$ *such that for all* $z^0 \in \mathbb{C}^n$ *the function* $g_{z^0}(t)$ *is of bounded* l_{z^0}*-index with* $N(g_{z^0}, l_{z^0}) \le M < +\infty$*, as a function of variable* $t \in \mathbb{C}$*. Thus,* $N_{\mathbf{b}}(F, L) = \max\{N(g_{z^0}, l_{z^0}) : z^0 \in \mathbb{C}^n\}$.

Theorem 2. *Let* $\mathbf{b} \in \mathbb{C}^n \setminus \{\mathbf{0}\}$ *be a given direction,* $A_0 \subset \mathbb{C}^n$ *such that* $\{z + t\mathbf{b} : t \in \mathbb{C}, z \in A_0\} = \mathbb{C}^n$*. A function* $F \in \widetilde{\mathcal{H}}^n_{\mathbf{b}}$ *has bounded L-index in the direction* $\mathbf{b}$ *if and only if there exists a number* $M > 0$ *such that for all* $z^0 \in A_0$ *the function* $g_{z^0}(t)$ *is of bounded* l_{z^0}*-index with* $N(g_{z^0}, l_{z^0}) \le M < +\infty$*, as a function of one variable* $t \in \mathbb{C}$ *and* $N_{\mathbf{b}}(F, L) = \max\{N(g_{z^0}, l_{z^0}) : z^0 \in A_0\}$.

Remark 1. *An arbitrary hyperplane* $A_0 = \{\tilde{z} \in \mathbb{C}^n : \langle \tilde{z}, c\rangle = 1\}$*, where* $\langle c, \mathbf{b}\rangle \ne 0$*, satisfies conditions of Theorem 2.*

Corollary 1. *If* $F \in \widetilde{\mathcal{H}}^n_{\mathbf{b}}$ *is of bounded L-index in the direction* $\mathbf{b}$ *and* j_0 *is chosen such that* $b_{j_0} \ne 0$*, then* $N_{\mathbf{b}}(F, L) = \max\{N(g_{z^0}, l_{z^0}) : z^0 \in \mathbb{C}^n, z^0_{j_0} = 0\}$*, and if* $\sum_{j=0}^{n} b_j \ne 0$*, then* $N_{\mathbf{b}}(F, L) = \max\{N(g_{z^0}, l_{z^0}) : z^0 \in \mathbb{C}^n, \sum_{j=0}^{n} z^0_j = 0\}$.

We note that for a given $z \in \mathbb{C}^n$ the choice of $z^0 \in \mathbb{C}^n$ and $t \in \mathbb{C}$ such that $\sum_{j=1}^{n} z^0_j = 0$ and $z = z^0 + t\mathbf{b}$, is unique.

Theorem 3 requires replacement of the space $\widetilde{\mathcal{H}}^n_{\mathbf{b}}$ by the space $\mathcal{H}^n_{\mathbf{b}}$. In other words, we use joint continuity in its proof.

Theorem 3. *Let* $\overline{A} = \mathbb{C}^n$*, i.e., A be an everywhere dense set in* $\mathbb{C}^n$ *and let a function* $F \in \mathcal{H}^n_{\mathbf{b}}$*. The function F is of bounded L-index in the direction* $\mathbf{b}$ *if and only if there exists* $M > 0$ *such that for all* $z^0 \in A$ *a function* $g_{z^0}(t)$ *is of bounded* l_{z^0}*-index* $N(g_{z^0}, l_{z^0}) \le M < +\infty$ *and* $N_{\mathbf{b}}(F, L) = \max\{N(g_{z^0}, l_{z^0}) : z^0 \in A\}$.

Proof. The necessity follows from Theorem 1.

Sufficiency. Since $\overline{A} = \mathbb{C}^n$, then for every $z^0 \in \mathbb{C}^n$ there exists a sequence $z^{(m)}$, that $z^{(m)} \to z^0$ as $m \to +\infty$ and $z^{(m)} \in A$ for all $m \in \mathbb{N}$. However, $F(z + t\mathbf{b})$ is of bounded l_z-index for all $z \in \overline{A}$ as a function of variable t. That is why in view the definition of bounded l_z-index there exists $M > 0$ that for all $z \in A$, $t \in \mathbb{C}$, $p \in \mathbb{Z}_+$ $\frac{|g_z^{(p)}(t)|}{p! l^p(t)} \le \max\left\{\frac{|g_z^{(k)}(t)|}{k! l_z^k(t)} : 0 \le k \le M\right\}$.

Substituting instead of z a sequence $z^{(m)} \in A$, $z^{(m)} \to z^0$, we obtain that for every $m \in \mathbb{N}$

$$\frac{|\partial^p_{\mathbf{b}} F(z^{(m)} + t\mathbf{b})|}{p! L^p(z^{(m)} + t\mathbf{b})} \le \max\left\{\frac{|\partial^k_{\mathbf{b}} F(z^{(m)} + t\mathbf{b})|}{k! L^k(z^{(m)} + t\mathbf{b})} : 0 \le k \le M\right\}.$$

However, F and $\partial^p_{\mathbf{b}} F$ are continuous in $\mathbb{C}^n$ for all $p \in \mathbb{N}$ and L is a positive continuous function. Thus, in the obtained expression the limiting transition is possible as $m \to +\infty$ ($z^{(m)} \to z^0$). Evaluating the limit as $m \to +\infty$ we obtain that for all $z^0 \in \mathbb{C}^n$, $t \in \mathbb{C}$, $m \in \mathbb{Z}_+$

$$\frac{|\partial^p_{\mathbf{b}} F(z^0+t\mathbf{b})|}{p!L^p(z^0+t\mathbf{b})} \le \max\left\{\frac{|\partial^k_{\mathbf{b}} F(z^0+t\mathbf{b})|}{k!L^k(z^0+t\mathbf{b})} : 0 \le k \le M\right\}.$$

This inequality implies that $F(z+t\mathbf{b})$ is of bounded $L(z+t\mathbf{b})$-index as a function of variable t for every given $z \in \mathbb{C}^n$. Applying Theorem 1 we obtain the desired conclusion. Theorem 3 is proved. □

Remark 1 and Theorem 3 imply the following corollary.

Corollary 2. *Let $\mathbf{b} \in \mathbb{C}^n \setminus \{\mathbf{0}\}$ be a given direction, $A_0 \subset \mathbb{C}^n$ such that its closure $\overline{A}_0 = \{z \in \mathbb{C}^n : \langle z, c\rangle = 1\}$, where $\langle c, \mathbf{b}\rangle \neq 0$. And let a function $F \in \mathcal{H}^n_{\mathbf{b}}$ and its derivatives $\partial^p_{\mathbf{b}} F \in \mathcal{H}^n_{\mathbf{b}}$ for all $p \in \mathbb{N}$. The function $F(z)$ is of bounded L-index in the direction $\mathbf{b}$ if and only if there exists a number $M > 0$ such that for all $z^0 \in A_0$ the function $g_{z^0}(t)$ is of bounded l_{z^0}-index with $N(g_{z^0}, l_{z^0}) \le M < +\infty$ and $N_{\mathbf{b}}(F, L) = \max\{N(g_{z^0}, l_{z^0}) : z^0 \in A_0\}$.*

3. Local Behavior of Directional Derivative

The following proposition is crucial in theory of functions of bounded index. It initializes series of propositions which are necessary to prove logarithmic criterion of index boundedness. It was first obtained by G. H. Fricke [17] for entire functions of bounded index. Later the proposition was generalized for entire functions of bounded l-index [18], analytic functions of bounded l-index [19], entire functions of bounded L-index in direction [7], functions analytic in a polydisc [4] or in a ball [5] with bounded **L**-index in joint variables,

Theorem 4. *Let $L \in Q^n_{\mathbf{b}}$. A function $F \in \widetilde{\mathcal{H}}^n_{\mathbf{b}}$ is of bounded L-index in the direction $\mathbf{b}$ if and only if for each $\eta > 0$ there exist $n_0 = n_0(\eta) \in \mathbb{Z}_+$ and $P_1 = P_1(\eta) \ge 1$ such that for every $z \in \mathbb{C}^n$ there exists $k_0 = k_0(z) \in \mathbb{Z}_+$, $0 \le k_0 \le n_0$, and*

$$\max\left\{\left|\partial^{k_0}_{\mathbf{b}} F(z+t\mathbf{b})\right| : |t| \le \frac{\eta}{L(z)}\right\} \le P_1 \left|\partial^{k_0}_{\mathbf{b}} F(z)\right|. \tag{3}$$

Proof. Our proof is based on the proof of appropriate theorem for entire functions of bounded L-index in direction [7].

Necessity. Let $N_{\mathbf{b}}(F; L) \equiv N < +\infty$. Let $[a]$, $a \in \mathbb{R}$, stands for the integer part of the number a in this proof. We denote

$$q(\eta) = [2\eta(N+1)(\lambda_{\mathbf{b}}(\eta))^{2N+1}] + 1.$$

For $z \in \mathbb{C}^n$ and $p \in \{0, 1, \ldots, q(\eta)\}$ we put

$$R^{\mathbf{b}}_p(z,\eta) = \max\left\{\frac{|\partial^k_{\mathbf{b}} F(z+t\mathbf{b})|}{k!L^k(z+t\mathbf{b})} : |t| \le \frac{p\eta}{q(\eta)L(z)}, 0 \le k \le N\right\},$$

$$\widetilde{R}^{\mathbf{b}}_p(z,\eta) = \max\left\{\frac{|\partial^k_{\mathbf{b}} F(z+t\mathbf{b})|}{k!L^k(z)} : |t| \le \frac{p\eta}{q(\eta)L(z)}, 0 \le k \le N\right\}.$$

However, $|t| \le \frac{p\eta}{q(\eta)L(z)} \le \frac{\eta}{L(z)}$, then $\lambda_{\mathbf{b}}\left(\frac{p\eta}{q(\eta)}\right) \le \lambda_{\mathbf{b}}(\eta)$. It is clear that $R_p^{\mathbf{b}}(z,\eta)$, $\widetilde{R}_p^{\mathbf{b}}(z,\eta)$ are well-defined. Moreover,

$$\begin{gathered} R_p^{\mathbf{b}}(z,\eta) = \\ = \max\left\{ \frac{|\partial_{\mathbf{b}}^k F(z+t\mathbf{b})|}{k!L^k(z)} \left(\frac{L(z)}{L(z+t\mathbf{b})}\right)^k : 0 \le k \le N, |t| \le \frac{p\eta}{q(\eta)L(z)} \right\} \le \\ \le \max\left\{ \frac{|\partial_{\mathbf{b}}^k F(z+t\mathbf{b})|}{k!L^k(z)} \left(\lambda_{\mathbf{b}}\left(\frac{p\eta}{q(\eta)}\right)\right)^k : |t| \le \frac{p\eta}{q(\eta)L(z)}, 0 \le k \le N \right\} \le \\ \le \max\left\{ \frac{|\partial_{\mathbf{b}}^k F(z+t\mathbf{b})|}{k!L^k(z)} (\lambda_{\mathbf{b}}(\eta))^k : |t| \le \frac{p\eta}{q(\eta)L(z)}, 0 \le k \le N \right\} \le \\ \le (\lambda_{\mathbf{b}}(\eta))^N \max\left\{ \frac{|\partial_{\mathbf{b}}^k F(z+t\mathbf{b})|}{k!L^k(z)} : |t| \le \frac{p\eta}{q(\eta)L(z)}, 0 \le k \le N \right\} = \\ = \widetilde{R}_p^{\mathbf{b}}(z,\eta)(\lambda_{\mathbf{b}}(\eta))^N, \end{gathered} \tag{4}$$

$$\begin{gathered} \widetilde{R}_p^{\mathbf{b}}(z,\eta) = \\ = \max\left\{ \frac{|\partial_{\mathbf{b}}^k F(z+t\mathbf{b})|}{k!L^k(z+t\mathbf{b})} \left(\frac{L(z+t\mathbf{b})}{L(z)}\right)^k : |t| \le \frac{p\eta}{q(\eta)L(z)}, 0 \le k \le N \right\} \le \\ \le \max\left\{ \frac{|\partial_{\mathbf{b}}^k F(z+t\mathbf{b})|}{k!L^k(z+t\mathbf{b})} \left(\lambda_{\mathbf{b}}\left(\frac{p\eta}{q(\eta)}\right)\right)^k : |t| \le \frac{p\eta}{q(\eta)L(z)}, 0 \le k \le N \right\} \le \\ \le \max\left\{ (\lambda_{\mathbf{b}}(\eta))^k \frac{|\partial_{\mathbf{b}}^k F(z+t\mathbf{b})|}{k!L^k(z+t\mathbf{b})} : |t| \le \frac{p\eta}{q(\eta)L(z)}, 0 \le k \le N \right\} \le \\ \le (\lambda_{\mathbf{b}}(\eta))^N \max\left\{ \frac{|\partial_{\mathbf{b}}^k F(z+t\mathbf{b})|}{k!L^k(z+t\mathbf{b})} : |t| \le \frac{p\eta}{q(\eta)L(z)}, 0 \le k \le N \right\} = \\ = R_p^{\mathbf{b}}(z,\eta)(\lambda_{\mathbf{b}}(\eta))^N. \end{gathered} \tag{5}$$

Let $k_p^z \in \mathbb{Z}, 0 \le k_p^z \le N$, and $t_p^z \in \mathbb{C}$, $|t_p^z| \le \frac{p\eta}{q(\eta)L(z)}$, be such that

$$\widetilde{R}_p^{\mathbf{b}}(z,\eta) = \frac{|\partial_{\mathbf{b}}^{k_p^z} F(z + t_p^z\mathbf{b})|}{k_p^z! L^{k_p^z}(z)}. \tag{6}$$

However, for every given $z \in \mathbb{C}^n$ the function $F(z + t\mathbf{b})$ and its derivative are entire as functions of variables t. Then by the maximum modulus principle, equality (6) holds for t_p^z such that $|t_p^z| = \frac{p\eta}{q(\eta)L(z)}$. We set $\widetilde{t}_p^z = \frac{p-1}{p} t_p^z$. Then

$$|\widetilde{t}_p^z| = \frac{(p-1)\eta}{q(\eta)L(z)}, \tag{7}$$

$$|\widetilde{t}_p^z - t_p^z| = \frac{|t_p^z|}{p} = \frac{\eta}{q(\eta)L(z)}. \tag{8}$$

It follows from (7) and the definition of $\widetilde{R}_{p-1}^{\mathbf{b}}(z,\eta)$ that

$$\widetilde{R}_{p-1}^{\mathbf{b}}(z,\eta) \ge \frac{|\partial_{\mathbf{b}}^{k_p^z} F(z + \widetilde{t}_p^z\mathbf{b})|}{k_p^z! L^{k_p^z}(z)}.$$

Therefore,

$$\begin{gathered} 0 \le \widetilde{R}_p^{\mathbf{b}}(z,\eta) - \widetilde{R}_{p-1}^{\mathbf{b}}(z,\eta) \le \frac{\left|\partial_{\mathbf{b}}^{k_p^z} F(z+t_p^z\mathbf{b})\right| - \left|\partial_{\mathbf{b}}^{k_p^z} F(z+\widetilde{t}_p^z\mathbf{b})\right|}{k_p^z! L^{k_p^z}(z)} = \\ = \frac{1}{k_p^z! L^{k_p^z}(z)} \int_0^1 \frac{d}{ds} \left|\partial_{\mathbf{b}}^{k_p^z} F(z + (\widetilde{t}_p^z + s(t_p^z - \widetilde{t}_p^z))\mathbf{b})\right| ds. \end{gathered} \tag{9}$$

For every analytic complex-valued function of real variable $\varphi(s)$, $s \in \mathbb{R}$, the inequality $\frac{d}{ds}|\varphi(s)| \le \left|\frac{d}{ds}\varphi(s)\right|$ holds, where $\varphi(s) \neq 0$. Applying this inequality to (9) and using the mean value theorem we obtain

$$\begin{gathered}\widetilde{R}_p^{\mathbf{b}}(z,t_0,\eta) - \widetilde{R}_{p-1}^{\mathbf{b}}(z,t_0,\eta) \le \\ \le \frac{|t_p^z - \widetilde{t}_p^z|}{k_p^z! L^{k_p^z}(z)} \int_0^1 \left|\partial_{\mathbf{b}}^{k_p^z+1} F(z + (\widetilde{t}_p^z + s(t_p^z - \widetilde{t}_p^z))\mathbf{b})\right| ds = \\ = \frac{|t_p^z - \widetilde{t}_p^z|}{k_p^z! L^{k_p^z}(z)} \left|\partial_{\mathbf{b}}^{k_p^z+1} F(z + (\widetilde{t}_p^z + s^*(t_p^z - \widetilde{t}_p^z))\mathbf{b})\right| = \\ = L(z)(k_p^z+1)|t_p^z - \widetilde{t}_p^z| \frac{|\partial_{\mathbf{b}}^{k_p^z+1} F(z + (\widetilde{t}_p^z + s^*(t_p^z - \widetilde{t}_p^z))\mathbf{b})|}{(k_p^z+1)! L^{k_p^z+1}(z)},\end{gathered}$$

where $s^* \in [0,1]$. The point $\widetilde{t}_p^z + s^*(t_p^z - \widetilde{t}_p^z)$ belongs to the set

$$\left\{t \in \mathbb{C} : |t| \le \frac{p\eta}{q(\eta)L(z)} \le \frac{\eta}{L(z)}\right\}.$$

Using the definition of boundedness of L-index in direction, the definition of $q(\eta)$, inequalities (4) and (8), for $k_p^z \le N$ we have

$$\begin{gathered}\widetilde{R}_p^{\mathbf{b}}(z,\eta) - \widetilde{R}_{p-1}^{\mathbf{b}}(z,\eta) \le \frac{|\partial_{\mathbf{b}}^{k_p^z+1} F(z + (\widetilde{t}_p^z + s^*(t_p^z - \widetilde{t}_p^z))\mathbf{b})|}{(k_p^z+1)! L^{k_p^z+1}(z + (\widetilde{t}_p^z + s^*(t_p^z - \widetilde{t}_p^z))\mathbf{b})} \times \\ \times \left(\frac{L(z + (\widetilde{t}_p^z + s^*(t_p^z - \widetilde{t}_p^z))\mathbf{b})}{L(z)}\right)^{k_p^z+1} L(z)(k_p^z+1)|t_p^z - \widetilde{t}_p^z| \le \eta \frac{N+1}{q(\eta)} (\lambda_{\mathbf{b}}(\eta))^{N+1} \times \\ \times \max\left\{\frac{|\partial_{\mathbf{b}}^k F(z + (\widetilde{t}_p^z + s^*(t_p^z - \widetilde{t}_p^z))\mathbf{b})|}{k! L^k(z + (\widetilde{t}_p^z + s^*(t_p^z - \widetilde{t}_p^z))\mathbf{b})} : 0 \le k \le N\right\} \le \eta \frac{N+1}{q(\eta)} (\lambda_{\mathbf{b}}(\eta))^{N+1} R_p^{\mathbf{b}}(z,\eta) \le \\ \le \frac{\eta(N+1)(\lambda_{\mathbf{b}}(\eta))^{2N+1}}{[2\eta(N+1)(\lambda_{\mathbf{b}}(\eta))^{2N+1}]+1} \widetilde{R}_p^{\mathbf{b}}(z,\eta) \le \frac{1}{2} \widetilde{R}_p^{\mathbf{b}}(z,\eta)\end{gathered}$$

It follows that $\widetilde{R}_p^{\mathbf{b}}(z,\eta) \le 2\widetilde{R}_{p-1}^{\mathbf{b}}(z,\eta)$. Using inequalities (4) and (5), we obtain for $R_p^{\mathbf{b}}(z,\eta)$

$$R_p^{\mathbf{b}}(z,\eta) \le 2(\lambda_{\mathbf{b}}(\eta))^N \widetilde{R}_{p-1}^{\mathbf{b}}(z,\eta) \le 2(\lambda_{\mathbf{b}}(\eta))^{2N} R_{p-1}^{\mathbf{b}}(z,\eta).$$

Hence,

$$\begin{gathered}\max\left\{\frac{|\partial_{\mathbf{b}}^k F(z+t\mathbf{b})|}{k! L^k(z+t\mathbf{b})} : |t| \le \frac{\eta}{L(z)}, 0 \le k \le N\right\} = R_{q(\eta)}^{\mathbf{b}}(z,\eta) \le \\ \le 2(\lambda_{\mathbf{b}}(\eta))^{2N} R_{q(\eta)-1}^{\mathbf{b}}(z,\eta) \le (2(\lambda_{\mathbf{b}}(\eta))^{2N})^2 R_{q(\eta)-2}^{\mathbf{b}}(z,\eta) \le \\ \le \cdots \le (2(\lambda_{\mathbf{b}}(\eta))^{2N})^{q(\eta)} R_0^{\mathbf{b}}(z,\eta) = \\ = (2(\lambda_{\mathbf{b}}(\eta))^{2N})^{q(\eta)} \max\left\{\frac{|\partial_{\mathbf{b}}^k F(z)|}{k! L^k(z)} : 0 \le k \le N\right\}.\end{gathered} \tag{10}$$

Let $k_z \in \mathbb{Z}$, $0 \le k_z \le N$, and $\widetilde{t}_z \in \mathbb{C}$, $|\widetilde{t}_z| = \frac{\eta}{L(z)}$ be such that

$$\frac{|\partial_{\mathbf{b}}^{k_z} F(z)|}{k_z! L^{k_z}(z)} = \max_{0 \le k \le N} \frac{|\partial_{\mathbf{b}}^k F(z)|}{k! L^k(z)},$$

and

$$|\partial_{\mathbf{b}}^{k_z} F(z + \widetilde{t}_z\mathbf{b})| = \max\{|\partial_{\mathbf{b}}^{k_z} F(z + t\mathbf{b})| : |t| \le \eta/L(z)\}.$$

Inequality (10) implies

$$\frac{|\partial_{\mathbf{b}}^{k_z}F(z+\widetilde{t_z}\mathbf{b})|}{k_z!L^{k_z}(z+\widetilde{t_z}\mathbf{b})} \leq \max\left\{\frac{|\partial_{\mathbf{b}}^{k_z}F(z+t\mathbf{b})|}{k_z!L^{k_z}(z+t\mathbf{b})} : |t| = \frac{\eta}{L(z)}\right\} \leq$$
$$\leq \max\left\{\frac{|\partial_{\mathbf{b}}^{k}F(z+t\mathbf{b})|}{k!L^{k}(z+t\mathbf{b})} : |t| = \frac{\eta}{L(z)}, 0 \leq k \leq N\right\} \leq$$
$$\leq (2(\lambda_{\mathbf{b}}(\eta))^{2N})^{q(\eta)}\frac{|\partial_{\mathbf{b}}^{k_z}F(z)|}{k_z!L^{k_z}(z)}.$$

Hence,

$$\max\left\{|\partial_{\mathbf{b}}^{k_z}F(z+t\mathbf{b})| : |t| \leq \eta/L(z)\right\} \leq$$
$$\leq (2(\lambda_{\mathbf{b}}(\eta))^{2N})^{q(\eta)}\frac{L^{k_z}(z+\widetilde{t_z}\mathbf{b})}{L^{k_z}(z)}|\partial_{\mathbf{b}}^{k_z}F(z)| \leq$$
$$\leq (2(\lambda_{\mathbf{b}}(\eta))^{2N})^{q(\eta)}(\lambda_{\mathbf{b}}(\eta))^{N}|\partial_{\mathbf{b}}^{k_z}F(z)| \leq$$
$$\leq (2(\lambda_{\mathbf{b}}(\eta))^{2N})^{q(\eta)}(\lambda_{\mathbf{b}}(\eta))^{N}|\partial_{\mathbf{b}}^{k_z}F(z)|.$$

Thus, we obtain (3) with $n_0 = N_{\mathbf{b}}(F, L)$ and

$$P_1(\eta) = (2(\lambda_{\mathbf{b}}(\eta))^{2N})^{q(\eta)}(\lambda_{\mathbf{b}}(\eta))^{N} > 1.$$

Sufficiency. Suppose that for each $\eta > 0$ there exist $n_0 = n_0(\eta) \in \mathbb{Z}_+$ and $P_1 = P_1(\eta) \geq 1$ such that for every $z \in \mathbb{C}^n$ there exists $k_0 = k_0(z) \in \mathbb{Z}_+, 0 \leq k_0 \leq n_0$, for which inequality (3) holds. We choose $\eta > 1$ and $j_0 \in \mathbb{N}$ such that $P_1 \leq \eta^{j_0}$. For given $z \in \mathbb{C}^n$, $k_0 = k_0(z)$ and $j \geq j_0$ by Cauchy's formula for $F(z+t\mathbf{b})$ as a function of one variable t

$$\partial_{\mathbf{b}}^{k_0+j}F(z) = \frac{j!}{2\pi i}\int_{|t|=\eta/L(z)}\frac{\partial_{\mathbf{b}}^{k_0}F(z+t\mathbf{b})}{t^{j+1}}dt.$$

Therefore, in view of (3) we have

$$\frac{|\partial_{\mathbf{b}}^{k_0+j}F(z)|}{j!} \leq \frac{L^j(z)}{\eta^j}\max\left\{|\partial_{\mathbf{b}}^{k_0}F(z+t\mathbf{b})| : |t| = \frac{\eta}{L(z)}\right\} \leq P_1\frac{L^j(z)}{\eta^j}|\partial_{\mathbf{b}}^{k_0}F(z)|,$$

Hence, for all $j \geq j_0$, $z \in \mathbb{C}^n$

$$\frac{|\partial_{\mathbf{b}}^{k_0+j}F(z)|}{(k_0+j)!L^{k_0+j}(z)} \leq \frac{j!k_0!}{(j+k_0)!}\frac{P_1}{\eta^j}\frac{|\partial_{\mathbf{b}}^{k_0}F(z)|}{k_0!L^{k_0}(z)} \leq \eta^{j_0-j}\frac{|\partial_{\mathbf{b}}^{k_0}F(z)|}{k_0!L^{k_0}(z)} \leq \frac{|\partial_{\mathbf{b}}^{k_0}F(z)|}{k_0!L^{k_0}(z)}.$$

Since $k_0 \leq n_0$, the numbers $n_0 = n_0(\eta)$ and $j_0 = j_0(\eta)$ are independent of z and t_0, this inequality means that a function F has bounded L-index in the direction $\mathbf{b}$ and $N_{\mathbf{b}}(F, L) \leq n_0 + j_0$. The proof of Theorem 4 is complete. □

Theorem 4 implies the next proposition that describes the boundedness of L-index in direction for an equivalent function to L. Let $L^*(z)$ be a positive continuous function in $\mathbb{C}^n$. We denote $L \asymp L^*$, if for some θ_1, θ_2, $0 < \theta_1 \leq \theta_2 < +\infty$, and for all $z \in \mathbb{C}^n$ the following inequalities hold $\theta_1 L(z) \leq L^*(z) \leq \theta_2 L(z)$.

Proposition 5. *Let $L \in Q_{\mathbf{b}}^n$, $L \asymp L^*$. A function $F \in \widetilde{\mathcal{H}}_{\mathbf{b}}^n$ has bounded L^*-index in the direction* **b** *if and only if F is of bounded L-index in the direction* **b**.

Proof. First, it is not easy to check that the function L^* also belongs to the class $Q^n_{\mathbf{b}}$. Let $N_{\mathbf{b}}(F, L^*) < +\infty$. Then by Theorem 4 for every $\eta^* > 0$ there exist $n_0(\eta^*) \in \mathbb{Z}_+$ and $P_1(\eta^*) \geq 1$ such that for every $z \in \mathbb{C}^n$ and some k_0, $0 \leq k_0 \leq n_0$, inequality (3) holds with L^* and η^* instead of L and η. But the condition $L \asymp L^*$ means that for some $\theta_1, \theta_2 \in \mathbb{R}_+, 0 < \theta_1 \leq \theta_2 < +\infty$ and for all $z \in \mathbb{C}^n$ the double inequality holds $\theta_1 L(z) \leq L^*(z) \leq \theta_2 L(z)$. Taking $\eta* = \theta_2 \eta$ we obtain

$$P_1|\partial^{k_0}_{\mathbf{b}} F(z)| \geq \max\left\{|\partial^{k_0}_{\mathbf{b}} F(z+t\mathbf{b})| : |t| \leq \eta^*/L^*(z)\right\} \geq$$
$$\geq \max\left\{|\partial^{k_0}_{\mathbf{b}} F(z+t\mathbf{b})| : \ |t| \leq \eta/L(z)\right\}.$$

Thus, by Theorem 4 in view of arbitrariness of η^* the function $F(z)$ has bounded L-index in the direction $\mathbf{b}$. We can obtain the converse proposition by replacing L with L^*. □

Please note that Proposition 5 can be slightly refined. The following proposition is easy deduced from (1).

Proposition 6. *Let $L_1(z)$, $L_2(z)$ be positive continuous functions, $F \in \widetilde{\mathcal{H}}^n_{\mathbf{b}}$ be a function of bounded L_1-index in the direction $\mathbf{b}$, for all $z \in \mathbb{C}^n$ the inequality $L_1(z) \leq L_2(z)$ holds. Then $N_{\mathbf{b}}(L_2, F) \leq N_{\mathbf{b}}(L_1, F)$.*

Using Fricke's idea [20], we obtain modification of Theorem 4.

Theorem 5. *Let $L \in Q^n_{\mathbf{b}}$. If there exist $\eta > 0$, $n_0 = n_0(\eta) \in \mathbb{Z}_+$ and $P_1 = P_1(\eta) \geq 1$ such that for all $z \in \mathbb{C}^n$ there exists $k_0 = k_0(z) \in \mathbb{Z}_+, 0 \leq k_0 \leq n_0$, for which the inequality holds*

$$\max\{|\partial^{k_0}_{\mathbf{b}} F(z+t\mathbf{b})| : |t| \leq \eta/L(z)\} \leq P_1|\partial^{k_0}_{\mathbf{b}} F(z)|,$$

then the function $F \in \widetilde{\mathcal{H}}^n_{\mathbf{b}}$ has bounded L-index in the direction $\mathbf{b} \in \mathbb{C}^n \setminus \{\mathbf{0}\}$.

Proof. Our proof is based on the proof of appropriate theorem for entire functions of bounded L-index in direction [21].

Assume that there exist $\eta > 0$, $n_0 = n_0(\eta) \in \mathbb{Z}_+$ and $P_1 = P_1(\eta) \geq 1$ such that for every $z \in \mathbb{C}^n$ there exists $k_0 = k_0(z) \in \mathbb{Z}_+, 0 \leq k_0 \leq n_0$, for which

$$\max\{|\partial^{k_0}_{\mathbf{b}} F(z+t\mathbf{b})| : |t| \leq \frac{\eta}{L(z)}\} \leq P_1|\partial^{k_0}_{\mathbf{b}} F(z)|. \tag{11}$$

If $\eta > 1$, then we choose $j_0 \in \mathbb{N}$ such that $P_1 \leq \eta^{j_0}$. And for $\eta \in (0;1]$ we choose $j_0 \in \mathbb{N}$ obeying the inequality $\frac{j_0!k_0!}{(j_0+k_0)!} P_1 < 1$. This j_0 exists because

$$\frac{j_0!k_0!}{(j_0+k_0)!} P_1 = \frac{k_0!}{(j_0+1)(j_0+2)\cdot \ldots \cdot (j_0+k_0)} P_1 \to 0, \ j_0 \to \infty.$$

Applying Cauchy's formula to the function $F(z+t\mathbf{b})$ as function of complex variable t for $j \geq j_0$ we obtain that for every $z \in \mathbb{C}^n$ there exists integer $k_0 = k_0(z)$, $0 \leq k_0 \leq n_0$, and

$$\partial^{k_0+j}_{\mathbf{b}} F(z) = \frac{j!}{2\pi i} \int\limits_{|t|=\frac{\eta}{L(z)}} \frac{\partial^{k_0}_{\mathbf{b}} F(z+t\mathbf{b})}{t^{j+1}} dt.$$

Taking into account (11), one has

$$\frac{|\partial^{k_0+j}_{\mathbf{b}} F(z)|}{j!} \leq \frac{L^j(z)}{\eta^j} \max\left\{|\partial^{k_0}_{\mathbf{b}} F(z+t\mathbf{b})| : \ |t| = \frac{\eta}{L(z)}\right\} \leq P_1 \frac{L^j(z)}{\eta^j} |\partial^{k_0}_{\mathbf{b}} F(z)|. \tag{12}$$

In view of choice j_0 for $\eta > 1$ and for all $j \geq j_0$ we deduce

$$\frac{|\partial_{\mathbf{b}}^{k_0+j}F(z)|}{(k_0+j)!L^{k_0+j}(z)} \leq \frac{j!k_0!}{(j+k_0)!}\frac{P_1}{\eta^j}\frac{|\partial_{\mathbf{b}}^{k_0}F(z)|}{k_0!L^{k_0}(z+t_0\mathbf{b})} \leq \eta^{j_0-j}\frac{|\partial_{\mathbf{b}}^{k_0}F(z)|}{k_0!L^{k_0}(z)} \leq \frac{|\partial_{\mathbf{b}}^{k_0}F(z)|}{k_0!L^{k_0}(z)}.$$

Since $k_0 \leq n_0$, the numbers $n_0 = n_0(\eta)$ and $j_0 = j_0(\eta)$ are independent of z, and $z \in \mathbb{C}^n$ is arbitrary, the last inequality means that the function F is of bounded L-index in the direction $\mathbf{b}$ and $N_{\mathbf{b}}(F,L) \leq n_0 + j_0$.

If $\eta \in (0,1)$, then (12) implies for all $j \geq j_0$

$$\frac{|\partial_{\mathbf{b}}^{k_0+j}F(z)|}{(k_0+j)!L^{k_0+j}(z)} \leq \frac{j!k_0!P_1}{(j+k_0)!}\frac{|\partial_{\mathbf{b}}^{k_0}F(z)|}{\eta^j k_0!L^{k_0}(z)} \leq \frac{|\partial_{\mathbf{b}}^{k_0}F(z)|}{\eta^j k_0!L^{k_0}(z)}$$

or in view of the choice of j_0

$$\frac{|\partial_{\mathbf{b}}^{k_0+j}F(z)|}{(k_0+j)!}\frac{\eta^{k_0+j}}{L^{k_0+j}(z)} \leq \frac{|\partial_{\mathbf{b}}^{k_0}F(z)|}{k_0!}\frac{\eta^{k_0}}{L^{k_0}(z)}.$$

Thus, the function F has bounded $\tilde{L}$-index in the direction $\mathbf{b}$, where $\tilde{L}(z) = \frac{L(z)}{\eta}$. Then by Proposition 5 the function F is of bounded L-index in the direction $\mathbf{b}$. Theorem is proved. □

4. Bounded Index in Direction in Bounded Domain

Let D be a bounded domain in $\mathbb{C}^n$. If inequality (1) is fulfilled for all $z \in D$ instead $\mathbb{C}^n$, then F is called *function of bounded L-index in the direction* $\mathbf{b}$ *in the domain D*. The least such integer m_0 is called *the L-index in the direction* $\mathbf{b}$ *in the domain D* and is denoted by $N_{\mathbf{b}}(F,L,D) = m_0$. By $\overline{D}$ we denote a closure of domain D.

Theorem 6. *Let D be an arbitrary bounded domain in $\mathbb{C}^n$, $\mathbf{b} \in \mathbb{C}^n \setminus \{\mathbf{0}\}$ be arbitrary direction. If $L\colon \mathbb{C}^n \to \mathbb{R}_+$ is continuous function, $F \in \mathcal{H}_{\mathbf{b}}^n$ and $(\forall p \in \mathbb{N})\ \partial_{\mathbf{b}}^p F \in \mathcal{H}_{\mathbf{b}}^n$ and $(\forall z^0 \in \overline{D})\colon\ F(z^0+t\mathbf{b}) \not\equiv 0$, then $N_{\mathbf{b}}(F,L,\overline{D}) < \infty$.*

Proof. Proof of this theorem is similar to proof of corresponding lemma in [13]. For every given $z^0 \in \overline{D}$ we develop the entire function $F(z^0+t\mathbf{b})$ in power series by powers t

$$F(z^0+t\mathbf{b}) = \sum_{m=0}^{\infty} \frac{\partial_{\mathbf{b}}^m F(z^0)}{m!}t^m \tag{13}$$

in the disc $\{t \in \mathbb{C} : |t| \leq \frac{1}{L(z^0)}\}$.

The quantity $\frac{|\partial_{\mathbf{b}}^m F(z^0)|}{m!}$ is the modulus of coefficient of power series (13) at the point $t = 0$. Substitute $t = \frac{1}{L(z^0)}$. Since $F \in \mathcal{H}_{\mathbf{b}}^n$, for every $z_0 \in \overline{D}$

$$\frac{|\partial_{\mathbf{b}}^m F(z^0)|}{m!L^m(z^0)} \to 0 \quad (m \to \infty),$$

i.e., there exists $m_0 = m(z^0,\mathbf{b})$ such that inequality (1) holds at the point $z = z^0$ for all $m \in \mathbb{Z}_+$.

We will show that $\sup\{m_0 \colon z^0 \in \overline{D}\} < +\infty$. On the contrary, we suppose that the set of all values m_0 is unbounded in z^0, that is $\sup\{m_0 \colon z^0 \in \overline{D}\} = +\infty$. Hence, for every $m \in \mathbb{Z}_+$ there exists $z^{(m)} \in \overline{D}$ and $p_m > m$

$$\frac{|\partial_{\mathbf{b}}^{p_m} F(z^{(m)})|}{p_m!L^{p_m}(z^{(m)})} > \max\left\{\frac{|\partial_{\mathbf{b}}^k F(z^{(m)})|}{k!L^k(z^{(m)})} \colon\ 0 \leq k \leq m\right\}. \tag{14}$$

Since $\{z^{(m)}\} \subset \overline{D}$, there exists subsequence $z'^{(m)} \to z' \in \overline{G}$ as $m \to +\infty$. By Cauchy's formula one has

$$\frac{\partial_{\mathbf{b}}^{p} F(z)}{p!} = \frac{1}{2\pi i} \int_{|t|=r} \frac{F(z+t\mathbf{b})}{t^{p+1}} dt$$

for any $p \in \mathbb{N}$, $z \in D$. Rewrite (14) in the form

$$\begin{array}{c} \max\left\{ \frac{|\partial_{\mathbf{b}}^{k} F(z^{(m)})|}{k! L^{k}(z^{(m)})} : \ 0 \le k \le m \right\} < \\ < \frac{1}{L^{p_m}(z^{(m)})} \int_{|t|=r/L(z^{(m)})} \frac{|F(z^{(m)}+t\mathbf{b})|}{|t|^{p_m+1}} |dt| \le \frac{1}{r^{p_m}} \max\{|F(z)| : z \in D_r\}, \end{array} \tag{15}$$

where $D_r = \bigcup_{z^* \in \overline{D}} \{z \in \mathbb{C}^n : |z - z^*| \le \frac{|b| r}{L(z^*)}\}$. We can choose $r > 1$, because $F \in \mathcal{H}_{\mathbf{b}}^{n}$. Evaluating limit for every directional derivative of fixed order in (15) as $m \to \infty$ we obtain

$$(\forall k \in \mathbb{Z}_+): \quad \frac{|\partial_{\mathbf{b}}^{k} F(z')|}{k! L^{k}(z')} \le \overline{\lim_{m \to \infty}} \frac{1}{r^{p_m}} \max\{|F(z)| : z \in D_r\} \le 0.$$

The passing to the limit is possible because $\partial_{\mathbf{b}}^{k} F$ is joint continuous. Thus, all derivatives in the direction $\mathbf{b}$ of the function F at the point z' equal 0 and $F(z') = 0$. In view of (13) $F(z' + t\mathbf{b}) \equiv 0$. This is a contradiction. □

5. Conclusions

The proposed approach can be applied in analytic theory of differential equations. It is known that concept of bounded index allows the investigation of properties of analytic solutions of linear higher-order differential equations with analytic coefficients. Therefore, it leads to the question of what the additional conditions are, providing index boundedness of every slice holomorphic solutions for linear higher-order directional derivative equations with slice holomorphic coefficients? In other words, is joint continuity a sufficient condition?

Since there are known analogs of Cauchy's formula for quaternionic variables and for Clifford algebras, the authors assume that The results in this paper can be generalized in these cases, i.e., in the case of slice holomorphic functions of quaternionic variable.

Author Contributions: These authors contributed equally to this work.

References

1. Bandura, A.; Skaskiv, O. Asymptotic estimates of entire functions of bounded **L**-index in joint variables. *Novi Sad J. Math.* **2018**, *48*, 103–116. [CrossRef]
2. Nuray, F.; Patterson, R.F. Multivalence of bivariate functions of bounded index. *Le Matematiche* **2015**, *70*, 225–233. [CrossRef]
3. Nuray, F.; Patterson, R.F. Vector-valued bivariate entire functions of bounded index satisfying a system of differential equations. *Mat. Stud.* **2018**, *49*, 67–74. [CrossRef]
4. Bandura, A.; Petrechko, N.; Skaskiv, O. Maximum modulus in a bidisc of analytic functions of bounded **L**-index and an analogue of Hayman's theorem. *Mat. Bohem.* **2018**, *143*, 339–354. [CrossRef]
5. Bandura, A.; Skaskiv, O. Sufficient conditions of boundedness of **L**-index and analog of Hayman's Theorem for analytic functions in a ball. *Stud. Univ. Babeş-Bolyai Math.* **2018**, *63*, 483–501. [CrossRef]
6. Bandura, A.I.; Skaskiv, O.B.; Tsvigun, V.L. Some characteristic properties of analytic functions in $\mathbb{D} \times \mathbb{C}$ of bounded **L**-index in joint variables. *Bukovyn. Mat. Zh.* **2018**, *6*, 21–31.
7. Bandura, A.I.; Skaskiv, O.B. Entire functions of bounded L-index in direction. *Mat. Stud.* **2007**, *27*, 30–52. (In Ukrainian)

8. Kuzyk, A.D.; Sheremeta, M.N. Entire functions of bounded l-distribution of values. *Math. Notes* **1986**, *39*, 3–8. [CrossRef]
9. Kuzyk, A.D.; Sheremeta, M.N. On entire functions, satisfying linear differential equations. *Differ. Equ.* **1990**, *26*, 1716–1722.
10. Lepson, B. Differential equations of infinite order, hyperdirichlet series and entire functions of bounded index. *Proc. Symp. Pure Math.* **1968**, *2*, 298–307.
11. Macdonnell, J.J. Some Convergence Theorems for Dirichlet-Type Series Whose Coefficients are Entire Functions of Bounded Index. Ph.D. Thesis, Catholic University of America, Washington, DC, USA, 1957.
12. Strelitz, S. Asymptotic properties of entire transcendental solutions of algebraic differential equations. *Contemp. Math.* **1983**, *25*, 171. [CrossRef]
13. Bandura, A.I. Sum of entire functions of bounded L-index in direction. *Mat. Stud.* **2016**, *45*, 149–158. [CrossRef]
14. Kosanyak, M. Slice Entire Functions of Bounded Index in Direction, Master's Thesis, Ivan Franko National University of Lviv, Lviv, Ukraine, 2018.
15. Sheremeta, M. *Analytic Functions of Bounded Index*; VNTL Publishers: Lviv, Ukraine, 1999; 141p.
16. Bandura, A.I.; Skaskiv, O.B. Sufficient sets for boundedness L-index in direction for entire functions. *Mat. Stud.* **2008**, *30*, 177–182.
17. Fricke, G.H. Functions of bounded index and their logarithmic derivatives. *Math. Ann.* **1973**, *206*, 215–223. [CrossRef]
18. Sheremeta, M.N.; Kuzyk, A.D. Logarithmic derivative and zeros of an entire function of bounded l-index. *Sib. Math. J.* **1992**, *33*, 304–312. [CrossRef]
19. Kushnir, V.O.; Sheremeta, M.M. Analytic functions of bounded l-index. *Mat. Stud.* **1999**, *12*, 59–66.
20. Fricke, G.H. Entire functions of locally slow growth. *J. Anal. Math.* **1975**, *28*, 101–122. [CrossRef]
21. Bandura, A.I. A modified criterion of boundedness of L-index in direction. *Mat. Stud.* **2013**, *39*, 99–102.

12

Some Identities for Euler and Bernoulli Polynomials and their Zeros

Taekyun Kim [1] and Cheon Seoung Ryoo [2,*]

1 Department of Mathematics, Kwangwoon University, Seoul 139-701, Korea; tkkim@kw.ac.kr
2 Department of Mathematics, Hannam University, Daejeon 306-791, Korea
* Correspondence: ryoocs@hnu.kr;

Abstract: In this paper, we study some special polynomials which are related to Euler and Bernoulli polynomials. In addition, we give some identities for these polynomials. Finally, we investigate the zeros of these polynomials by using the computer.

Keywords: Appell sequence; Appell numbers and polynomials; Bernoulli and Euler polynomials; cosine–Bernoulli and cosine–Euler polynomials; sine–Bernoulli and sine–Euler polynomials

---

1. Introduction

Many mathematicians have studied in the area of the Bernoulli numbers and polynomials, Euler numbers and polynomials, Genocchi numbers and polynomials, and tangent numbers and polynomials. The class of Appell polynomial sequences is one of the important classes of polynomial sequences. The Appell polynomial sequences arise in numerous problems of applied mathematics, mathematical physics and several other mathematical branches (see [1–14]). The Appell polynomials can be defined by considering the following generating function:

$$\begin{aligned} A(t)e^{xt} &= A_0(x) + A_1(x)\frac{t}{1!} + A_2(x)\frac{t^2}{2!} + \cdots + A_n(x)\frac{t^n}{n!} + \cdots \\ &= \sum_{n=0}^{\infty} A_n(x)\frac{t^n}{n!}, \text{ (see [5,7,8])}, \end{aligned} \tag{1}$$

where

$$A(t) = A_0 + A_1\frac{t}{1!} + A_2\frac{t^2}{2!} + \cdots + A_n\frac{t^n}{n!} + \cdots, A_0 \neq 0.$$

Alternatively, the sequence $A_n(x)$ is Appell sequence for $(g(t), t)$ if and only if

$$\frac{1}{g(t)}e^{xt} = \sum_{n=0}^{\infty} A_n(x)\frac{t^n}{n!}, \text{ (see [5,7,8])},$$

where

$$g(t) = \sum_{n=0}^{\infty} g_n\frac{t^n}{n!}, g_0 \neq 0.$$

Differentiating generating Equation (1) with respect to x and equating coefficients of $\frac{t^n}{n!}$, we have

$$\frac{d}{dx}A_n(x) = nA_{n-1}(x), n = 0, 1, 2, 3, \cdots.$$

The typical examples of Appell polynomials are the Bernoulli and Euler polynomials (see [1–14]). It is well known that the Bernoulli polynomials are defined by the generating function to be

$$\frac{t}{e^t-1}e^{xt}=\sum_{n=0}^{\infty}B_n(x)\frac{t^n}{n!}. \tag{2}$$

When $x=0$, $B_n=B_n(0)$ are called the Bernoulli numbers. The Euler polynomials are given by the generating function to be

$$\frac{2}{e^t+1}e^{xt}=\sum_{n=0}^{\infty}E_n(x)\frac{t^n}{n!}. \tag{3}$$

When $x=0$, $E_n=E_n(0)$ are called the Euler numbers.

The Bernoulli polynomials $\mathbf{B}_n^{(r)}(x)$ of order r are defined by the following generating function

$$\left(\frac{t}{e^t-1}\right)^r e^{xt}=\sum_{n=0}^{\infty}\mathbf{B}_n^{(r)}(x)\frac{t^n}{n!},\quad (|t|<2\pi). \tag{4}$$

The Frobenius–Euler polynomials of order r, denoted by $\mathbf{H}_n^{(r)}(u,x)$, are defined as

$$\left(\frac{1-u}{e^t-u}\right)^r e^{xt}=\sum_{n=0}^{\infty}\mathbf{H}_n^{(r)}(u,x)\frac{t^n}{n!}. \tag{5}$$

The values at $x=0$ are called Frobenius–Euler numbers of order r; when $r=1$, the polynomials or numbers are called ordinary Frobenius–Euler polynomials or numbers.

In this paper, we study some special polynomials which are related to Euler and Bernoulli polynomials. In addition, we give some identities for these polynomials. Finally, we investigate the zeros of these polynomials by using the computer.

2. Cosine–Bernoulli, Sine–Bernoulli, Cosine–Euler and Sine–Euler Polynomials

In this section, we define the cosine–Bernoulli, sine–Bernoulli, cosine–Euler and sine–Euler polynomials. Now, we consider the Euler polynomials that are given by the generating function to be

$$\frac{2}{e^t+1}e^{(x+iy)t}=\sum_{n=0}^{\infty}E_n(x+iy)\frac{t^n}{n!}. \tag{6}$$

On the other hand, we observe that

$$e^{(x+iy)t}=e^{xt}e^{iyt}=e^{xt}(\cos yt+i\sin yt). \tag{7}$$

From Equations (6) and (7), we have

$$\sum_{n=0}^{\infty}E_n(x+iy)\frac{t^n}{n!}=\frac{2}{e^t+1}e^{(x+iy)t}=\frac{2}{e^t+1}e^{xt}(\cos yt+i\sin yt), \tag{8}$$

and

$$\sum_{n=0}^{\infty}E_n(x-iy)\frac{t^n}{n!}=\frac{2}{e^t+1}e^{(x-iy)t}=\frac{2}{e^t+1}e^{xt}(\cos yt-i\sin yt). \tag{9}$$

Thus, by (8) and (9), we can derive

$$\frac{2}{e^t+1}e^{xt}\cos yt=\sum_{n=0}^{\infty}\left(\frac{E_n(x+iy)+E_n(x-iy)}{2}\right)\frac{t^n}{n!}, \tag{10}$$

and

$$\frac{2}{e^t+1}e^{xt}\sin yt = \sum_{n=0}^{\infty}\left(\frac{E_n(x+iy)+E_n(x-iy)}{2i}\right)\frac{t^n}{n!}. \tag{11}$$

It follows that we define the following cosine–Euler polynomials and sine–Euler polynomials.

Definition 1. *The cosine–Euler polynomials $E_n^{(C)}(x,y)$ and sine–Euler polynomials $E_n^{(S)}(x,y)$ are defined by means of the generating functions*

$$\sum_{n=0}^{\infty} E_n^{(C)}(x,y)\frac{t^n}{n!} = \frac{2}{e^t+1}e^{xt}\cos yt, \tag{12}$$

and

$$\sum_{n=0}^{\infty} E_n^{(S)}(x,y)\frac{t^n}{n!} = \frac{2}{e^t+1}e^{xt}\sin yt, \tag{13}$$

respectively.

Note that $E_n^{(C)}(x,0) = E_n(x), E_n^{(S)}(x,0) = 0, (n \geq 0)$. The cosine–Euler and sine–Euler polynomials can be determined explicitly. A few of them are

$$\begin{aligned}
&E_0^{(C)}(x,y) = 1, \quad E_1^{(C)}(x,y) = -\frac{1}{2}+x,\\
&E_2^{(C)}(x,y) = -x+x^2-y^2,\\
&E_3^{(C)}(x,y) = \frac{1}{4}-\frac{3x^2}{2}+x^3+\frac{3y^2}{2}-3xy^2,\\
&E_4^{(C)}(x,y) = x-2x^3+x^4+6xy^2-6x^2y^2+y^4,
\end{aligned}$$

and

$$\begin{aligned}
&E_0^{(S)}(x,y) = 0, \quad E_1^{(S)}(x,y) = y,\\
&E_2^{(S)}(x,y) = -y+2xy,\\
&E_3^{(S)}(x,y) = -3xy+3x^2y-y^3,\\
&E_4^{(S)}(x,y) = y-6x^2y+4x^3y+2y^3-4xy^3.
\end{aligned}$$

By (10)–(13), we have

$$\begin{aligned}
E_n^{(C)}(x,y) &= \frac{E_n(x+iy)+E_n(x-iy)}{2},\\
E_n^{(S)}(x,y) &= \frac{E_n(x+iy)-E_n(x-iy)}{2i}.
\end{aligned}$$

Clearly, we can get the following explicit representations of $E_n(x+iy)$

$$\begin{aligned}
E_n(x+iy) &= \sum_{k=0}^{n}\binom{n}{k}(x+iy)^{n-k}E_k,\\
E_n(x+iy) &= \sum_{k=0}^{n}\binom{n}{k}(iy)^{n-k}E_k(x).
\end{aligned}$$

Let

$$e^{xt}\cos yt = \sum_{k=0}^{\infty} C_k(x,y)\frac{t^k}{k!}, \qquad e^{xt}\sin yt = \sum_{k=0}^{\infty} S_k(x,y)\frac{t^k}{k!}. \tag{14}$$

Then, by Taylor expansions of $e^{xt}\cos yt$ and $e^{xt}\sin yt$, we get

$$e^{xt}\cos yt=\sum_{k=0}^{\infty}\left(\sum_{m=0}^{[\frac{k}{2}]}\binom{k}{2m}(-1)^{m}x^{k-2m}y^{2m}\right)\frac{t^{k}}{k!} \tag{15}$$

and

$$e^{xt}\sin yt=\sum_{k=0}^{\infty}\left(\sum_{m=0}^{[\frac{k-1}{2}]}\binom{k}{2m+1}(-1)^{m}x^{k-2m-1}y^{2m+1}\right)\frac{t^{k}}{k!}, \tag{16}$$

where [ ] denotes taking the integer part. By (14)–(16), we get

$$C_k(x,y)=\sum_{m=0}^{[\frac{k}{2}]}\binom{k}{2m}(-1)^{m}x^{k-2m}y^{2m},$$

and

$$S_k(x,y)=\sum_{m=0}^{[\frac{k-1}{2}]}\binom{k}{2m+1}(-1)^{m}x^{k-2m-1}y^{2m+1},\ (k\geq 0).$$

The two polynomials can be determined explicitly. A few of them are

$$\begin{aligned}&C_0(x,y)=1,\quad C_1(x,y)=x,\quad C_2(x,y)=x^2-y^2,\\&C_3(x,y)=x^3-3xy^2,\quad C_4(x,y)=x^4-6x^2y^2+y^4,\\&C_5(x,y)=x^5-10x^3y^2+5xy^4,\quad C_6(x,y)=x^6-15x^4y^2+15x^2y^4-y^6,\end{aligned}$$

and

$$\begin{aligned}&S_0(x,y)=0,\quad S_1(x,y)=y,\quad S_2(x,y)=2xy,\\&S_3(x,y)=3x^2y-y^3,\quad S_4(x,y)=4x^3y-4xy^3,\\&S_5(x,y)=5x^4y-10x^2y^3+y^5,\quad S_6(x,y)=6x^5y-20x^3y^3+6xy^5.\end{aligned}$$

Now, we observe that

$$\begin{aligned}\frac{2}{e^t+1}e^{xt}\cos yt&=\left(\sum_{l=0}^{\infty}E_l\frac{t^l}{l!}\right)\left(\sum_{m=0}^{\infty}C_m(x,y)\frac{t^m}{m!}\right)\\&=\sum_{n=0}^{\infty}\left(\sum_{l=0}^{n}\binom{n}{l}E_lC_{n-l}(x,y)\right)\frac{t^n}{n!}.\end{aligned} \tag{17}$$

Therefore, we obtain the following theorem:

Theorem 1. *For $n\geq 0$, we have*

$$E_n^{(C)}(x,y)=\sum_{l=0}^{n}\binom{n}{l}E_lC_{n-l}(x,y)$$

and

$$E_n^{(S)}(x,y)=\sum_{l=0}^{n}\binom{n}{l}E_lS_{n-l}(x,y).$$

From (12), we have

$$\begin{aligned} 2e^{xt}\cos yt &= \left(\sum_{l=0}^{\infty} E_l^{(C)}(x,y)\frac{t^l}{l!}\right)(e^t+1) \\ &= \sum_{n=0}^{\infty}\left(\sum_{l=0}^{n}\binom{n}{l}E_l^{(C)}(x,y)+E_n^{(C)}(x,y)\right)\frac{t^n}{n!}. \end{aligned} \tag{18}$$

By (14) and (18), we get

$$C_n(x,y)=\frac{1}{2}\left(\sum_{l=0}^{n}\binom{n}{l}E_l^{(C)}(x,y)+E_n^{(C)}(x,y)\right). \tag{19}$$

Therefore, we obtain the following theorem:

Theorem 2. *For $n \geq 0$, we have*

$$C_n(x,y)=\frac{1}{2}\left(\sum_{l=0}^{n}\binom{n}{l}E_l^{(C)}(x,y)+E_n^{(C)}(x,y)\right),$$

and

$$S_n(x,y)=\frac{1}{2}\left(\sum_{l=0}^{n}\binom{n}{l}E_l^{(S)}(x,y)+E_n^{(S)}(x,y)\right).$$

From (12), we note that

$$\begin{aligned} \sum_{n=0}^{\infty} E_n^{(C)}(1-x,y)\frac{t^n}{n!} &= \frac{2}{e^t+1}e^{(1-x)t}\cos yt \\ &= \frac{2}{e^{-t}+1}e^{-xt}\cos(-yt) \\ &= \left(\sum_{l=0}^{\infty}(-1)^l E_l\frac{t^l}{l!}\right)\left(\sum_{m=0}^{\infty}(-1)^m C_{m,}(x,y)\frac{t^m}{m!}\right) \\ &= \sum_{n=0}^{\infty}\left(\sum_{l=0}^{n}\binom{n}{l}E_l C_{n-l}(x,y)\right)\frac{(-1)^n}{n!}t^n. \end{aligned} \tag{20}$$

Therefore, we obtain the following theorem:

Theorem 3. *For $n \geq 0$, we have*

$$\begin{aligned} E_n^{(C)}(1-x,y) &= (-1)^n\sum_{l=0}^{n}\binom{n}{l}E_l C_{n-l}(x,y) \\ &= (-1)^n E_n^{(C)}(x,y), \end{aligned}$$

and

$$\begin{aligned} E_n^{(S)}(1-x,y) &= (-1)^{n+1}E_n^{(S)}(x,y) \\ &= (-1)^{n+1}\sum_{l=0}^{n}\binom{n}{l}E_l S_{n-l}(x,y). \end{aligned}$$

Now, we observe that

$$\begin{aligned}\sum_{n=0}^{\infty} E_n^{(C)}(x+1,y)\frac{t^n}{n!} &= \frac{2}{e^t+1}e^{(x+1)t}\cos yt \\ &= \frac{2}{e^t+1}e^{xt}(e^t-1+1)\cos yt \\ &= 2e^{xt}\cos yt - \frac{2}{e^t+1}e^{xt}\cos yt \\ &= \sum_{n=0}^{\infty}\left(2C_n(x,y)-E_n^{(C)}(x,y)\right)\frac{t^n}{n!}.\end{aligned} \tag{21}$$

By comparing the coefficients on the both sides, we get

$$E_n^{(C)}(x+1,q)+E_n^{(C)}(x,y)=2C_n(x,y),\ (n\geq 0). \tag{22}$$

Therefore, we obtain the following theorem:

Theorem 4. *For $n \geq 0$, we have*

$$E_n^{(C)}(x+1,y)+E_n^{(C)}(x,y)=2C_n(x,y),$$

and

$$E_n^{(S)}(x+1,y)+E_n^{(S)}(x,y)=2S_n(x,y).$$

From (14) and (15), we have

$$\sum_{k=0}^{\infty} C_k(0,y)\frac{t^k}{k!}=\sum_{m=0}^{\infty}(-1)^m y^{2m}\frac{t^{2m}}{(2m)!}. \tag{23}$$

Therefore, by Theorem 4 and (23), we obtain the following corollary:

Corollary 1. *For $n \geq 0$, we have*

$$E_{2n}^{(C)}(1,y)+E_{2n}^{(C)}(0,y)=2(-1)^n y^{2n},$$

and

$$E_{2n+1}^{(S)}(1,y)+E_{2n+1}^{(S)}(0,y)=2(-1)^n y^{2n+1}.$$

By (12), we get

$$\begin{aligned}\sum_{n=0}^{\infty} E_n^{(C)}(x+r,y)\frac{t^n}{n!} &= \left(\frac{2e^{xt}}{e^t+1}\cos yt\right)e^{rt} \\ &= \left(\sum_{l=0}^{\infty}E_l^{(C)}(x,y)\frac{t^l}{l!}\right)\left(\sum_{k=0}^{\infty}r^k\frac{t^k}{k!}\right) \\ &= \sum_{n=0}^{\infty}\left(\sum_{k=0}^{n}\binom{n}{k}E_k^{(C)}(x,y)r^{n-k}\right)\frac{t^n}{n!}.\end{aligned} \tag{24}$$

Therefore, by comparing the coefficients on the both sides, we obtain the following theorem:

Theorem 5. *For $n \geq 0, r \in \mathbb{N}$, we have*

$$E_n^{(C)}(x+r,y) = \sum_{k=0}^{n} \binom{n}{k} E_k^{(C)}(x,y) r^{n-k},$$

and

$$E_n^{(S)}(x+r,y) = \sum_{k=0}^{n} \binom{n}{k} E_k^{(S)}(x,y) r^{n-k}.$$

Taking $r = 1$ in Theorem 5, we obtain the following corollary:

Corollary 2. *For $n \geq 0$, we have*

$$2C_n(x,y) = E_n^{(C)}(x,y) + \sum_{k=0}^{n} \binom{n}{k} E_k^{(C)}(x,y),$$

and

$$2S_n(x,y) = E_n^{(S)}(x,y) + \sum_{k=0}^{n} \binom{n}{k} E_k^{(S)}(x,y).$$

From Corollary 2, we note that

$$E_n^{(C)}(0,y) + \sum_{k=0}^{n} \binom{n}{k} E_k^{(C)}(0,y) = \begin{cases} 0, & \text{if } n = 2m+1, \\ 2(-1)^m y^{2m}, & \text{if } n = 2m, \end{cases} \tag{25}$$

and

$$E_n^{(S)}(0,y) + \sum_{k=0}^{n} \binom{n}{k} E_k^{(S)}(0,y) = \begin{cases} 2(-1)^m y^{2m+1}, & \text{if } n = 2m+1, \\ 0, & \text{if } n = 2m. \end{cases} \tag{26}$$

By (12), we get

$$\begin{aligned} \sum_{n=1}^{\infty} \frac{\partial}{\partial x} E_n^{(C)}(x,y) \frac{t^n}{n!} &= \frac{\partial}{\partial x} \left(\frac{2}{e^t+1} e^{xt} \cos yt \right) \\ &= \frac{2}{e^t+1} t e^{xt} \cos yt \\ &= \sum_{n=1}^{\infty} \left(n E_{n-1}^{(C)}(x,y) \right) \frac{t^n}{n!}. \end{aligned} \tag{27}$$

Comparing the coefficients on the both sides of (27), we have

$$\frac{\partial}{\partial x} E_n^{(C)}(x,y) = n E_{n-1}^{(C)}(x,y).$$

Similarly, for $n \geq 1$, we have

$$\begin{aligned} \frac{\partial}{\partial x} E_n^{(S)}(x,y) &= n E_{n-1}^{(S)}(x,y), \\ \frac{\partial}{\partial y} E_n^{(C)}(x,y) &= -n E_{n-1}^{(S)}(x,y), \\ \frac{\partial}{\partial y} E_n^{(S)}(x,y) &= n E_{n-1}^{(C)}(x,y). \end{aligned}$$

Now, we consider the Bernoulli polynomials that are given by the generating function to be

$$\frac{t}{e^t-1}e^{(x+iy)t} = \sum_{n=0}^{\infty} B_n(x+iy)\frac{t^n}{n!}.$$

We also have

$$\sum_{n=0}^{\infty} B_n(x+iy)\frac{t^n}{n!} = \frac{t}{e^t-1}e^{(x+iy)t} = \frac{t}{e^t-1}e^{xt}(\cos yt + i\sin yt), \tag{28}$$

and

$$\sum_{n=0}^{\infty} B_n(x-iy)\frac{t^n}{n!} = \frac{t}{e^t-1}e^{(x-iy)t} = \frac{t}{e^t-1}e^{xt}(\cos yt - i\sin yt). \tag{29}$$

Thus, by (28) and (29), we can derive

$$\frac{t}{e^t-1}e^{xt}\cos yt = \sum_{n=0}^{\infty}\left(\frac{B_n(x+iy)+B_n(x-iy)}{2}\right)\frac{t^n}{n!}, \tag{30}$$

and

$$\frac{t}{e^t-1}e^{xt}\sin yt = \sum_{n=0}^{\infty}\left(\frac{B_n(x+iy)+B_n(x-iy)}{2i}\right)\frac{t^n}{n!}. \tag{31}$$

It follows that we define the following cosine–Bernoulli and sine–Bernoulli polynomials.

Definition 2. *The cosine–Bernoulli polynomials $B_n^{(C)}(x,y)$ and sine–Bernoulli polynomials $B_n^{(S)}(x,y)$ are defined by means of the generating functions*

$$\sum_{n=0}^{\infty} B_n^{(C)}(x,y)\frac{t^n}{n!} = \frac{t}{e^t-1}e^{xt}\cos yt, \tag{32}$$

and

$$\sum_{n=0}^{\infty} B_n^{(S)}(x,y)\frac{t^n}{n!} = \frac{t}{e^t-1}e^{xt}\sin yt, \tag{33}$$

respectively.

By (30), (31), (32), and (33), we have

$$B_n^{(C)}(x,y) = \frac{B_n(x+iy)+B_n(x-iy)}{2},$$
$$B_n^{(S)}(x,y) = \frac{B_n(x+iy)-B_n(x-iy)}{2i}.$$

Note that $B_n^{(C)}(x,0) = B_n(x)$ are the Bernoulli polynomials. The cosine–Bernoulli and sine–Bernoulli polynomials can be determined explicitly. A few of them are

$$B_0^{(C)}(x,y) = 1, \quad B_1^{(C)}(x,y) = -\frac{1}{2}+x,$$
$$B_2^{(C)}(x,y) = \frac{1}{6}-x+x^2-y^2,$$
$$B_3^{(C)}(x,y) = \frac{x}{2}-\frac{3x^2}{2}+x^3+\frac{3y^2}{2}-3xy^2,$$
$$B_4^{(C)}(x,y) = -\frac{1}{30}+x^2-2x^3+x^4-y^2+6xy^2-6x^2y^2+y^4,$$

and

$$\begin{aligned} &B_0^{(S)}(x,y) = 0, \quad B_1^{(S)}(x,y) = y, \quad B_2^{(S)}(x,y) = -y + 2xy, \\ &B_3^{(S)}(x,y) = \frac{y}{2} - 3xy + 3x^2y - y^3, \\ &B_4^{(S)}(x,y) = 2xy - 6x^2y + 4x^3y + 2y^3 - 4xy^3. \end{aligned}$$

From (32), we have

$$\begin{aligned} \sum_{n=0}^{\infty} B_n^{(C)}(x,y)\frac{t^n}{n!} &= \frac{t}{e^t - 1}e^{xt}\cos yt, \\ &= \left(\sum_{l=0}^{\infty} B_n \frac{t^l}{l!}\right)\left(\sum_{m=0}^{\infty} C_m(x,y)\frac{t^m}{m!}\right) \\ &= \sum_{n=0}^{\infty}\left(\sum_{l=0}^{n}\binom{n}{l} B_l C_{n-l}(x,y)\right)\frac{t^n}{n!}. \end{aligned} \tag{34}$$

Comparing the coefficients on the both sides of (34), we obtain the following theorem:

Theorem 6. *For $n \geq 0$, we have*

$$B_n^{(C)}(x,y) = \sum_{l=0}^{n}\binom{n}{l} B_l C_{n-l}(x,y),$$

and

$$B_n^{(S)}(x,y) = \sum_{l=0}^{n}\binom{n}{l} B_l S_{n-l}(x,y).$$

By replacing x by $1-x$ in (32), we get

$$\begin{aligned} \sum_{n=0}^{\infty} B_n^{(C)}(1-x,y)\frac{t^n}{n!} &= \frac{t}{e^t - 1}e^{(1-x)t}\cos yt \\ &= \frac{t}{1-e^{-t}}e^{-xt}\cos yt \\ &= \sum_{n=0}^{\infty}(-1)^n B_n(x,y)\frac{t^n}{n!}. \end{aligned} \tag{35}$$

Therefore, we obtain the following theorem:

Theorem 7. *For $n \geq 0$, we have*

$$B_n^{(C)}(1-x,y) = (-1)^n B_n^{(C)}(x,y),$$

and

$$B_n^{(S)}(1-x,y) = (-1)^{n+1} B_n^{(S)}(x,y).$$

Now, we observe that

$$\begin{aligned}\sum_{n=0}^{\infty} B_n^{(C)}(x+1,q)\frac{t^n}{n!} &= \frac{t}{e^t-1}e^{(x+1)t}\cos yt \\ &= te^{xt}\cos yt + \frac{t}{e^t-1}e^{xt}\cos yt \\ &= \sum_{n=1}^{\infty} nC_{n-1}(x,y)\frac{t^n}{n!} + \sum_{n=0}^{\infty} B_n^{(C)}(x,y)\frac{t^n}{n!} \\ &= \sum_{n=0}^{\infty}\left(nC_{n-1}(x,y) + B_n^{(C)}(x,y)\right)\frac{t^n}{n!}.\end{aligned} \tag{36}$$

Thus, by (36), we get

$$B_n^{(C)}(x+1,y) = nC_{n-1}(x,y) + B_n^{(C)}(x,y), (n \geq 1). \tag{37}$$

Therefore, by (37), we obtain the following theorem:

Theorem 8. *For $n \geq 1$, we have*

$$B_n^{(C)}(x+1,y) - B_n^{(C)}(x,y) = nC_{n-1}(x,y),$$

and

$$B_n^{(S)}(x+1,y) - B_n^{(S)}(x,y) = nS_{n-1}(x,y).$$

Now, we define the new type polynomials that are given by the generating functions to be

$$\frac{2}{e^t+1}\cos yt = \sum_{n=0}^{\infty} E_n^{(C)}(y)\frac{t^n}{n!}, \tag{38}$$

and

$$\frac{2}{e^t+1}\sin yt = \sum_{n=0}^{\infty} E_n^{(S)}(y)\frac{t^n}{n!}, \tag{39}$$

respectively.

Note that $E_n^{(C)}(0) = E_n$, $E_n^{(S)}(0) = 0$, $E_n^{(C)}(0,y) = E_n^{(C)}(y)$, $E_n^{(S)}(0,y) = E_n^{(S)}(y)$, $(n \geq 0)$. The new type polynomials can be determined explicitly. A few of them are

$$E_0^{(C)}(y) = 1, \quad E_1^{(C)}(x,y) = -\frac{1}{2}, \quad E_2^{(C)}(x,y) = -y^2, \quad E_3^{(C)}(y) = \frac{1}{4} + \frac{3y^2}{2},$$
$$E_4^{(C)}(y) = y^4, \quad E_5^{(C)}(y) = -\frac{1}{2} - \frac{5y^2}{2} - \frac{5y^4}{2}, \quad E_6^{(C)}(y) = -y^6,$$

and

$$E_0^{(S)}(x,y) = 0, \quad E_1^{(S)}(x,y) = y, \quad E_2^{(S)}(x,y) = -y, \quad E_3^{(S)}(x,y) = -y^3,$$
$$E_4^{(S)}(x,y) = y + 2y^3 \quad E_5^{(S)}(x,y) = y^5, \quad E_6^{(S)}(x,y) = -3y - 5y^3 - 3y^5.$$

From (38) and (39), we derive the following equations:

$$\frac{2}{e^t+1}\cos yt = \sum_{k=0}^{\infty}\left(\sum_{m=0}^{[\frac{k}{2}]}\binom{k}{2m}(-1)^m E_{k-2m}y^{2m}\right)\frac{t^k}{k!} \tag{40}$$

and

$$\frac{2}{e^t+1}\sin yt=\sum_{k=0}^{\infty}\left(\sum_{m=0}^{[\frac{k-1}{2}]}\binom{k}{2m+1}(-1)^m E_{k-2m-1}y^{2m+1}\right)\frac{t^k}{k!}. \tag{41}$$

By (38)–(41), we get

$$E_n^{(C)}(y)=\sum_{m=0}^{[\frac{n}{2}]}\binom{n}{2m}(-1)^m y^{2m}E_{n-2m}, \tag{42}$$

and

$$E_n^{(S)}(y)=\sum_{m=0}^{[\frac{n-1}{2}]}\binom{n}{2m+1}(-1)^m y^{2m+1}E_{n-2m-1}, (k\geq 0). \tag{43}$$

From (12), (13), (38) and (39), we derive the following theorem:

Theorem 9. *For $n\geq 0$, we have*

$$E_n^{(C)}(x,y)=\sum_{k=0}^{n}\binom{n}{k}x^{n-k}E_k^{(C)}(y),$$

and

$$E_n^{(S)}(x,y)=\sum_{k=0}^{n}\binom{n}{k}x^{n-k}E_k^{(S)}(y).$$

Now, we define the new type polynomials that are given by the generating functions to be

$$\frac{t}{e^t-1}\cos yt=\sum_{n=0}^{\infty}B_n^{(C)}(y)\frac{t^n}{n!}, \tag{44}$$

and

$$\frac{t}{e^t-1}\sin yt=\sum_{n=0}^{\infty}B_n^{(S)}(y)\frac{t^n}{n!}, \tag{45}$$

respectively.

Note that $B_n^{(C)}(0)=B_n$, $B_n^{(S)}(0)=0$, $B_n^{(C)}(0,y)=B_n^{(C)}(y)$, $B_n^{(S)}(0,y)=B_n^{(S)}(y)$, $(n\geq 0)$. The new type polynomials can be determined explicitly. A few of them are

$$B_0^{(C)}(x,y)=1,\quad B_1^{(C)}(x,y)=-\frac{1}{2},\quad B_2^{(C)}(x,y)=\frac{1}{6}-y^2,$$
$$B_3^{(C)}(x,y)=\frac{3y^2}{2},\quad B_4^{(C)}(x,y)=-\frac{1}{30}-y^2+y^4,\quad B_5^{(C)}(x,y)=\frac{5y^4}{2},$$

and

$$B_0^{(S)}(x,y)=0,\quad B_1^{(S)}(x,y)=y,\quad B_2^{(S)}(x,y)=-y,$$
$$B_3^{(S)}(x,y)=\frac{y}{2}-y^3,\quad B_4^{(S)}(x,y)=2y^3,\quad B_5^{(S)}(x,y)=-\frac{y}{6}-\frac{5y^3}{3}+y^5.$$

From (44) and (45), we derive the following equations:

$$\frac{t}{e^t-1}\cos yt=\sum_{k=0}^{\infty}\left(\sum_{m=0}^{[\frac{k}{2}]}\binom{k}{2m}(-1)^m B_{k-2m}y^{2m}\right)\frac{t^k}{k!} \tag{46}$$

and

$$\frac{t}{e^t-1}\sin yt=\sum_{k=0}^{\infty}\left(\sum_{m=0}^{[\frac{k-1}{2}]}\binom{k}{2m+1}(-1)^m B_{k-2m-1}y^{2m+1}\right)\frac{t^k}{k!}. \tag{47}$$

By (44)–(47), we get

$$B_n^{(C)}(y) = \sum_{m=0}^{[\frac{n}{2}]} \binom{n}{2m}(-1)^m y^{2m} B_{n-2m}, \tag{48}$$

and

$$B_n^{(S)}(y) = \sum_{m=0}^{[\frac{n-1}{2}]} \binom{n}{2m+1}(-1)^m y^{2m+1} B_{n-2m-1}, (k \geq 0). \tag{49}$$

From (32), (33), (44) and (45), we derive the following theorem:

Theorem 10. *For $n \geq 0$, we have*

$$B_n^{(C)}(x,y) = \sum_{k=0}^{n} \binom{n}{k} x^{n-k} B_k^{(C)}(y),$$

and

$$B_n^{(S)}(x,y) = \sum_{k=0}^{n} \binom{n}{k} x^{n-k} B_k^{(S)}(y).$$

We remember that the classical Stirling numbers of the first kind $S_1(n,k)$ and $S_2(n,k)$ are defined by the relations (see [12])

$$(x)_n = \sum_{k=0}^{n} S_1(n,k) x^k \text{ and } x^n = \sum_{k=0}^{n} S_2(n,k)(x)_k, \tag{50}$$

respectively. Here, $(x)_n = x(x-1)\cdots(x-n+1)$ denotes the falling factorial polynomial of order n. The numbers $S_2(n,m)$ also admit a representation in terms of a generating function

$$(e^t - 1)^m = m! \sum_{n=m}^{\infty} S_2(n,m) \frac{t^n}{n!}. \tag{51}$$

By (12), (51) and by using Cauchy product, we get

$$\begin{aligned}
\sum_{n=0}^{\infty} E_n^{(C)}(x,y)\frac{t^n}{n!} &= \left(\frac{2}{e^t+1}\right)(1-(1-e^{-t}))^{-x}\cos yt \\
&= \left(\frac{2}{e^t+1}\right)\cos yt \sum_{l=0}^{\infty}\binom{x+l-1}{l}(1-e^{-t})^l \\
&= \sum_{l=0}^{\infty} < x >_l \frac{(e^t-1)^l}{l!}\left(\frac{2}{e^t+1}\right)e^{-lt}\cos yt \\
&= \sum_{l=0}^{\infty} < x >_l \sum_{n=0}^{\infty} S_2(n,l)\frac{t^n}{n!}\sum_{n=0}^{\infty} E_n^{(C)}(-l,y)\frac{t^n}{n!} \\
&= \sum_{n=0}^{\infty}\left(\sum_{l=0}^{\infty}\sum_{i=l}^{n}\binom{n}{i} S_2(i,l) E_{n-i}^{(C)}(-l,y) < x >_l\right)\frac{t^n}{n!},
\end{aligned} \tag{52}$$

where $< x >_l = x(x+1)\cdots(x+l-1)(l \geq 1)$ with $< x >_0 = 1$.

By comparing the coefficients on both sides of (52), we have the following theorem:

Theorem 11. *For $n \in \mathbb{Z}_+$, we have*

$$E_n^{(C)}(x,y) = \sum_{l=0}^{\infty}\sum_{i=l}^{n}\binom{n}{i} S_2(i,l) E_{n-i}^{(C)}(-l,y) < x >_l,$$
$$E_n^{(S)}(x,y) = \sum_{l=0}^{\infty}\sum_{i=l}^{n}\binom{n}{i} S_2(i,l) E_{n-i}^{(S)}(-l,y) < x >_l .$$

By (12), (38), (50), (51) and by using Cauchy product, we have

$$\begin{aligned}
\sum_{n=0}^{\infty} E_n^{(C)}(x,y)\frac{t^n}{n!} &= \left(\frac{2}{e^t+1}\right)((e^t-1)+1)^x \cos(yt) \\
&= \frac{2}{e^t+1}\cos(yt)\sum_{l=0}^{\infty}\binom{x}{l}(e^t-1)^l \\
&= \sum_{l=0}^{\infty}(x)_l \frac{(e^t-1)^l}{l!}\left(\frac{2}{e^t+1}\cos(yt)\right) \\
&= \sum_{l=0}^{\infty}(x)_l \sum_{n=0}^{\infty} S_2(n,l)\frac{t^n}{n!}\sum_{n=0}^{\infty} E_n^{(C)}(y)\frac{t^n}{n!} \\
&= \sum_{n=0}^{\infty}\left(\sum_{l=0}^{\infty}\sum_{i=l}^{n}\binom{n}{i}(x)_l S_2(i,l) E_{n-i}^{(C)}(y)\right)\frac{t^n}{n!}.
\end{aligned} \tag{53}$$

By comparing the coefficients on both sides of (53), we have the following theorem:

Theorem 12. *For $n \in \mathbb{Z}_+$, we have*

$$E_n^{(C)}(x,y) = \sum_{l=0}^{\infty}\sum_{i=l}^{n}\binom{n}{i}(x)_l S_2(i,l) E_{n-i}^{(C)}(y),$$
$$E_n^{(S)}(x,y) = \sum_{l=0}^{\infty}\sum_{i=l}^{n}\binom{n}{i}(x)_l S_2(i,l) E_{n-i}^{(S)}(y).$$

By (4), (12), (38), (50), (51) and by using Cauchy product, we have

$$\begin{aligned}
&\sum_{n=0}^{\infty} E_n^{(C)}(x,y)\frac{t^n}{n!} \\
&= \left(\frac{2}{e^t+1}\right)e^{xt}\cos(yt) \\
&= \frac{(e^t-1)^r}{r!}\frac{r!}{t^r}\left(\frac{t}{e^t-1}\right)^r e^{xt}\sum_{n=0}^{\infty} E_n^{(C)}(y)\frac{t^n}{n!} \\
&= \frac{(e^t-1)^r}{r!}\left(\sum_{n=0}^{\infty}\mathbf{B}_n^{(r)}(x)\frac{t^n}{n!}\right)\left(\sum_{n=0}^{\infty} E_n^{(C)}(y)\frac{t^n}{n!}\right)\frac{r!}{t^r} \\
&= \sum_{n=0}^{\infty}\left(\sum_{l=0}^{n}\frac{\binom{n}{l}}{\binom{l+r}{r}} S_2(l+r,r)\sum_{i=0}^{n-l}\binom{n-l}{i}\mathbf{B}_i^{(r)}(x) E_{n-l-i}^{(C)}(y)\right)\frac{t^n}{n!}.
\end{aligned}$$

By comparing the coefficients on both sides, we have the following theorem:

Theorem 13. *For $n \in \mathbb{Z}_+$ and $r \in \mathbb{N}$, we have*

$$E_n^{(C)}(x,y) = \sum_{l=0}^{n} \frac{\binom{n}{l}}{\binom{l+r}{r}} S_2(l+r,r) \sum_{i=0}^{n-l} \binom{n-l}{i} E_{n-l-i}^{(C)}(y) \mathbf{B}_i^{(r)}(x),$$
$$E_n^{(S)}(x,y) = \sum_{l=0}^{n} \frac{\binom{n}{l}}{\binom{l+r}{r}} S_2(l+r,r) \sum_{i=0}^{n-l} \binom{n-l}{i} E_{n-l-i}^{(S)}(y) \mathbf{B}_i^{(r)}(x).$$

By (5), (12), (38), (50), (51) and by using the Cauchy product, we get

$$\begin{aligned}
\sum_{n=0}^{\infty} E_n^{(C)}(x,y) \frac{t^n}{n!} &= \left(\frac{2}{e^t+1}\right) e^{xt} \cos(yt) \\
&= \frac{(e^t-u)^r}{(1-u)^r} \left(\frac{1-u}{e^t-u}\right)^r e^{xt} \left(\frac{2}{e^t+1}\right) \cos(yt) \\
&= \sum_{n=0}^{\infty} \mathbf{H}_n^{(r)}(u,x) \frac{t^n}{n!} \sum_{i=0}^{r} \binom{r}{i} e^{it} (-u)^{r-i} \frac{1}{(1-u)^r} \left(\frac{2}{e^t+1}\right) \cos(yt) \\
&= \frac{1}{(1-u)^r} \sum_{i=0}^{r} \binom{r}{i} (-u)^{r-i} \sum_{n=0}^{\infty} \mathbf{H}_n^{(r)}(u,x) \frac{t^n}{n!} \sum_{n=0}^{\infty} E_n^{(C)}(i,y) \frac{t^n}{n!} \\
&= \sum_{n=0}^{\infty} \left(\frac{1}{(1-u)^r} \sum_{i=0}^{r} \binom{r}{i} (-u)^{r-i} \sum_{l=0}^{n} \binom{n}{l} \mathbf{H}_l^{(r)}(u,x) E_{n-l}^{(C)}(i,y) \right) \frac{t^n}{n!}.
\end{aligned}$$

By comparing the coefficients on both sides, we have the following theorem:

Theorem 14. *For $n \in \mathbb{Z}_+$ and $r \in \mathbb{N}$, we have*

$$E_n^{(C)}(x,y) = \frac{1}{(1-u)^r} \sum_{i=0}^{r} \sum_{l=0}^{n} \binom{r}{i} \binom{n}{l} (-u)^{r-i} \mathbf{H}_l^{(r)}(u,x) E_{n-l}^{(C)}(i,y),$$
$$E_n^{(S)}(x,y) = \frac{1}{(1-u)^r} \sum_{i=0}^{r} \sum_{l=0}^{n} \binom{r}{i} \binom{n}{l} (-u)^{r-i} \mathbf{H}_l^{(r)}(u,x) E_{n-l}^{(S)}(i,y).$$

By Theorems 12–14, we have the following corollary.

Corollary 3. *For $n \in \mathbb{Z}_+$ and $r \in \mathbb{N}$, we have*

$$\begin{aligned}
&\sum_{l=0}^{\infty} \sum_{i=l}^{n} \binom{n}{i} (x)_l S_2(i,l) E_{n-i}^{(C)}(y) \\
&= \frac{1}{(1-u)^r} \sum_{i=0}^{r} \sum_{l=0}^{n} \binom{r}{i} \binom{n}{l} (-u)^{r-i} \mathbf{H}_l^{(r)}(u,x) E_{n-l}^{(C)}(i,y) \\
&= \sum_{l=0}^{n} \frac{\binom{n}{l}}{\binom{l+r}{r}} S_2(l+r,r) \sum_{i=0}^{n-l} \binom{n-l}{i} E_{n-l-i}^{(C)}(y) \mathbf{B}_i^{(r)}(x).
\end{aligned}$$

3. Distribution of Zeros of the Cosine–Euler and Sine–Euler Polynomials

This section aims to demonstrate the benefit of using numerical investigation to support theoretical prediction and to discover a new interesting pattern of the zeros of the cosine–Euler and sine–Euler polynomials. Using a computer, a realistic study for the cosine–Euler polynomials $E_n^{(C)}(x,y)$ and sine–Euler polynomials $E_n^{(S)}(x,y)$ is very interesting. It is the aim of this paper to observe an interesting phenomenon of "scattering" of the zeros of the the cosine–Euler polynomials $E_n^{(C)}(x,y)$ and sine–Euler polynomials $E_n^{(S)}(x,y)$ in a complex plane. We investigate the beautiful zeros of the cosine–Euler and sine–Euler polynomials by using a computer. We plot the zeros of the cosine–Euler polynomials $E_n^{(C)}(x,y)$ (Figure 1).

In Figure 1 (top-left), we choose $n = 30$ and $y = -3$. In Figure 1 (top-right), we choose $n = 30$ and $y = 0$. In Figure 1 (bottom-left), we choose $n = 30$ and $y = 1/2$. In Figure 1 (bottom-right), we choose $n = 30$ and $y = 3$.

We plot the zeros of the sine–Euler polynomials $E_n^{(S)}(x,y)$ (Figure 2).

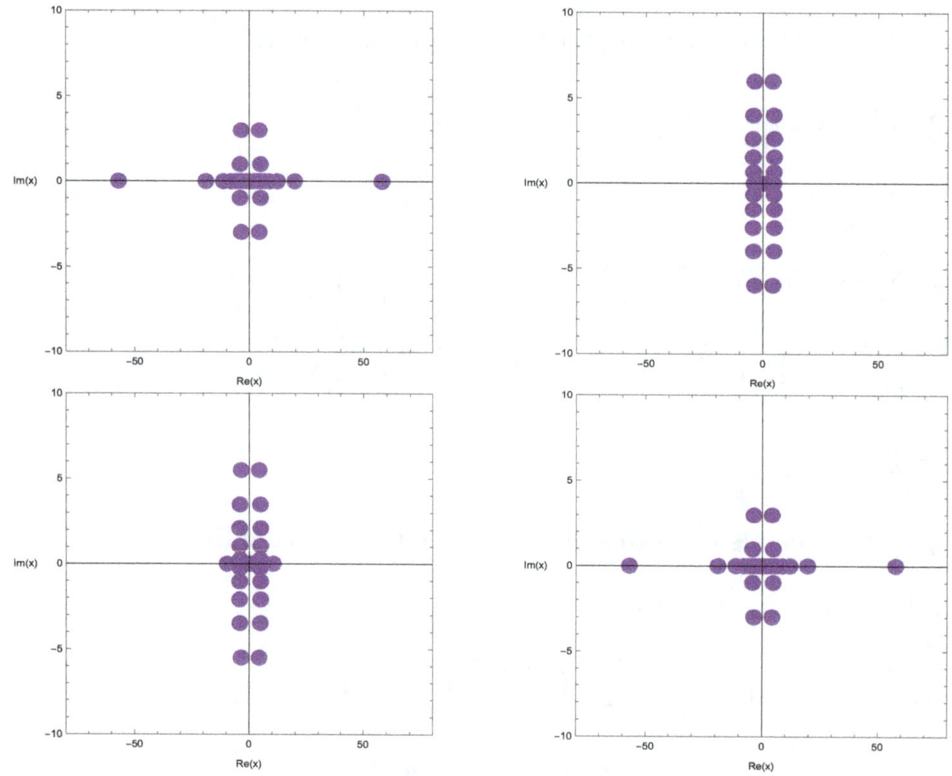

Figure 1. Zeros of $E_n^{(C)}(x,y)$.

In Figure 2 (top-left), we choose $n = 30$ and $x = -3$. In Figure 2 (top-right), we choose $n = 30$ and $x = -1$. In Figure 2 (bottom-left), we choose $n = 30$ and $x = 1$. In Figure 2 (bottom-right), we choose $n = 30$ and $x = 3$.

We observe that $E_n^{(C)}(x,a), x \in \mathbb{C}$ has $Re(x) = \frac{1}{2}$ reflection symmetry in addition to the usual $Im(x) = 0$ reflection symmetry analytic complex functions, where $a \in \mathbb{R}$(Figures 1 and 2).

Since

$$\sum_{n=0}^{\infty} E_n^{(C)}(1-x,-y)\frac{(-1)^n t^n}{n!} = \frac{2}{e^{-t}+1}e^{(1-x)(-t)}\cos yt$$
$$= \frac{2}{e^t+1}e^{xt}\cos yt = \sum_{n=0}^{\infty} E_n^{(C)}(x,y)\frac{t^n}{n!},$$

we obtain

$$E_n^{(C)}(x,y) = (-1)^n E_n^{(C)}(1-x,-y), \quad E_n^{(C)}(x,y) = (-1)^n E_n^{(C)}(1-x,y),$$
$$E_n^{(S)}(x,y) = (-1)^n E_n^{(S)}(1-x,-y), \quad E_n^{(S)}(x,y) = (-1)^{n+1} E_n^{(S)}(1-x,y).$$

Hence, we have the following theorem:

Theorem 15. *If* $n \equiv 1 \pmod 2$, *then*

$$E_n^{(C)}(1/2, y) = 0, \quad B_n^{(C)}(1/2, y) = 0, \text{ for } n \in \mathbb{N}.$$

If $n \equiv 0 \pmod 2$, *then*

$$E_n^{(S)}(1/2, y) = 0, \quad B_n^{(S)}(1/2, y) = 0, \text{ for } n \in \mathbb{N}.$$

Our numerical results for numbers of real and complex zeros of the cosine–Euler polynomials $E_n^{(C)}(x, y) = 0$ are displayed (Table 1).

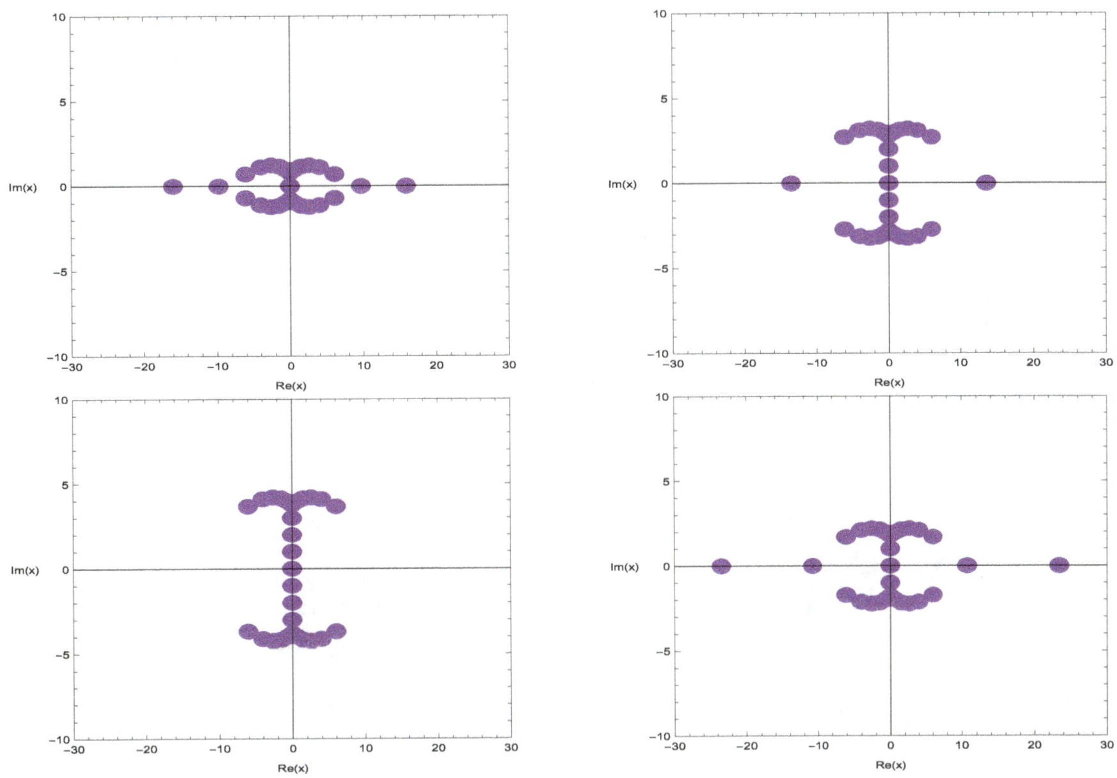

Figure 2. Zeros of $E_n^{(S)}(x, y)$.

Table 1. Numbers of real and complex zeros of $E_n^{(C)}(x, y)$.

Degree n	$y = -3$		$y = 2$	
	Real Zeros	**Complex Zeros**	**Real Zeros**	**Complex Zeros**
1	1	0	1	0
2	2	0	2	0
3	3	0	3	0
4	4	0	4	0
5	5	0	5	0
6	6	0	6	0
7	7	0	7	0
8	8	0	8	0
9	9	0	9	0
10	10	0	10	0

Our numerical results for numbers of real and complex zeros of the sine–Euler polynomials $E_n^{(S)}(x, y) = 0$ are displayed (Table 2).

Stacks of zeros of the cosine–Euler polynomials $E_n^{(C)}(x,y)$ for $1 \leq n \leq 40$ from a 3D structure are presented (Figure 3).

In Figure 3 (left), we choose $y = -3$. In Figure 3 (right), we choose $y = 1/2$. The plot of real zeros of the cosine–Euler polynomials $E_n^{(C)}(x,y)$ for $1 \leq n \leq 40$ structure are presented (Figure 4).

In Figure 4 (left), we choose $y = -3$. In Figure 4 (right), we choose $y = 1/2$. Stacks of zeros of the sine–Euler polynomials $E_n^{(S)}(x,y)$ for $1 \leq n \leq 40$ from a 3D structure are presented (Figure 5).

Table 2. Numbers of real and complex zeros of $E_n^{(S)}(x,y)$.

Degree n	$x = -3$		$x = 1$	
	Real Zeros	**Complex Zeros**	**Real Zeros**	**Complex Zeros**
1	1	0	1	0
2	1	0	1	0
3	3	0	3	0
4	3	0	3	0
5	5	0	5	0
6	5	0	1	4
7	7	0	7	0
8	7	0	1	6
9	9	0	9	0
10	9	0	1	8

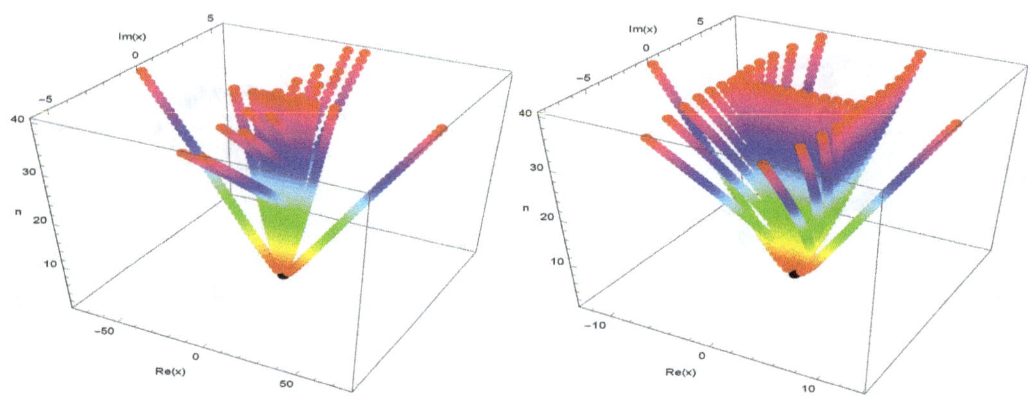

Figure 3. Stacks of zeros of $E_n^{(C)}(x,y), 1 \leq n \leq 40$.

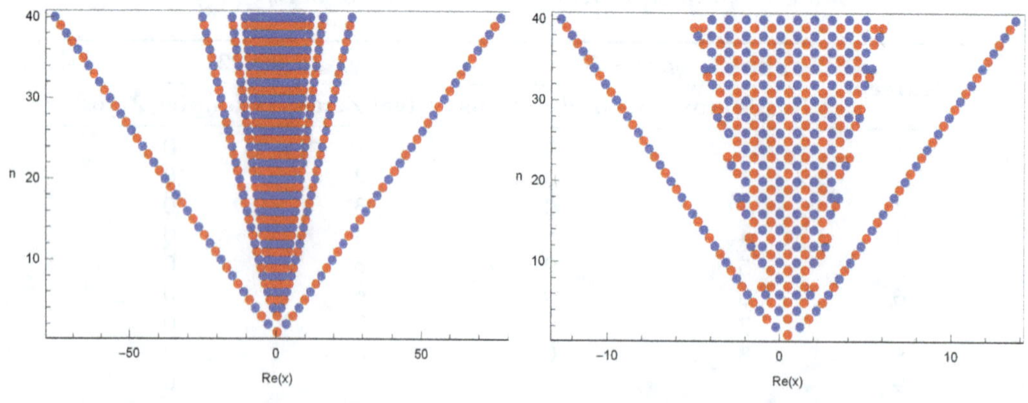

Figure 4. Real zeros of $E_n^{(C)}(x,y)$.

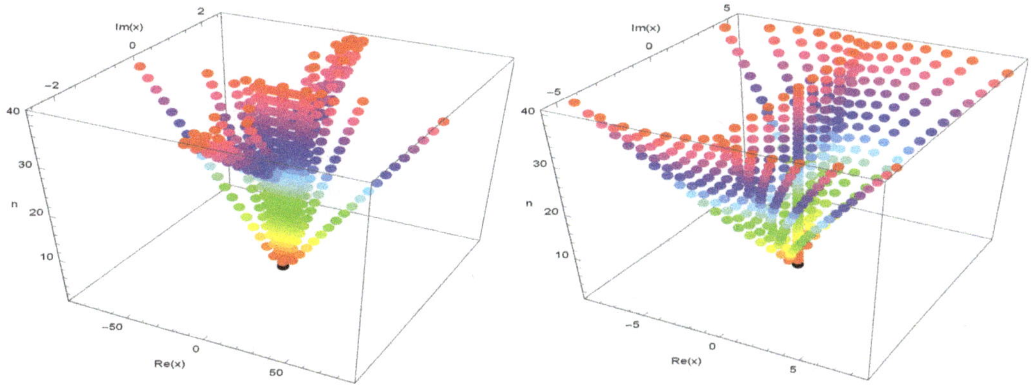

Figure 5. Stacks of zeros of $E_n^{(S)}(x,y), 1 \le n \le 40$.

In Figure 5 (left), we choose $x = -3$. In Figure 3 (right), we choose $x = 1$. The plot of real zeros of the sine–Euler polynomials $E_n^{(S)}(x,y)$ for $1 \le n \le 40$ structure are presented (Figure 6).

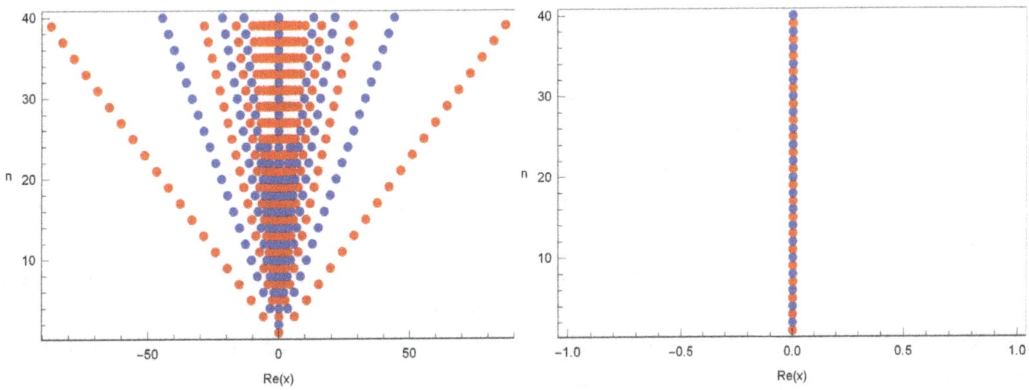

Figure 6. Real zeros of $E_n^{(S)}(x,y)$.

In Figure 6 (left), we choose $x = -3$. In Figure 6 (right), we choose $x = 1$.

We observe a remarkable regular structure of the complex roots of the cosine–Euler polynomials $E_n^{(C)}(x,y)$. We also hope to verify a remarkable regular structure of the complex roots of the cosine–Euler polynomials $E_n^{(C)}(x,y)$. Next, we calculated an approximate solution satisfying $E_n^{(C)}(x,y) = 0, x \in \mathbb{R}$. The results are given in Table 3.

Table 3. Approximate solutions of $E_n^{(C)}(x,-3) = 0, x \in \mathbb{R}$.

Degree n				x			
1				0.50000			
2			−2.5414,	3.5414			
3			−4.7678,	0.50000,	5.7678		
4		−6.8305,	−0.82832,	1.8283,	7.8305		
5		−8.8303,	−1.8336,	0.50000,	2.8336,	9.8303	
6		−10.799,	−2.7017,	−0.40666,	1.4067,	3.7017,	11.799
7	−12.751,	−3.4960,	−1.1389,	0.50000,	2.1389,	4.4960,	13.751

Next, we calculated an approximate solution satisfying $E_n^{(S)}(x,y) = 0, y \in \mathbb{R}$. The results are given in Table 4.

Table 4. Approximate solutionsof $E_n^{(S)}(-3, y) = 0, y \in \mathbb{R}$.

Degree n	y
1	0.00000
2	0.00000
3	−6.0000, 0, 6.0000
4	−3.3912, 0, 3.3912
5	−10.687, −2.4038, 0, 2.4038, 10.687
6	−5.9045, −1.8630, 0, 1.8630, 5.9045
7	−15.241, −4.1727, −1.5184, 0, 1.5184, 4.1727, 15.241

Author Contributions: T.K. and C.S.R. wrote and checked the results of the paper; C.S.R. conducted numerical experiments of this paper; T.K. completed the revision of the article.

Acknowledgments: The authors would like to thank the referees for their valuable comments.

References

1. Ağyüz, E.; Acikgoz, M.; Araci, S. A symmetric identity on the q-Genocchi polynomials of higher-order under third dihedral group D3. *Proc. Jangjeon Math. Soc.* **2015**, *18*, 177–187
2. Bayad, A.; Chikhi, J. Non linear recurrences for Apostol-Bernoulli-Euler numbers of higher order. *Adv. Stud. Contemp. Math. (Kyungshang)* **2012** , *22*, 1–6.
3. Carlitz, L. Recurrences for the Bernoulli and Euler numbers. II. *Math. Nachr.* **1965**, *29*, 151–160. [CrossRef]
4. He, Y.; Kim, T. A higher-order convolution for Bernoulli polynomials of the second kind. *Appl. Math. Comput.* **2018**, *324*, 51–58. [CrossRef]
5. Kim, D.S.; Kim, T. Some identities of Bernoulli and Euler polynomials arising from umbral calculus. *Adv. Stud. Contemp. Math. (Kyungshang)* **2013**, *23*, 159–171. [CrossRef]
6. Kim, D.S.; Kim, T.; Kim, Y.H.; Lee, S.H. Some arithmetic properties of Bernoulli and Euler numbers. *Adv. Stud. Contemp. Math. (Kyungshang)* **2012**, *22*, 467–480.
7. Kim, D.S.; Kim, T.; Kim, Y.-H.; Dolgy, D.V. A note on Eulerian polynomials associated with Bernoulli and Euler numbers and polynomials. *Adv. Stud. Contemp. Math. (Kyungshang)* **2012**, *22*, 379–389.
8. Roman, S. *The Umbral Calculu, Pure and Applied Mathematics, 111*; Academic Press, Inc.: New York, NY, USA; Harcourt Brace Jovanovich Publishers: San Diego, CA, USA, 1984; ISBN 0-12-594380-6.
9. Ryoo, C.S.; Kim, Y.H. A numerical investigation on the structure of the roots of the twisted q-Euler polynomials. *Adv. Stud. Contemp. Math. (Kyungshang)* **2009**, *19*, 131–141.
10. Ryoo, C.S.; Agarwal, R.P. Some identities involving q-poly-tangent numbers and polynomials and distribution of their zeros. *Adv. Differ. Equ.* **2017**, *213*. [CrossRef]
11. Simsek, Y. Identities on the Changhee numbers and Apostol-type Daehee polynomials. *Adv. Stud. Contemp. Math. (Kyungshang)* **2017**, *27*, 199–212.
12. Srivastava, H.M.; Pintér, Á. Remarks on some relationships between the Bernoulli and Euler polynomials. *Appl. Math. Lett.* **2004**, *17*, 375–380. [CrossRef]
13. Srivastava, H.M.; Pintér, Á. Addition theorems for the Appell polynomials and the associated classes of polynomial expansions. *Aequ. Math.* **2013**, *85*, 483–495.
14. Young, P.T. Degenerate Bernoulli polynomials, generalized factorial sums, and their applications. *J. Number Theor.* **2008**, *128*, 738–758 [CrossRef]

13

Common Fixed Point Theorems for Generalized Geraghty (α, ψ, ϕ)-Quasi Contraction type Mapping in Partially Ordered Metric-Like Spaces

Haitham Qawaqneh [1,*,†], Mohd Noorani [1,†], Wasfi Shatanawi [2,3,†] and Habes Alsamir [1,†]

1 Department of Mathematics, University Kebangsaan Malaysia, Bangi 43600, Malaysia; msn@ukm.my (M.N.); h.alsamer@gmail.com (H.A.)
2 Department of Mathematics, Hashemite University, Zarqa 1315, Jordan; swasfi@hu.edu.jo
3 Department of Mathematics and General Courses, Prince Sultan University, Riyadh 11586, Saudi Arabia; wshatanawi@psu.edu.sa
* Correspondence: Haitham.math77@gmail.com;
† These authors contributed equally to this work.

Abstract: The aim of this paper is to establish the existence of some common fixed point results for generalized Geraghty (α, ψ, ϕ)-quasi contraction self-mapping in partially ordered metric-like spaces. We display an example and an application to show the superiority of our results. The obtained results progress some well-known fixed (common fixed) point results in the literature. Our main results cannot be specifically attained from the corresponding metric space versions. This paper is scientifically novel because we take Geraghty contraction self-mapping in partially ordered metric-like spaces via $\alpha-$admissible mapping. This opens the door to other possible fixed (common fixed) point results for non-self-mapping and in other generalizing metric spaces.

Keywords: common fixed point; metric-like space; α-Geraghty contraction; triangular α-admissible mapping

---

1. Introduction

Fixed point theory occupies a central role in the study of solving nonlinear equations of kinds $Sx = x$, where the function S is characterized on abstract space X. It is outstanding that the Banach contraction principle is a standout amongst essential and principal results in the fixed point theorem. It ensures the existence of fixed points for certain self-maps in a complete metric space and provides a helpful technique to find those fixed points. Many authors studied and extended it in many generalizations of metric spaces with new contractive mappings, for example, see References [1–3] and the references therein.

Otherwise, Hitzler and Seda [4] introduce the notation of metric-like (dislocated) metric space as a generalization of a metric space, they introduced variants of the Banach fixed point theorem in such space. Metric like spaces were revealed by Amini-Harandi [5] who proved the existence of fixed point results. This interesting subject has been mediated by certain authors, for example, see References [6–8]. In partial metric spaces and partially ordered metric-like spaces, the usual contractive condition is weakened and many researchers apply their results to problems of existence and uniqueness of solutions for some boundary value problems of differential and Integral equations, for example, see References [9–22] and the references therein.

Additionally, Geraghty [23] characterized a kind of the set of functions $\mathfrak{S}$ to be classified as the functions β:$[0, \infty) \to [0, 1)$ such that if $\{t_n\}$ is a sequence in $[0, +\infty)$ with $\beta(t_n) \to 1$, then $t_n \to 0$.

By using the function $\beta \in \mathfrak{S}$, Geraghty [23] presented the following exceptional theorem

Theorem 1. *Suppose (Y,d) is a complete metric space. Assume that $T{:}Y \to Y$ and $\beta{:}[0,\infty) \to [0,1)$ are functions such that for all $u,v \in Y$,*

$$d(Tu,Tv) \leq \beta(d(u,v))d(u,v), \tag{1}$$

where $\beta \in \mathcal{S}$, then T has a fixed point and has to be unique.

The main results of Geraghty have engaged many of authors, see References [24–26] and the references therein.

Recently, Amini-Harandi and Emami [27] reconsidered Theorem 1 as the framework of partially ordered metric spaces and they presented taking into account existence theorem.

Theorem 2. *Let (Y,d) be a partially ordered complete metric space. Assume $S{:}Y \to Y$ is a mapping such that there exists $u_0 \in Y$ with $u_0 \preceq Su_0$ and $\alpha \in \mathcal{F}$ such that*

$$d(Su,Sv) \leq \alpha(d(u,v))d(u,v), \; for\; any\; u,v \in Y\; with\; u \succeq v. \tag{2}$$

Hence, S has a fixed point supported that either S is continuous or Y is such that if an increasing sequence $\{u_n\} \to u$, then $u_n \leq u$ for all n.

In 2015, Karapinar [28] demonstrated the following specific results:

Theorem 3. *[28] Let (Y,σ) be a complete metric-like space. Assume that $S{:}Y \to Y$ is a mapping. If there exists $\beta \in \mathfrak{S}$ such that*

$$\sigma(Su,Sv) \leq \beta(\sigma(u,v))\sigma(u,v) \tag{3}$$

for all $u,v \in Y$, then S has a unique fixed point $u^* \in Y$ with $\sigma(u^*,u^*) = 0$.

The notion of quasi-contraction presented by Reference [29], is known as one of the foremost common contractive self-mappings.

A mapping $S{:}Y \to Y$ is expressed to be a quasi contraction if there exists $0 \leq \lambda < 1$ such that

$$d(Su,Sv) \leq \lambda \max\{d(u,v).d(u,fv),d(u,fv),d(fu,v),d(u,fv)\}, \tag{4}$$

for any $u,v \in Y$.

In this paper, we show the generalized Geraghty (α,ψ,ϕ)-quasi contraction type mapping in partially ordered metric like space, then we present some fixed and common fixed point theorems for such mappings in an ordered complete metric-like space. We investigate this new contractive mapping as a generalized weakly contractive mapping in our main results, then we display an example and an application to support our obtained results.

2. Preliminaries

In this section, we review a few valuable definitions and assistant results that will be required within the following sections.

Definition 1. *[5] Let Y be a nonempty set. A function $\sigma{:}Y \times Y \to [0,\infty)$ is expressed to be a metric-like space on X if for any $u,v,z \in Y$, the accompanying stipulations satisfied:*

(σ_1) $\sigma(u,v) = 0 \Rightarrow u = v$,
(σ_2) $\sigma(u,v) = \sigma(v,u)$,
(σ_3) $\sigma(u,z) \leq \sigma(u,v) + \sigma(v,z)$.

The pair (Y,σ) is called a metric-like space.

Obviously, we can consider that every metric space and partial metric space could be a metric-like space. However, this assertion isn't valid.

Example 1. *[5] Let $Y = \{0, 1\}$ and*

$$\sigma(u,v) = \begin{cases} 2, & \text{if } u = v = 0; \\ 1, & \text{otherwise}. \end{cases} \tag{5}$$

We note that $\sigma(0,0) \not\leq \sigma(0,1)$. So, (Y, σ) is a metric-like space and at the same time it is not a partial metric space.

Additonally, each metric-like σ on Y create a topology τ_σ on Y whose use as a basis of the group of open σ-balls

$$B_\sigma(Y, \epsilon) = \{u \in Y : \mid \sigma(u,v) - \sigma(u,u) \mid < \epsilon\}, \text{ for all } u, v \in Y \text{ and } \epsilon > 0.$$

Let (Y, σ) be a metric-like space and $f{:}Y \to Y$ be a continuous mapping. Then

$$\lim_{n\to\infty} u_n = u \Rightarrow \lim_{n\to\infty} fu_n = fu.$$

A sequence $\{u_n\}$ of elements of Y is considered σ-Cauchy if the limit $\lim_{n,m\to\infty} \sigma(u_n, u_m)$ exists as a finite number. The metric-like space (Y, σ) is considered complete if for each σ-Cauchy sequence $\{u_n\}$, there is some $u \in Y$ such that

$$\lim_{n\to\infty} \sigma(u_n, u) = \sigma(u,u) = \lim_{n,m\to\infty} \sigma(u_n, u_m).$$

Remark 1. *[30] Let $Y = \{0, 1\}$, and $\sigma(u,v) = 1$ for each $u, v \in Y$ and $u_n = 1$ for each $n \in \mathbb{N}$. Then, it is easy to see that $u_n \to 0$ and $u_n \to 1$ and so in metric-like spaces the limit of a convergent sequence is not necessarily unique.*

Lemma 1. *[30] Let (Y, σ) be a metric-like space. Let $\{u_n\}$ be a sequence in Y such that $u_n \to u$ where $u \in Y$ and $\sigma(u,u) = 0$. Then, for all $u, v \in Y$, we have $\lim_{n\to\infty} \sigma(u_n, v) = \sigma(u,v)$.*

Example 2. *[5] Let $Y = \mathbb{R}$ and $\sigma{:}Y \times Y \to [0, +\infty)$ be defined by*

$$\sigma(u,v) = \begin{cases} 2n, & \text{if } u = v = 0; \\ n, & \text{otherwise}. \end{cases}$$

Then, we can consider (Y, σ) to be a metric-like space, but it does not satisfy the conditions of the partial metric space, as $\sigma(0,0) \not\leq \sigma(0,1)$.

Samet et al. [31] displayed the definition of α-admissible mapping as followings:

Definition 2. *[31] Let $S{:}X \to X$ and $\alpha{:}X \times X \to [0, \infty)$ are two functions. Then, S is called α-admissible if $\forall u, v \in X$ with $\alpha(u,v) \geq 1$ implies $\alpha(fu, fv) \geq 1$.*

Definition 3. *[32] Let $S, T{:}X \to X$ be two mappings and $\alpha{:}X \times X \to \mathbb{R}$ be a function. We consider that the pair (S, T) is α-admissible if*

$$u, v \in X, \ \alpha(u,v) \geq 1 \Rightarrow \alpha(Su, Tv) \geq 1 \text{ and } \alpha(Tu, Sv) \geq 1$$

Definition 4. *[33] Let $S{:}X \to X$ and $\alpha{:}X \times X \to [0, \infty)$. Then, S is called a triangular α-admissible mapping if*

(1) *S is α-admissible,*
(2) *$\alpha(u,z) \geq 1$ and $\alpha(z,v) \geq 1$ imply $\alpha(u,v) \geq 1$.*

Definition 5. *[32] Let $S,T:X \to X$ and $\alpha:X \times X \to [0,\infty)$. Then, (S,T) is called a triangular α-admissible mapping if*

(1) *The pair (S,T) is α-admissible,*
(2) *$\alpha(u,z) \geq 1$ and $\alpha(z,v) \geq 1$ imply $\alpha(u,v) \geq 1$.*

Let Ψ indicate the set of functions $\psi:[0,\infty) \to [0,\infty)$ that approve the following stipulations:

(1) ψ is strictly continuous increasing,
(2) $\psi(t) = 0 \Leftrightarrow t = 0$.

and Φ indicates the set of all continuous functions $\phi:[0,\infty) \to [0,\infty)$ with $\phi(t) > \psi(t)$ for all $t > 0$ and $\phi(0) = 0$.

Definition 6. *[12] Let $(X,d,\preceq)$ be a partially ordered metric space. Assume $f,g:X \to X$ are two mappings. Then:*

(1) *For all $x,y \in X$ are said to be comparable if $x \preceq y$ or $y \preceq x$ holds,*
(2) *f is said to be nondecreasing if $x \preceq y$ implies $fx \preceq fy$,*
(3) *f,g are called weakly increasing if $fx \preceq gfx$ and $gx \preceq fgx$ for all $x \in X$,*
(4) *f is called weakly increasing if f and I are weakly increasing, where I is denoted to the identity mapping on X.*

3. Main Results

In this section, we present the notation of generalized Geraghty (α,ψ,ϕ)-quasi contraction self-mappings in partially ordered metric-like space. Then, we present some fixed and common fixed point theorems for such self-mappings. We investigate this new contractive self-mapping as a generalized weakly contractive self-mapping which is a generalization of the results of Reference [34]. Results of this kind are amongst the most useful in fixed point theory and it's applications.

Definition 7. *Let (X,σ) be a partially ordered metric-like space and $S,T:X \to X$ be two mappings. Then, we consider that the pair (S,T) is generalized Geraghty (α,ψ,ϕ)-quasi contraction self-mapping if there exist $\alpha:X \times X \to [0,\infty)$, $\beta \in \mathfrak{S}$, $\psi \in \Psi$ and $\phi:[0,\infty) \to [0,\infty)$ are continuous functions with $\phi(t) \leq \psi(t)$ for all $t > 0$ such that*

$$\alpha(x,y)\psi(\sigma(Sx,Ty)) \leq \lambda\beta(\psi(M_{x,y}))\phi(M_{x,y}), \tag{6}$$

holds for all elements $x,y \in X$ and $0 \leq \lambda < 1$, where

$$M_{x,y} = \max\{\sigma(x,y), \sigma(x,Sx), \sigma(y,Ty), \sigma(Sx,y), \sigma(x,Ty)\}.$$

The following two lemmas will be utilized proficiently within the verification of our fundamental result.

Lemma 2. *If $\psi \in \Psi$ and $\phi:[0,\infty) \to [0,\infty)$ are continuous function that satisfy the condition $\psi(t) > \phi(t)$ for all $t > 0$, then $\phi(0) = 0$.*

Proof. From the assumption $\phi(t) < \psi(t)$, since ψ and ϕ are continuous, we have

$$0 \leq \phi(0) = \lim_{t\to 0} \phi(t) \leq \lim_{t\to 0} \psi(t) = \psi(0) = 0.$$

□

Lemma 3. *Let $S, T{:}X \to X$ be two mappings and $\alpha{:}X \times X \to [0,\infty)$ be a function such that S, T are triangular $\alpha-$admissible. Suppose that there exists $x_0 \in X$ such that $\alpha(x_0, Sx_0) \geq 1$. Define a sequence $\{x_n\}$ in X by $Sx_{2n} = x_{2n+1}$ and $Tx_{2n+1} = x_{2n+2}$. Then $\alpha(x_n, x_m) \geq 1$ for all $m, n \in \mathbb{N}$ with $n < m$.*

Proof. Since $\alpha(x_0, Sx_0) \geq 1$ and S, T are $\alpha-$admissible, we get

$$\alpha(x_0, x_1) = \alpha(x_0, Sx_0) \geq 1.$$

By triangular $\alpha-$admissibility, we get

$$\alpha(Sx_0, Tx_1) = \alpha(x_1, x_2) \geq 1$$

and

$$\alpha(TSx_0, STx_1) = \alpha(x_2, x_3) \geq 1.$$

Again, since $\alpha(x_2, x_3) \geq 1$, then

$$\alpha(Sx_2, Tx_3) = \alpha(x_3, x_4) \geq 1$$

and

$$\alpha(TSx_2, STx_3) = \alpha(x_4, Sx_5) \geq 1.$$

By proceeding the above process, we conclude that $\alpha(x_n, x_{n+1}) \geq 1$ for all $n \in \mathbb{N} \cup \{0\}$. Now, we prove that $\alpha(x_n, x_m) \geq 1$, for all$m, n \in \mathbb{N}$ with $n < m$. Since

$$\begin{cases} \alpha(x_n, x_{n+1}) \geq 1, \\ \alpha(x_{n+1}, x_{n+2}) \geq 1, \end{cases}$$

then, we have

$$\alpha(x_n, x_{n+2}) \geq 1.$$

Again, since

$$\begin{cases} \alpha(x_n, x_{n+2}) \geq 1 \\ \alpha(x_{n+2}, x_{n+3}) \geq 1, \end{cases}$$

we deduce that

$$\alpha(x_n, x_{n+3}) \geq 1.$$

By continuing this process, we have

$$\alpha(x_n, x_m) \geq 1$$

for all $n \in \mathbb{N}$ with $m > n$. □

Lemma 4. *Let $(X, \preceq, \sigma)$ be a partially ordered metric-like space. Assume S, T are two self-mappings of X which the pair (S, T) is generalized (α, ψ, ϕ)-quasi contraction self-mappings. Fix $x_1 \in X$ and define a sequence $\{x_n\}$ by $x_{2n+1} = Sx_{2n}$ and $x_{2n+2} = Tx_{2n+1}$ for all $n \in \mathbb{N}$. If $\lim_{n\to\infty} \sigma(x_n, x_{n+1}) = 0$ and the sequence $\{x_n\}$ is nondecreasing, then $\{x_n\}$ is a Cauchy sequence.*

Proof. Since S, T are a generalized (α, ψ, ϕ)-quasi contraction non-self mapping, then there exist $\psi \in \Psi, \phi \in \Phi$ such that

$$\alpha(x,y)\psi(\sigma(Sx,Ty)) \leq \lambda\beta(\psi(M_{x,y}))\phi(M_{x,y}), \tag{7}$$

holds for all elements $x, y \in X$ and $0 \leq \lambda < 1$, where

$$M_{x,y} = \max\{\sigma(x,y), \sigma(x,Sx), \sigma(y,Ty), \sigma(Sx,y), \sigma(x,Ty)\}.$$

Now, we show that the sequence $\{x_n\}$ is Cauchy sequence. Assume, for contradiction's sake, that $\{x_n\}$ isn't Cauchy sequence. Therefore, there exist $\epsilon > 0$ and two subsequences $\{n_k\}$ and $\{m_k\}$ of the sequence $\{x_n\}$ such that $\sigma(x_{2n_k}, x_{2m_k}), \sigma(x_{2n_k-1}, x_{2m_k})$ and $\sigma(x_{2n_k}, x_{2m_k+1})$ converge to ϵ^+ when $k \to \infty$.

$$n_k > m_k > k,\ \sigma(x_{2n_k}, x_{2m_k-2}) < \epsilon,\ \sigma(x_{2n_k}, x_{2m_k}) \geq \epsilon. \tag{8}$$

By the above inequalities and triangle inequality property, we imply that

$$\begin{aligned} \epsilon &\leq \sigma(x_{2n_k}, x_{2m_k}) \leq \sigma(x_{2n_k}, x_{2m_k-2}) + \sigma(x_{2m_k-2}, x_{2m_k-1}) + \sigma(x_{2m_k-1}, x_{2m_k}) \\ &< \epsilon + \sigma(x_{2m_k-2}, x_{2m_k-1}) + \sigma(x_{2m_k-1}, x_{2m_k}). \end{aligned}$$

In view of $\lim_{n\to\infty} \sigma(x_n, x_{n+1}) = 0$ and letting $k \to \infty$ in the above inequalities, we obtain

$$\lim_{k\to\infty} \sigma(x_{2n_k}, x_{2m_k}) = \epsilon. \tag{9}$$

By the triangle inequality, we have

$$\begin{aligned} \sigma(x_{2n_k}, x_{2m_k}) &\leq \sigma(x_{2n_k}, x_{2n_k+1}) + \sigma(x_{2n_k+1}, x_{2m_k}) \\ &\leq \sigma(x_{2n_k}, x_{2n_k+1}) + \sigma(x_{2n_k+1}, x_{2m_k+1}) + \sigma(x_{2m_k+1}, x_{2m_k}) \\ &\leq \sigma(x_{2n_k}, x_{2n_k+1}) + \sigma(x_{2n_k+1}, x_{2n_k+2}) + \sigma(x_{2n_k+2}, x_{2m_k}) + 2\sigma(x_{2m_k}, x_{2m_k+1}) \\ &\leq 2\sigma(x_{2n_k}, x_{2n_k+1}) + 2\sigma(x_{2m_k+2}, x_{2m_k+1}) + \sigma(x_{2n_k}, x_{2m_k}) + 2\sigma(x_{2m_k}, x_{2m_k+1}). \end{aligned}$$

Taking the limit as $k \to \infty$ in the above inequalities and using Equation (9), we get

$$\lim_{k\to\infty} \sigma(x_{2n_k}, x_{2m_k}) = \lim_{k\to\infty} \sigma(x_{2n_k+1}, x_{2m_k}) = \lim_{k\to\infty} \sigma(x_{2n_k+1}, x_{2m_k+1}) = \epsilon. \tag{10}$$

Since $x_{n_k+1} \preceq x_{m_k}$ and $\alpha(x_{n_k+1}, x_{m_k}) \geq 1$ for all $k \in \mathbb{N}$, so by substituting x with x_{n_k+1} and y with x_{m_k} in Equation (7), it follows that

$$\psi(\sigma(x_{n_k+1}, x_{m_k})) \leq \alpha(x_{n_k+1}, x_{m_k})\psi(\sigma(Sx_{n_k}, Tx_{m_k-1})) \leq \lambda\beta(\psi(M_{x,y}))\phi(M_{x,y}), \tag{11}$$

holds for all elements $x, y \in X$ and $0 \leq \lambda < 1$, where

$$\begin{aligned} M_{x_{n_k}, x_{m_k-1}} &= \max\{\sigma(x_{n_k}, x_{m_k-1}), \sigma(x_{n_k}, Sx_{n_k}), \sigma(x_{m_k-1}, Tx_{m_k-1}), \\ &\quad \sigma(Sx_{n_k}, x_{m_k-1}), \sigma(x_{n_k}, Tx_{m_k-1})\} \\ &= \max\{\sigma(x_{n_k}, x_{m_k-1}), \sigma(x_{n_k}, x_{n_k+1}), \sigma(x_{m_k-1}, x_{m_k}), \\ &\quad \sigma(x_{n_k+1}, x_{m_k-1}), \sigma(x_{n_k}, x_{m_k})\}. \end{aligned}$$

Taking the limit as $k \to \infty$ of the above inequality and applying Equations (9), (10), we get

$$\lim_{k\to\infty} M_{x_{2n_k}, x_{2m_k}} = \epsilon. \tag{12}$$

Letting $k \to \infty$ in Equation (11) and using $\phi \in \Phi$, $\beta \in \mathcal{S}$ and Equation (12), we deduce that

$$\begin{aligned}\psi(\epsilon) &\leq \lambda\beta(\psi(\epsilon))\phi(\epsilon)\\ &< \lambda\phi(\epsilon)\\ &< \lambda\psi(\epsilon).\end{aligned}$$

This is possible only if $\epsilon = 0$. Which contradicts the positivity of ϵ. Therefore, we get the desired result. □

Theorem 4. *Let (X, σ) be a partially ordered metric like space. Assume that $S, T: X \to X$ are two self-mappings fulfilling the following conditions:*

(1) *(S, T) is triangular α-admissible and there exists an $x_0 \in X$ such that $\alpha(x_0, Sx_0) \geq 1$,*
(2) *the pair (S, T) is weakly increasing,*
(3) *the pair (S, T) is a generalized Geraghty (α, ψ, ϕ)-quasi contraction non-self mapping,*
(4) *S and T are σ-continuous mappings.*

Then, the pair (S, T) has a common fixed point $z \in X$ with $\sigma(z, z) = 0$. Moreover, assume that if $x_1, x_2 \in X$ such $\sigma(x_1, x_1) = \sigma(x_2, x_2) = 0$ implies that x_1 and x_2 are comparable elements. Then the common fixed point of the pair (S, T) is unique.

Proof. Let $x_0 \in X$ such that $\alpha(x_0, Sx_0) \geq 1$. Define the sequence $\{x_n\}$in X as follows:

$$x_{2n+1} = Sx_{2n} \;\; x_{2n+2} = Tx_{2n+1} \;\; for\, all \;\; n \geq 0. \tag{13}$$

Suppose that $x_{2n} \neq x_{2n+1}$ for all $n \in \mathbb{N}_0$. Then, $\sigma(x_{2n}, x_{2n+1}) > 0$ for all $n \in \mathbb{N}_0$. Indeed, if $x_{2n} \neq x_{2n+1}$, which is a contradiction. By using the assumption of Equations (1), (2), and Lemma 3, we have

$$\alpha(x_n, x_{n+1}) \geq 1 \tag{14}$$

for all $n \in \mathbb{N} \cup \{0\}$.

Since the pair (S, T) is weakly increasing, we have

$$x_1 = Sx_0 \preceq TSx_0 = x_2 = Sx_1 = \preceq ...x_{2n} \preceq TSx_{2n} = x_{2n+2} \preceq ....$$

Thus, $x_n \preceq x_{n+1}$, for all $n \in \mathbb{N}$. Since $\alpha(x_{2n}, x_{2n+1}) \geq 1$, by applying Equation (6), we obtain

$$\begin{aligned}\psi(\sigma(x_{2n+1}, x_{2n+2})) &= \psi(\sigma(Sx_{2n}, Tx_{2n+1}))\\ &\leq \alpha(x_{2n}, x_{2n+1})\psi(\sigma(Sx_{2n}, Tx_{2n+1}))\\ &\leq \lambda\beta(\psi(M_{x_{2n},x_{2n+1}}))\phi(M_{x_{2n},x_{2n+1}}).\end{aligned} \tag{15}$$

Set $\sigma_n = \sigma(x_{2n+1}, x_{2n+2})$. We have

$$\psi(\sigma_n) = \psi(\sigma(x_{2n+1}, x_{2n+2})) \tag{16}$$

$$\leq \lambda\beta(\psi(M_{x_{2n},x_{2n+1}}))\phi(M_{x_{2n},x_{2n+1}}). \tag{17}$$

For the rest, for each n assume that $(\sigma_n \neq 0)$.

$$\begin{aligned}M_{x_{2n},x_{2n+1}} &= \max\{\sigma(x_{2n}, x_{2n+1}), \sigma(x_{2n}, Sx_{2n}), \sigma(x_{2n+1}, Tx_{2n+1}), \sigma(Sx_{2n}, x_{2n+1}), \sigma(x_{2n}, Tx_{2n+1})\}\\ &= \max\{\sigma(x_{2n}, x_{2n+1}), \sigma(x_{2n}, x_{2n+1}), \sigma(x_{2n+1}, x_{2n+2}), \sigma(x_{2n+1}, x_{2n+1}), \sigma(x_{2n}, x_{2n+2})\}\\ &= \max\{\sigma(x_{2n}, x_{2n+1}), \sigma(x_{2n+1}, x_{2n+2}), \sigma(x_{2n}, x_{2n+2})\}\\ &= \max\{\sigma_{n-1}, \sigma_n, \sigma_{n-1} + \sigma_n\}\end{aligned}$$

If for some $n \in \mathbb{N}$, $\max\{\sigma_{n-1}, \sigma_n, \sigma_{n-1}+\sigma_n\} = \sigma_n$ then from Equation (16), we find that $\psi(\sigma_n) < \lambda\psi(\sigma_n)$ which is a contradiction with respect to $0 \leq \lambda < 1$. We deduce $\max\{\sigma_{n-1}, \sigma_n, \sigma_{n-1}+\sigma_n\} = \max\{\sigma_{n-1}, \sigma_{n-1}+\sigma_n\}$. Therefore Equation (16) becomes

$$\psi(\sigma_n) < \lambda\psi(\max\{\sigma_{n-1}, \sigma_{n-1}+\sigma_n\}).$$

Put

$$\gamma = \max\{\lambda, \frac{\lambda}{1-\lambda}\}.$$

Thus,

$$\psi(\sigma_n) \leq \gamma\beta(\psi(\sigma_{n-1}))\phi(\sigma_{n-1}), for\ all\ n \in \mathbb{N}_0. \tag{18}$$

It is clear that $\gamma < 1$. Therefore, the sequence $\{\sigma(x_n, x_{n+1})\}$ is a decreasing sequence. Thus, there exists $r \geq 0$ such that

$$\lim_{n\to\infty} \sigma(x_n, x_{n+1}) = r.$$

Now, we show that $r = 0$. Presume to the contrary, that is $r > 0$. Since $\beta \in \mathcal{S}$ and by using the condition of Theorem 4 and taking the limit as $k \to \infty$ in Equation (18), we conclude

$$\psi(r) \leq \lambda\beta(\psi(r))\phi(r) < \lambda\phi(r) < \lambda\psi(r),$$

which could be a contradiction. So $r = 0$. Then,

$$\lim_{n\to\infty} \sigma(x_n, x_{n+1}) = 0.$$

Lemma 4 implies that $\{x_n\}$ is a Cauchy sequence and from the completeness of (X, σ), then there exists a $x^* \in X$ in order that

$$\lim_{n\to\infty} \sigma(x_n, x^*) = \sigma(x^*, x^*) = \lim_{n,m\to\infty} \sigma(x_n, x_m). \tag{19}$$

Whereas, S and T are continuous, we conclude

$$\lim_{n\to\infty} \sigma(x_{n+1}, Tx^*) = \lim_{n\to\infty} \sigma(Sx_n, Tx^*) = \sigma(Sx^*, Tx^*), \tag{20}$$

$$\lim_{n\to\infty} \sigma(Sx^*, x_{n+1}) = \lim_{n\to\infty} \sigma(Sx^*, Tx_n) = \sigma(Sx^*, Tx^*). \tag{21}$$

By Lemma 1 and Equation (19), we obtain that

$$\lim_{n\to\infty} \sigma(x_{n+1}, Tx^*) = \sigma(x^*, Tx^*) \tag{22}$$

and

$$\lim_{n\to\infty} \sigma(Sx^*, x_{n+1}) = \sigma(Sx^*, x^*). \tag{23}$$

By merging Equations (20) and (22), we deduce that $\sigma(x^*, Tx^*) = \sigma(Sx^*, x^*)$. In addition, by Equations (21) and (23), we deduce that $\sigma(Sx^*, x^*) = \sigma(Sx^*, Tx^*)$. So

$$\sigma(x^*, Tx^*) = \sigma(Sx^*, x^*) = \sigma(Sx^*, Tx^*). \tag{24}$$

Presently, we display that $\sigma(x^*, Tx^*) = 0$. Assume the opposite, that is, $\sigma(x^*, Tx^*) > 0$, we get

$$\begin{aligned} \psi(\sigma(x^*, Tx^*)) &= \psi(\sigma(Sx^*, Tx^*)) \\ &\leq \lambda\beta(\psi(M_{x^*,x^*}))\phi(M_{x^*,x^*}), \end{aligned} \tag{25}$$

where

$$\begin{aligned} M_{x^*,x^*} &= \max\{\sigma(x^*,x^*),\sigma(x^*,Sx^*),\sigma(x^*,Tx^*),\sigma(Sx^*,x^*),\sigma(x^*,Tx^*),\} \\ &= \max\{\sigma(x^*,Tx^*),\sigma(x^*,Sx^*)\} \\ &= \max\{\sigma(x^*,Tx^*),\sigma(x^*,Tx^*)\}. \end{aligned}$$

Therefore, from Equation (25), we get

$$\begin{aligned} \psi(\sigma(x^*,Tx^*)) &\le \beta(\psi(\sigma(x^*,Tx^*)))\phi(\sigma(x^*,Tx^*)) \\ &< \lambda\phi(\sigma(x^*,Tx^*) \\ &< \lambda\psi(\sigma(x^*,Tx^*)) \end{aligned} \tag{26}$$

Since $\psi \in \Psi$, we have $\sigma(x^*,Tx^*) < \lambda\sigma(x^*,Tx^*)$ which is a discrepancy. Thus, we have $\sigma(x^*,Tx^*) = 0$. Hence, $Tx^* = x^*$. From Equation (24), we deduce that $\sigma(x^*,Sx^*) = 0$. Therefore, $Sx^* = x^*$. Hence, x^* is a common fixed point of S and T. To demonstrate the uniqueness of the common fixed point, we suppose that $\bar{x}$ is another fixed point of S and T. Directly, we prove that $\sigma(\bar{x},\bar{x}) = 0$. Assume the antithesis, that is, $\sigma(\bar{x},\bar{x}) > 0$. Since $\bar{x} \preceq \bar{x}$, we get

$$\begin{aligned} \psi(\sigma(\bar{x},\bar{x})) &= \psi(\sigma(S\bar{x},T\bar{x})) \\ &\le \lambda\beta(\psi(\sigma(\bar{x},\bar{x})))\phi(\sigma(\bar{x},\bar{x})) \\ &< \lambda\phi(\sigma(\bar{x},\bar{x})) \\ &< \lambda\psi(\sigma(\bar{x},\bar{x})) \end{aligned}$$

which is a discrepancy. Thus, $\sigma(\bar{x},\bar{x}) = 0$. Therefore, by the further conditions on X, we deduce that x^* and $\bar{x}$ are comparable. Presently, suppose that $\sigma(x^*,\bar{x}) \neq 0$. Then

$$\begin{aligned} \psi(\sigma(x^*,\bar{x})) &= \psi(\sigma(Sx^*,T\bar{x})) \\ &\le \lambda(\psi(\sigma(x^*,\bar{x})))\phi(\sigma(x^*,\bar{x})) \\ &< \lambda\phi(\sigma(x^*,\bar{x})) \end{aligned}$$

which is a discrepancy with the condition of Theorem 4. Therefore, $\sigma(x^*,\bar{x}) = 0$. Hence, $x^* = \bar{x}$. Thus, S and T have a unique common fixed point. □

It is additionally conceivable to expel the continuity of S and T by exchanging a weaker condition. (C) If $\{x_n\}$ is a nondecreasing sequence in X such that $\alpha(x_n,x_{n+1}) \ge 1$ for all $n \in \mathbb{N}\cup\{0\}$ and $x_n \to u \in X$ as $n \to \infty$, then there exists a subsequence $\{x_{n_l}\}$ of $\{x_n\}$ such that $x_{n_l} \preceq u$ for all l.

Theorem 5. *Let (X,σ) be a partially ordered metric-like space. Assume that $S,T{:}X \to X$ are two self-mappings fulfilling the following conditions:*

1. *the pair (S,T) is triangular α-admissible,*
2. *there exists an $x_0 \in X$ such that $\alpha(x_0,Sx_0) \ge 1$,*
3. *the pair (S,T) is a generalized Geraghty (α,ψ,ϕ)-quasi contraction non-self mapping,*
4. *the pair (S,T) is weakly increasing,*
5. *(C) holds.*

Then, the pair (S,T) has a common fixed point $v \in X$ with $\sigma(v,v) = 0$. Moreover, suppose that if $x_1,x_2 \in X$ such $\sigma(x_1,x_1) = \sigma(x_2,x_2) = 0$ implies that x_1 and x_2 are comparable. Then, the common fixed point of the pair (S,T) is unique.

Proof. Here, we define $\{x_n\}$ as in the proof of Theorem 4. Clearly $\{x_n\}$ is a Cauchy sequence in X, then there exists $v \in X$ in order that

$$\lim_{n\to\infty} x_n = v \tag{27}$$

As a result of the condition of Equation (5), there exists a subsequence $\{x_{n_l}\}$ of $\{x_n\}$ in order that $x_{n_l} \preceq v$ for all l. Therefore, x_{n_l} and v are comparable. In addition, from Equation (13) on taking limit as $n \to \infty$ and using Equation (27), we get

$$\lim_{n\to\to\infty} x_n = v.$$

$$\lim_{n\to\to\infty} Sx_{2n_l} = \lim_{n\to\to\infty} x_{2n_l+1} = v, \lim_{n\to\to\infty} Tx_{2n_l+1} = \lim_{n\to\to\infty} x_{2n_l+2} = v. \tag{28}$$

From the definition of α yields that $\alpha(x_{n_l}, v) \geq 1$ for all l. Now by applying Equation (6), we have

$$\begin{aligned} & \psi(\sigma(x_{2n_l+1}, Tv)) \\ = & \psi(\sigma(Sx_{2n}, Tv)) \\ \leq & \lambda\beta(\psi(M_{x_{2n_l},v})\phi(M_{x_{2n_l},v}) \\ < & \lambda\phi(M_{x_{2n_l},v}) \\ < & \lambda\psi(M_{x_{2n_l},v}) \end{aligned} \tag{29}$$

where

$$M_{x_{2n_l},v} = \max\{\sigma(x_{2n}, v), \sigma(x_{2n}, Sx_{2n}), \sigma(v, Tv), \sigma(Sx_{2n}, v), \sigma(x_{2n}, Tv)\}$$

Letting $l \to +\infty$ and using Equations (27) and (28), we have

$$\lim_{l\to\infty} M_{x_{2n_l},v} = \max\{\sigma(v, Sv), \sigma(v, Tv)\} \tag{30}$$

Case **I**: Assume that $\lim_{l\to\infty} M_{x_{2n_l},v} = \sigma(v, Tv)$.
From Equation (30) and letting $l \to \infty$ in Equation (29). Then, we have

$$\psi(\sigma(v, Tv)) < \lambda\psi(\sigma(v, Tv)).$$

Regarding the concept of ψ, we deduce that $\sigma(v, Tv) < \lambda\sigma(v, Tv)$ which is a discrepancy. Hence, we get that $\sigma(v, Tv) = 0$. As a result of (σ_1), we have $v = Tv$.

Case **II**: Assume that $\lim_{l\to\infty} M_{x_{2n_l},v} = \sigma(v, Sv)$. Then, arguing like above, we get $v = Sv$. Thus, $v = Sv = Tv$. Uniqueness of the fixed point is follows from the Theorem 4. This completes the proof. □

If we set $S = T$ and $M(x, y) = \max\{\sigma(x, y), \sigma(x, Tx), \sigma(y, Ty), \sigma(Tx, y), \sigma(x, Ty)\}$ in Theorems 4 and 5, then we obtain the following corollaries.

Corollary 1. *Let (X, σ) be a partially ordered metric-like space and $\alpha: X \times X \to [0, \infty)$ a function. Assume that $S: X \to X$ holds the following:*

(1) *there exists $\psi \in \Psi, \beta \in \mathfrak{S}$ and a continuous function $\phi : [0, \infty) \to [0, \infty)$ are continuous functions with $\phi(t) < \psi(t)$ for all $t > 0$ such that*

$$\alpha(x, y)\psi(\sigma(Sx, Sy)) \leq \lambda\beta(\psi(M_{x,y}))\phi(M_{x,y}), \tag{31}$$

holds for all comparable elements $x, y \in X$ and $0 \leq \lambda < 1$,

(2) *S is triangular α-admissible and there exists an $x_0 \in X$ such that $\alpha(x_0, Sx_0) \geq 1$,*

(3) *$Sx \preceq S(Sx)$ for all $x, y \in X$,*

(4) *T is σ-continuous mappings.*

Then, S has an unique fixed point $v \in X$ with $\sigma(v,v) = 0$.

Corollary 2. *Let (X,σ) be a partially ordered metric-like space and $\alpha : X \times X \to [0,\infty)$ a function. Assume that $S : X \to X$ holds the following:*

(1) *there exists $\psi \in \Psi, \beta \in \mathfrak{S}$ and a continuous function $\phi : [0,\infty) \to [0,\infty)$ are continuous functions with $\phi(t) < \psi(t)$ for all $t > 0$ such that*

$$\alpha(x,y)\psi(\sigma(Sx,Sy)) \leq \lambda\beta(\psi(M_{x,y}))\phi(M_{x,y}), \tag{32}$$

holds for all comparable elements $x,y \in X$ and $0 \leq \lambda < 1$,

(2) *S is triangular α-admissible and there exists an $x_0 \in X$ such that $\alpha(x_0, Sx_0) \geq 1$,*

(3) *$Sx \preceq S(Sx)$ for all $x,y \in X$,*

(4) *$(\mathcal{C})$ holds.*

Then, S has an unique fixed point $v \in X$ with $\sigma(v,v) = 0$.

If we take $\alpha(x,y) = 1$ in Theorems 4 and 5, we have the following corollaries.

Corollary 3. *Let (X,σ) be a partially ordered metric-like space. Assume $S,T : X \to X$ are two mappings holding the following:*

(1) *there exists $\psi \in \Psi, \beta \in \mathfrak{S}$ and a continuous function $\phi : [0,\infty) \to [0,\infty)$ are continuous functions with $\phi(t) < \psi(t)$ for all $t > 0$ such that*

$$\psi(\sigma(Sx,Ty)) \leq \lambda\beta(\psi(M_{x,y}))\phi(M_{x,y}), \tag{33}$$

holds for all comparable elements $x,y \in X$ and $0 \leq \lambda < 1$, where

$$M_{x,y} = \max\{\sigma(x,y), \sigma(x,Sx), \sigma(y,Ty), \sigma(Sx,y), \sigma(x,Ty)\}.$$

(2) *the pair (S,T) is weakly increasing,*

(3) *S and T are σ-continuous mappings.*

Then, the pair S,T has an unique common fixed point $v \in X$ with $\sigma(v,v) = 0$.

Corollary 4. *Let (X,σ) be a partially ordered metric-like space, Assume $S,T : X \to X$ are two mappings holding the following:*

(1) *there exists $\psi \in \Psi, \beta \in \mathfrak{S}$ and a continuous function $\phi : [0,\infty) \to [0,\infty)$ are continuous functions with $\phi(t) < \psi(t)$ for all $t > 0$ such that*

$$\psi(\sigma(Tx,Ty)) \leq \lambda\beta(\psi(M_{x,y}))\phi(M_{x,y}), \tag{34}$$

holds for all comparable elements $x,y \in X$ and $0 \leq \lambda < 1$, where

$$M_{x,y} = \max\{\sigma(x,y), \sigma(x,Sx), \sigma(y,Ty), \sigma(Sx,y), \sigma(x,Ty)\},$$

(2) *the pair (S,T) is weakly increasing,*

(3) *the pair (S,T) is a generalized (α,ψ,ϕ)-quasi contraction non-self,*

(4) *$(\mathcal{C})$ holds.*

Then, the pair S,T has an unique common fixed point $v \in X$ with $\sigma(v,v) = 0$.

4. Consequences

If we put $M_{x,y} = \sigma(x,y)$, then, by Theorems 4 and 5, we get the following corollaries as an expansion of results from the literature.

Corollary 5. *Let (X,σ) be a partially ordered metric like space and $\alpha{:}X \times X \to [0,\infty)$ be a function. Suppose that $S,T{:}X \to X$ are two self-mappings holding the following:*

(1) *(S,T) is triangular α-admissible and there exists an $x_0 \in X$ such that $\alpha(x_0,Sx_0) \geq 1$,*
(2) *there exists $\psi \in \Psi, \beta \in \mathfrak{S}$ and a continuous function $\phi : [0,\infty) \to [0,\infty)$ are continuous functions with $\phi(t) < \psi(t)$ for all $t > 0$ in order that*

$$\psi(\sigma(Sx,Ty)) \leq \lambda\beta(\psi(\sigma(x,y))\phi(\sigma(x,y)), \tag{35}$$

satisfies for $x,y \in X$ and $0 \leq \lambda < 1$,
(3) *the pair (S,T) is weakly increasing,*
(4) *the pair (S,T) is σ-continuous mappings.*

Then, the pair (S,T) has an unique common fixed point $v \in X$ with $\sigma(v,v) = 0$.

Corollary 6. *Let (X,σ) be a partially ordered metric-like space. Assume $S,T{:}X \to X$ are two mappings holding the following:*

(1) *(S,T) is triangular α-admissible and there exists an $x_0 \in X$ such that $\alpha(x_0,Sx_0) \geq 1$,*
(2) *there exists $\psi \in \Psi, \beta \in \mathfrak{S}$ and a continuous function $\phi : [0,\infty) \to [0,\infty)$ are continuous functions with $\phi(t) \leq \psi(t)$ for all $t > 0$ in order that*

$$\psi(\sigma(Sx,Ty)) \leq \lambda\beta(\psi(\sigma(x,y))\phi(\sigma(x,y)), \tag{36}$$

satisfies for $x,y \in X$ and $0 \leq \lambda \leq 1$,
(3) *the pair (S,T) is weakly increasing,*
(4) *(C) holds.*

Then, the pair (S,T) has an unique common fixed point $v \in X$ with $\sigma(v,v) = 0$.

Corollary 7. *Let (X,σ) be a partially ordered metric-like space. Assume $\alpha{:}X \times X \to [0,\infty)$ is a function and $S{:}X \to X$ is a mapping holding the following:*

(1) *S is triangular α-admissible and there exists an $x_0 \in X$ such that $\alpha(x_0,Sx_0) \geq 1$.*
(2) *there exists $\psi \in \Psi, \beta \in \mathfrak{S}$ and a continuous function $\phi : [0,\infty) \to [0,\infty)$ are continuous functions with $\phi(t) < \psi(t)$ for all $t > 0$ in order that*

$$\alpha(x,y)\psi(\sigma(Sx,Sy)) \leq \lambda\beta(\psi(\sigma(x,y))\phi(\sigma(x,y)), \tag{37}$$

holds for all comparable elements $x,y \in X$ and $0 \leq \lambda < 1$,
(3) *$S \preceq S(Sx)$,*
(4) *the pair (S,T) is σ-continuous mappings.*

Then, S has an unique fixed point $v \in X$ with $\sigma(v,v) = 0$.

Corollary 8. *Let (X,σ) be a partially ordered metric-like space. Assume $\alpha{:}X \times X \to [0,\infty)$ is a function and $S{:}X \to X$ is a mapping holding the following:*

(1) *S is triangular α-admissible and there exists an $x_0 \in X$ such that $\alpha(x_0,Sx_0) \geq 1$,*

(2) *there exists* $\psi \in \Psi, \beta \in \mathfrak{S}$ *and a continuous function* $\phi : [0,\infty) \to [0,\infty)$ *are continuous functions with* $\phi(t) < \psi(t)$ *for all* $t > 0$ *in order that*

$$\alpha(x,y)\psi(\sigma(Sx,Sy)) \leq \lambda\beta(\psi(\sigma(x,y))\phi(\sigma(x,y)), \tag{38}$$

satisfies for $x, y \in X$ *and* $0 \leq \lambda < 1$,

(3) $S \preceq S(Sx)$,

(4) (C) *holds.*

Then S has an unique fixed point $v \in X$ *with* $\sigma(v,v) = 0$.

Example 3. *Let* $X = \{0,1,2\}$ *and specify the partial order* $\preceq$ *on X in order that*

$$\preceq := \{(0,0),(1,1),(2,2),(0,2),(2,1),(0,1)\}.$$

Take into consideration that the function $S : X \to X$ *specified as*

$$S = \begin{pmatrix} 0 & 1 & 2 \\ 1 & 1 & 0 \end{pmatrix}, \tag{39}$$

which increasing with respect to $\preceq$. *Let* $x_0 = 0$. *Hence,* $S(x_0) = 1$ *and* $S(S(X_0)) = S(1) = 1$. *Characterize to begin with the metric like space* σ *on X by* $\sigma(0,1) = 1, \sigma(0,2) = \frac{5}{2}, \sigma(1,2) = \frac{3}{2}$ *and* $\sigma(x,x) = 0$. *Then,* (X,σ) *is a complete metric-like space. Let* $\beta \in \mathcal{S}$ *is given by* $\beta(t) = \frac{e^t}{2}, \psi(t) = t, \lambda = \frac{1}{2}$ *and* $\phi(t) = \frac{2}{3}t$. *Define a function* α:$X \times X \to [0,\infty)$ *in order that*

$$\alpha(x,y) = \begin{cases} 1 \text{ if } x \in \{0,1,2\} \\ 0 \text{ if } \text{ otherwise.} \end{cases}$$

Note that $S \in X$ *and is continuous. S is* α*-admissible mapping. Indeed,* $\alpha(Sx,Sy) = 1$. *If* $(x,y) = (0,1)$, *then* $\alpha(0,1) = 1$ *and*

$$\begin{aligned} M_{0,1} &= \max\{\sigma(0,1),\sigma(0,S0),\sigma(1,S1),\sigma(S0,1),\sigma(0,S1)\} \\ &= \max\{\sigma(0,1),\sigma(0,1),\sigma(1,1),\sigma(1,1),\sigma(0,1)\} \\ &= \max\{1,1,0,1,0\} = 1. \end{aligned}$$

$\sigma(S0,S1) = \sigma(1,1) = 0$. *Now*

$$0 = \alpha(0,1)\psi(\sigma(\sigma(S0,S1))) \leq \beta(\psi(M_{0,1}))\phi(M_{0,1}) = \frac{1}{2}\beta(1) \simeq \phi(1) = \frac{1}{2} \times \frac{e}{2} \times \frac{2}{3} = \frac{e}{6}$$

holds.

If $(x,y) = (0,2)$, *then* $\alpha(0,2) = 1$ *and*

$$\begin{aligned} M_{0,2} &= \max\{\sigma(0,2),\sigma(0,S0),\sigma(2,S2),\sigma(S0,2),\sigma(0,S2)\} \\ &= \max\{\sigma(0,2),\sigma(0,1),\sigma(2,0),\sigma(1,2),\sigma(0,0)\} \\ &= \max\{\frac{5}{2},1,\frac{5}{2},\frac{3}{2},0\} = \frac{5}{2}. \end{aligned}$$

$\sigma(S0,S2) = \sigma(1,0) = \frac{5}{2}$. *Now*

$$\frac{5}{2} = \alpha(0,2)\psi(\sigma(\sigma(S0,S2))) \leq \beta(\psi(M_{0,2}))\phi(M_{0,2}) = \frac{1}{2}\beta(\frac{e^{\frac{5}{2}}}{2}) \times \frac{2}{3} \times \frac{5}{2} = \frac{5e^{\frac{5}{2}}}{12}$$

holds. Similarly, for the case $(x = 1, y = 2)$, *it is simple to examine that the contractive condition in Corollary 1 is satisfied.*

All conditions (1)–(4) of Corollary 1 are satisfied. Hence S has a unique fixed point $x = 1$.

5. Application

The aim of this section is to give the existence of fixed points of an integral equation, where we can apply the obtained result of Corollary 1 to get a common solution.

We consider X with the partial order $\preceq$ presented by:

$$x \preceq y \Leftrightarrow x(t) \preceq y(t) \ for\ all\ t \in [0,1].$$

Let $X = C(I,\mathbb{R})$ be the set of continuous functions specified on $I = [0,1]$. The metric-like space $\sigma : X \times X \to [0,\infty)$ presented by

$$\sigma(x,y) = \sup_{t\in[0,1]} \mid x(t) - y(t) \mid,$$

for all $x, y \in X$. Since (X,σ) is a complete metric-like space. We consider the integral equation

$$x(t) = g(t) + \int_0^1 P(t,r)f(r,x(r))dr;\ t \in [0,1] \tag{40}$$

for all $x \in X$.

We suppose that f:$[0,1] \times \mathbb{R} \to \mathbb{R}$ and g:$[0,1] \to \mathbb{R}$ are two continuous functions. Suppose that P:$[0,1] \times [0,1] \to [0,\infty)$ in order that

$$Sx(t) = g(t) + \int_0^1 P(t,r)f(r,x(r))dr;\ t \in [0,1] \tag{41}$$

for all $x \in X$. Then, a solution of Equation (40) is a fixed point of S.

Now, We will prove the following Theorem with our obtained results.

Theorem 6. *Assume that the following conditions are satisfied:*

(*i*) *There exists* ζ:$X \times X \to [0,1)$ *such that for all* $r \in [0,1]$ *and for all* $x, y \in X$

$$0 \leq \mid f(r,x(r)) - f(r,y(r)) \mid \leq \zeta(x,y) \mid x(r) - y(r) \mid,$$

(*ii*) *there exists* β:$[0,\infty) \to [0,1)$ *such that*

$$\lim_{n\to\infty} \beta(t_n) = 1 \Rightarrow \lim_{n\to\infty} t_n = 0,$$

and

$$\| \int_0^1 P(t,r)\zeta(x,y)dr \|_\infty \leq (\frac{1}{4}\beta(\| x - y \|_\infty)).$$

Then the integral Equation (41) has a unique solution in X.

Proof. By conditions (i) and (ii), we get

$$\begin{aligned}
\mid S(x)(t) - S(y)(t) \mid &= \left| \int_0^1 P(t,r)[f(r,x(r)) - f(r,x(r))]dr \right| \\
&\leq \int_0^1 P(t,r) \mid f(r,x(r)) - f(r,y(r)) \mid dr \\
&\leq \int_0^1 P(t,r)\zeta(x,y) \mid f(r,x(r)) - f(r,y(r)) \mid dr \\
&\leq \int_0^1 P(t,r)\zeta(x,y) \parallel x - y \parallel_\infty dr \\
&\leq \sigma(x,y) \int_0^1 P(t,r)\zeta(x,y)dr \\
&\leq \frac{1}{4}\beta(\sigma(x,y))\sigma(x,y) \\
&= \frac{1}{2}\beta(\sigma(x,y))\frac{1}{2}\sigma(x,y) \\
&= \frac{1}{2}\beta(\sigma(x,y))\phi(\sigma(x,y)).
\end{aligned}$$

At that point, we have

$$\parallel S(x)(t) - S(y)(t) \parallel_\infty \leq \frac{1}{2}\beta(\sigma(x,y))\phi(\sigma(x,y)).$$

for all $x, y \in X$.

Thus, we obtain

$$\sigma(Sx, Sy) \leq \frac{1}{2}\beta(\sigma(x,y))\phi(\sigma(x,y)),\ forall x, y \in X.$$

Lastly, we specify $\beta{:}X \times X \to [0,\infty)$ such that

$$\alpha(x,y) = \begin{cases} 1 & \text{if } x,y \in X, \\ 0 & \text{if } otherwise. \end{cases}$$

Then, we have

$$\alpha(x,y)\sigma(Sx,Sy) \leq \frac{1}{2}\beta(\sigma(x,y))\sigma(x,y).$$

Obviously, $\alpha(x,y) = 1$ and $\alpha(Sx,Sy) = 1$ for all $x,y,z \in X$. Therefore, S is triangular $\alpha-$admissible mapping.

Hence, the hypotheses of Corollary 1 hold with $\psi(t) = t, \lambda = \frac{1}{2}$ and $\phi(t) = \frac{t}{2}$. Thus, S has a unique fixed point, that is, the integral Equation (40) has a unique solution in X. □

6. Conclusions

We have introduced some common fixed point results for generalized (α,ψ,ϕ)-quasi contraction self-mapping in partially ordered metric-like spaces. We have generalized weakly contractive mapping as we used quasi contraction self-mapping, α-admissible mapping, triangular α-admissible mapping and ψ,ϕ as strictly increasing and continuous functions. We have provided an example and application to show the superiority of our results over corresponding (common) fixed point results. Alternatively, we suggest finding new results by replacing the single-valued mapping with multi-valued mapping. Furthermore, we suggest generalizing more results in other spaces like b-metric space, metric-like space, and others. Otherwise, we suggest using our main results for non-self-mapping to establish the existence of an optimal approximate solution.

Author Contributions: All authors contributed equally to the main text.

Acknowledgments: The author would like to thank the anonymous reviewers and editor for their valuable comments.

References

1. Qawaqneh, H.; Noorani, M.S.M.; Shatanawi, W.; Alsamir, H. Common fixed points for pairs of triangular α-admissible mappings. *J. Nonlinear Sci. Appl.* **2017**, *10*, 6192–6204. [CrossRef]
2. Qawaqneh, H.; Noorani, M.S.M.; Shatanawi, W.; Abodayeh, K.; Alsamir, H. Fixed point for mappings under contractive condition based on simulation functions and cyclic (α, β)-admissibility. *J. Math. Anal.* **2018**, *9*, 38–59. [CrossRef]
3. Qawaqneh, H.; Noorani, M.S.M.; Shatanawi, W. Fixed point for mappings under contractive condition based on simulation functions and cyclic (α, β)-admissibility. *J. Math. Anal.* **2018**, *11*, 702–716. [CrossRef]
4. Hitzler, P.; Seda, A.K. Dislocated topologies. *J. Electr. Eng.* **2000**, *12*, 4223–4229.
5. Amini-Harandi, A. Metric-like spaces, partial metric spaces and fixed points. *Fixed Point Theory Appl.* **2012**, *2012*, 204. [CrossRef]
6. Chen, C.; Dong, J.; Zhu, C. Some fixed point theorems in b-metric-like spaces. *Fixed Point Theory Appl.* **2015**, *2015*. [CrossRef]
7. Roshan, J.R.; Parvaneh, V.; Sedghi, S.; Shbkolaei, N.; Shatanawi, W. Common fixed points of almost generalized contractive mappings in ordered b-metric spaces. *Fixed Point Theory Appl.* **2013**, *2013*, 159. [CrossRef]
8. Zoto, K. Some generalizations for (α, ψ, ϕ)-contractions in b-metric-like spaces and an application. *Fixed Point Theory Appl.* **2017**, *2017*, 26. [CrossRef]
9. Aydi, H.; Felhi, A. Best proximity points for cyclic Kannan-Chatterjea- Ciric non-self contractions on metric-like spaces. *J. Nonlinear Sci. Appl.* **2016**, *9*, 2458–2466. [CrossRef]
10. Aydi, H.; Felhi, A.; Sahmim, S. On common fixed points for (α, ψ)-contractions and generalized cyclic contractions in b-metric-like spaces and consequences. *J. Nonlinear Sci. Appl.* **2016**, *9*, 2492–2510. [CrossRef]
11. Aydi, H.; Felhi, A. On best proximity points for various α-proximal contractions on metric-like spaces. *J. Nonlinear Sci. Appl.* **2016**, *9*, 5202–5218. [CrossRef]
12. Aydi, H.; Felhi, A.; Afshari, H. New Geraghty non-self contractions on metric-like spaces. *J. Nonlinear Sci. Appl.* **2017**, *10*, 780–788. [CrossRef]
13. Aydi, H.; Felhi, A.; Sahmim, S. Common fixed points via implicit contractions on b-metric-like spaces. *J. Nonlinear Sci. Appl.* **2017**, *10*, 1524–1537. [CrossRef]
14. Alsamir, H; Noorani, M.S.M.; Shatanawi, W. On fixed points of (η, θ)-quasicontraction mappings in generalized metric spaces. *J. Nonlinear Sci. Appl.* **2016**, *9*, 4651–4658. [CrossRef]
15. Shatanawi, W.; Nashine, H.K. A generalization of Banachs contraction principle for nonlinear contraction in a partial metric space. *J. Nonlinear Sci. Appl.* **2012**, *5*, 37–43. [CrossRef]
16. Aydi, H.; Shatanawi, W.; Vetro, C. On generalized weak G-contraction mapping in G-metric spaces. *Comput. Math. Appl.* **2011**, *62*, 4223–4229. [CrossRef]
17. Cho, S. Fixed point theorems for generalized weakly contractive mappings in metric spaces with application. *Fixed Point Theory Appl.* **2018**, *2018*, 3. [CrossRef]
18. Zoto, K.; Radenovic, S.; Ansari, A.H. On some fixed point results for (s, p, α)-contractive mappings in b-metric-like spaces and applications to integral equations. *Open Math.* **2018**, *16*, 235–249. [CrossRef]
19. Mohammadi, B.; Golkarmanesh, F.; Parvaneh, V. Common fixed point results via implicit contractions for multi-valued mappings on b-metric like spaces. *Cogent Math. Stat.* **2018**, *5*, 1493761. [CrossRef]
20. Karapinar, E.; Czerwik, S.; Aydi, H. Meir-Keeler Contraction Mappings in Generalized-Metric Spaces. *J. Funct. Spaces* **2018**, *2018*, 3264620. [CrossRef]
21. Liang, M.; Zhu, C.; Wu, Z.; Chen, C. Some New Coupled Coincidence Point and Coupled Fixed Point Results in Partially Ordered Metric-Like Spaces and an Application. *J. Funct. Spaces* **2018**, *2018*, 1378379. [CrossRef]
22. Hussain, N.; Roshan, J.R.; Parvaneh, V.; Kadelburg, Z. Fixed Points of Contractive Mappings in-Metric-Like Spaces. *Sci. World J.* **2014**, *2014*, 471827. [CrossRef] [PubMed]
23. Geraghty, M. On contractive mappings. *Proc. Am. Math. Soc.* **1973**, *40*, 604–608. [CrossRef]
24. Karapinar, E. A Discussion on α-ψ-Geraghty Contraction Non-self Mappings. *Filomat* **2014**, *28*, 761–766. [CrossRef]

25. Karapinar, E. α-ψ-Geraghty Contraction Non-self Mappings and Some Related Fixed Point Results. *Filomat* **2014**, *28*, 37–48. [CrossRef]
26. Karapinar, E.; Pitea, A. On alpha-psi-Geraghty contraction type mappings on quasi-Branciari metric spaces. *J. Nonlinear Convex Anal.* **2016**, *17*, 1291–1301.
27. Amini-Harandi, A.; Emami, H. A fixed point theorem for contraction non-self maps in partially ordered metric spaces and application to ordinary differential equations. *Nonlinear Anal.* **2010**, *72*, 2238–2242. [CrossRef]
28. karapinar, E.; Alsulami, H.; Noorwali, M. Some extensions for Geragthy non-self contractive mappings. *Fixed Point Theory Appl.* **2015**, *2015*, 303. [CrossRef]
29. Ćiriċ, L. A generalization of Banach contraction principle. *Proc. Am. Math. Soc.* **1974**, *45*, 267–273. [CrossRef]
30. Isik, H.; Turkoglu, D. Fixed point theorems for weakly contractive mappings in partially ordered metric-like spaces. *Fixed Point Theory Appl.* **2013**, *2013*, 51. [CrossRef]
31. Samet, B.; Vetro, C.; Vetro, P. Fixed point theorems for a α-ψ-contractive non-self mappings. *Nonlinear Anal.* **2012**, *75*, 2154–2165. [CrossRef]
32. Abdeljawad, T. Meir-Keeler a-contractive fixed and common fixed point theorems. *Fixed Point Theory Appl.* **2013**, *2013*, 19. [CrossRef]
33. Karapinar, E.; Kumam, P.; Salimi, P. On α-ψ-Meir-Keeler contractive mappings. *Fixed Point Theory Appl.* **2013**, *2013*, 94. [CrossRef]
34. Alsamir, H.; Noorani, M.S.M.; Shatanawi, W.; Abodayeh, K. Common Fixed Point Results for Generalized (ψ, β)-Geraghty Contraction Non-self Mapping in Partially Ordered Metric-like Spaces with Application. *Filomat* **2017**, *31*, 5497–5509. [CrossRef]

14

Some Exact Solutions to Non-Fourier Heat Equations with Substantial Derivative

Konstantin Zhukovsky *, Dmitrii Oskolkov and Nadezhda Gubina

Faculty of Physics, M.V. Lomonosov Moscow State University, Leninskie Gory, 119991 Moscow, Russia; oskolkov.di15@physics.msu.ru (D.O.); gubina_nadya@mail.ru (N.G.)
* Correspondence: zhukovsk@physics.msu.ru;

Abstract: One-dimensional equations of telegrapher's-type (TE) and Guyer–Krumhansl-type (GK-type) with substantial derivative considered and operational solutions to them are given. The role of the exponential differential operators is discussed. The examples of their action on some initial functions are explored. Proper solutions are constructed in the integral form and some examples are studied with solutions in elementary functions. A system of hyperbolic-type inhomogeneous differential equations (DE), describing non-Fourier heat transfer with substantial derivative thin films, is considered. Exact harmonic solutions to these equations are obtained for the Cauchy and the Dirichlet conditions. The application to the ballistic heat transport in thin films is studied; the ballistic properties are accounted for by the Knudsen number. Two-speed heat propagation process is demonstrated—fast evolution of the ballistic quasi-temperature component in low-dimensional systems is elucidated and compared with slow diffusive heat-exchange process. The comparative analysis of the obtained solutions is performed.

Keywords: exponential operator; differential operator; Guyer–Krumhansl equation; moving media; non–Fourier; heat conduction; Knudsen number

---

1. Introduction

Recent progress in technology and science has driven interest to studies of heat conduction beyond common Fourier law [1]: $\partial_t T = k\partial_x^2 T$, where k is the thermal diffusivity, T is the temperature; Fourier law describes heat conduction in homogeneous matter at normal conditions well. New heat sources, such as lasers, microwaves etc, are employed in medicine, science and material processing for melting, welding, cutting, drilling, etc. Often heat source and treated media are in motion. Some of modern materials and media have one or two dimensions: ultra-thin films, layers and nano-wires. Highly inhomogeneous porous and multilayered media are also common in industry. First major deviations from Fourier law were found in liquid Helium and in some solid crystal dielectrics at low temperatures <25 ° K [2–5]. Proper phenomenon was called Second Sound [6]; its satisfactory qualitative mathematical description was proposed by Cattaneo and Vernotte [7], who supposed phonon heat transport in addition to Fourier heat diffusion. In terms of temperature it can be put as follows:

$$(\tau\partial_t^2 + \partial_t)T = k_F \nabla^2 T \quad (1)$$

where k_F is the Fourier thermal diffusivity, $\tau = k_F/C^2$ is the relaxation time of the heat waves propagation, which relates the moments of the temperature change and of the respective heat flux change. Cattaneo-Vernotte constitution implies a phase lag between the heat flux vector and the temperature gradient; in addition to heat diffusion the temperature perturbation propagates in matter like damped sound-wave at finite speed $C = \sqrt{k_F/\tau}$. The relaxation time τ is associated with th phonon–phonon interaction time; at normal conditions it is very small: $\tau \approx 10^{-13}$ s. The Cattaneo Equation (1) is the particular case of the telegrapher's equation (TE).

$$\left(\partial_t^2 + \varepsilon\partial_t\right)F(x,t) = \left(\alpha\partial_x^2 + \kappa\right)F(x,t),\ \varepsilon,\ \alpha,\ \kappa = \text{const} \tag{2}$$

which takes its name because it describes the electric signal propagation in long electric lines without radiation [8]. For heat conduction we have $\tau = 1/\varepsilon$, $k_F = \alpha/\varepsilon$, $\mu = \kappa/\varepsilon$ and $\mu = 0$ in Equation (1); the source term, $\kappa \neq 0$, describes heat exchange with the environment of low excess temperatures. However, precise quantitative description of the Second Sound with Equation (1) was not successful. There are reports on non-Fourier heat transport in highly inhomogeneous matter even at normal conditions [9–17], in fuel droplets [18], in biomaterials [19], in energy saving and insulating materials [20], in graphene, nanofibers, carbon nanotubes, silicon wires, etc. [21–27]. Heat transport in these cases was found close to that described by Guyer and Krumhansl (GK) in [28,29]. In the one-dimensional case in terms of temperature alone it has the following form:

$$\left(\partial_t^2 + \varepsilon\partial_t - \delta\partial_{t,x,x}^3\right)T(x,t) = \left(\alpha\partial_x^2 + \kappa\right)T(x,t),\ \alpha,\ \varepsilon,\ \delta,\ \kappa = \text{const.} \tag{3}$$

In its pure form the Guyer–Krumhansl law has $\kappa = 0$. The following formulation:

$$\left(\tau\partial_t^2 + \partial_t\right)T(x,t) = \left(D_b\partial_{t,x,x}^3 + k_F\partial_x^2 + \mu\right)T(x,t), \tag{4}$$

involves the parameters $\tau = 1/\varepsilon$, $\mu = \kappa/\varepsilon$, $k_F = \alpha/\varepsilon$, and $D_b = \delta/\varepsilon$, where the latter, k_b, is the ballistic type heat conductivity. The parameters in the above Equations (3) and (4) have the following dimensions: $[\tau] = s$, $[k_F] = \frac{m^2}{s}$, $[D_b] = m^2$, $[\mu] = \frac{1}{s}$, $[\kappa] = \frac{1}{s^2}$, $[\alpha] = \frac{m^2}{s^2}$, $[\delta] = \frac{m^2}{s}$, and $[\varepsilon] = \frac{1}{s}$.

Despite GK equation is usually associated with ballistic properties, which manifest when the characteristic system scale L is comparable or less than the mean free path l of the phonons $L < l$, contradictory opinions on it exist [30–32]. In particular, the term "ballistic" is questioned in the context of GK type heat conduction in macroscopic inhomogeneous materials at room temperature [10,11]. Moreover, more complicated system of equations was proposed for the ballistic heat transport in [30], while heat propagation in highly inhomogeneous materials was reported to be close to GK law [9,11].

Solutions to Hyperbolic Heat Conduction Equation (HHCE) can be obtained both analytically and numerically [33–40], although numerical methods seem to be more commonly used [41–44]. Analytical study gives deeper insight in the problem; operational analytical approach and solutions to HHE were developed in [45–50]. This method easily handles also other linear DE of high order and fractional DE [51–58]. Use of the exponential differential operators, such as the heat operator $S = e^{\partial_x^2}$ [59] allows operational solution of GK-type Equation (4) as demonstrated in [47–50].

In what follows we will obtain some exact solutions to non-Fourier heat transfer in GK-type equation with the substantial derivative, where the speed of the media is constant, and will obtain some particular operational solutions to modified GK-type equation in this case. Moreover, we will study the important case of periodic initial conditions, which occur, for example, in polymer electrolyte fuel cells (PEFCs) [60–68], as well as in thin membranes and highly inhomogeneous porous periodic structures, in printed wired heating boards [69] etc. We will analyze the solutions for the heat transport, described by the system of inhomogeneous partial derivative equation, and use the Knudsen number to account for the ballistic conditions in 1- and 2-dimensional structures [70].

2. Ballistic Heat Transport Equations with Substantial Derivative

For ultra-thin films and wires neither pure GK nor Cattaneo laws describe exactly heat transport phenomena, but their combination with Fourier law; it forms the following inhomogeneous system of differential equations for the ballistic $\theta_b(x,t)$ and diffusive component $\theta_d(x,t)$ of the complete dimensionless quasi-temperature $\theta = \theta_d + \theta_b$ (see [30]):

$$\left\{\partial_t^2 + 2\partial_t - \frac{10Kn_b^2}{3}\partial_x^2 - 3Kn_b^2\partial_{t,x,x}^3 + 1\right\}\theta_b(x,t) = 0, \tag{5}$$

$$\left(\partial_t^2 + \frac{Kn_b^2}{Kn_d^2}\partial_t - \frac{Kn_b^4}{3Kn_d^2}\partial_x^2\right)\theta_d(x,t) = \left(\partial_t + \frac{Kn_b^2}{Kn_d^2}\right)\theta_b(x,t), \tag{6}$$

where indices b and d stand for "ballistic" and "diffusive", respectively. The dimensionless quasi-temperature θ can be described as the non-dimensional energy, associated with the internal energy; the quantities θ_d and θ_b must therefore be understood to quasi-temperatures, defined as a measure of corresponding internal energy components u_d and u_b, respectively to which they are related by the simple relations $\theta_d = u_d/c$ $\theta_b = u_b/c$ (see [30] for details), where c is the heat capacity.

It should be understood though, that the ballistic heat transport is effectively described by both contributions in $\theta = \theta_d + \theta_b$, $\theta_d(x,t)$ and $\theta_b(x,t)$ are distinguished for convenience. It is easy to see that the above Equations (5) and (6) are of GK-type (5) and of telegrapher's type (6); the inhomogeneous right hand side (r.h.s.) of Equation (6) is given by the solution of (5). In the above equations and in all following equations the space- and time-coordinates are obviously dimensionless as well as the proper coefficients in the equations are. The dimensionless form of equations for the temperature is useful at least because of the space-time temperature distribution is expressed in our work explicitly in terms of x and t. Proper renormalization in SI units or else how is elementary (see, for example, [30] for details) and good, especially in view of heavy notations in the analytical solutions, which we will obtain in what follows.

Low-dimensional system is characterized by the dimensionless Knudsen number $Kn = l/L$, which for the phonon heat transport describes the ratio of the free mean path of the phonons l to the characteristic scale of the system L. Knudsen number usually arises in problems, where the scale of the system and the characteristic scale of the processes in it compare with each other, such as for gas flow in ultra-narrow channels [71], etc. In distinct ballistic case in (5), $Kn = 1$, we have $\alpha = 3.333$, $\delta = 3$, $\varepsilon = 2$, $\kappa = -1$ in (3) and $k_F = 5/3$, $D_b = 3/2$, $\tau = 1/2$, $\mu = -1/2$ in (4). We assume dimensionless equations here and in what follows. In the weak ballistic case, $Kn = 0.1$, we get $\alpha = 0.03333$, $\delta = 0.03$, $\varepsilon = 2$, $\kappa = -1$ in (3) and $k_F = 0.01666$, $k_b = 0.015$, $\tau = 1/2$, $\mu = -1/2$ in (4). In the distinct ballistic case, $Kn = 1$, all heat transport terms in GK-type equation contribute more or less equally, in the weak ballistic case, $Kn = 0.1$, the Cattaneo wave-term prevails. The inhomogeneous system of PDE (5) and (6) for ballistic heat transport thin film was studied numerically in [30]; based on the periodic analytical solutions to GK-type equation [72] some solutions to (5) and (6) were obtained in [73,74].

Above Equations (5) and (6), as well as Equations (2)–(4), describe non-Fourier heat transport in stationary media. In the case when the observer moves relatively to the media with constant speed $\vec{v}$, the substantial derivative $D/Dt = \partial/\partial t + \vec{v}\vec{\nabla}$ should be used instead of the time-derivative $\partial/\partial t$ to describe the variation of quantity F along a path. Generally speaking, both the material equations and the energy, mass and momentum conservation equations in case of moving observer should be considered with account for the substantial derivative, resulting in a proper heat conduction equation for the temperature alone. This is relatively simple in the case of Fourier heat diffusion and even for relativistic heat equation [75,76], some more complicated for GK-type equation. While not intending to rigorously derive a set of equations for the heat transport in case of moving observer, accounting for the relativistic heat flux, the energy balance, etc., we will solve the extended form of the known GK-type equation with the simple substantial derivative. Derivation of a physically comprehensive one-dimensional analogue of GK-equation in case of moving observer remains beyond the scope of the present study. The following modified form of GK-type equation arises upon the use of the substantial time derivative:

$$\left(\partial_t^2 + \varepsilon\partial_t + 2v\partial_{t,x}^2 - \delta\partial_{t,x,x}^3\right)T(x,t) = \left(\left(\alpha - v^2\right)\partial_x^2 - \varepsilon v\partial_x + \delta v\partial_x^3 + \kappa\right)T(x,t). \tag{7}$$

This simply presumes that the temperature as a function of time is recorded by a floating instrument in a flow, such as a weather balloon in meteorology and oceanography, which implies the substantial derivative along the pathline traveled. With this said, we notice that the above Equation (7) differs from GK-type Equation (4) by some additional terms: the second mixed time-space derivative in the left hand side (l.h.s.) and the third- and first-order space-derivatives in the r.h.s. We need to emphasize that this equation is related to a rigid domain meanwhile the observer is moving with speed $\vec{v}$. Moreover, the coefficient for the second-order space-derivative in the r.h.s. now contains the speed v, which can compensate α, eliminating the second-order space-derivative, if $v = \sqrt{\alpha}$, or invert the sign of this term, if $v > \sqrt{\alpha}$. Telegrapher's equation accordingly modifies as follows:

$$\left(\partial_t^2 + \varepsilon\partial_t + 2v\partial_{t,x}^2\right)T(x,t) = \left(\left(\alpha - v^2\right)\partial_x^2 - \varepsilon v\partial_x + \kappa\right)T(x,t). \tag{8}$$

Differently from GK-type equation with substantial derivative (7), TE (8) has only second-order differential operators; both GK-type and TE can be solved by the operational method, employing exponential differential operators $e^{\hat{D}}$ and involving the heat operator $\hat{S} = e^{\partial_x^2}$ [59]. The operational solution to the third-order PDE (7) apparently involves exponential differential operator of the third order $e^{a\partial_x^3}$. However, in what follows we will demonstrate as this solution effectively reduces to the shift of the solution in a stationary media. Considering thin films, we have obvious relations between the coefficients in (7) and (5): $\alpha_b \leftrightarrow 10Kn_b^2/3$, $\varepsilon_b \leftrightarrow 2$, $\delta \leftrightarrow 3Kn_b^2$, $\kappa_b \leftrightarrow -1$, while for Equation (6) we see that $\alpha_d \leftrightarrow Kn_b^4/3Kn_d^2$, $\varepsilon_d \leftrightarrow Kn_b^2/Kn_d^2$. For heat transport in moving thin film, the system of DE (5) and (6) for the complete quasi-temperature $\theta = \theta_d + \theta_b$ takes the following form:

$$\left(\partial_t^2 + 2\partial_t + 2v\partial_{t,x}^2 - \left(\frac{10}{3}Kn_b^2 - v^2\right)\partial_x^2 - 3Kn_b^2\partial_{t,x,x}^3 + 2v\partial_x - 3vKn_b^2\partial_x^3 + 1\right)\theta_b(x,t) = 0, \tag{9}$$

$$\left(\partial_t^2 + \frac{Kn_b^2}{Kn_d^2}\partial_t + 2v\partial_{t,x}^2 - \left(\frac{Kn_b^4}{3Kn_d^2} - v^2\right)\partial_x^2 + v\frac{Kn_b^2}{Kn_d^2}\partial_x\right)\theta_d(x,t) = \left(\partial_t + v\partial_x + \frac{Kn_b^2}{Kn_d^2}\right)\theta_b(x,t). \tag{10}$$

In what follows we will approach the GK-type equation with substantial derivative (7) operationally, and will also provide exact analytical harmonic solution to the system of PDE (9) and (10), describing ballistic heat transfer with substantial derivative.

3. Operational Approach to Transport Equations

The above Equations (7) and (8) are the particular cases of the following DE with the coordinate-dependent operators $\hat{\varepsilon}(x)$ and $\hat{D}(x)$:

$$\left(\partial_t^2 + \hat{\varepsilon}(x)\partial_t\right)F(x,t) = \hat{D}(x)F(x,t), \tag{11}$$

which can be solved using the operational method. Equation (11) models a broad spectrum of physical phenomena. It becomes telegrapher's equation for $\hat{D}(x) = \alpha\partial_x^2 + \kappa$ and constant ε term and further reduces to Cattaneo heat equation for $\hat{D}(x) = \partial_x^2$; GK-type equation appears when $\hat{\varepsilon} = \varepsilon - \delta\partial_x^2$. Some of these equations were studied in [45–50]. Other second-order PDE and fractional DE were explored with the help of the operational approach in [51–58]. The formal converging particular operational solution to (11) reads as follows:

$$F(x,t) = e^{-\frac{t\hat{\varepsilon}(x)}{2}} e^{-\frac{t}{2}\sqrt{\hat{\varepsilon}^2(x) + 4\hat{D}(x)}} C(x), \tag{12}$$

where $C(x)$ can be obtained from the initial condition $F(x,0) = f(x)$. The particular form of the initial function will be chosen below. The other branch of the solution contains the positive argument in the exponential, $e^{\frac{t}{2}\sqrt{\hat{\varepsilon}^2(x)+4\hat{D}(x)}}$. No Laplace transforms exist for it. However, symmetry with respect to inversion, $t \to -t$, $\varepsilon \to -\varepsilon$, $\delta \to -\delta$, $v \to -v$, allows writing the other solution to Equation (7), based on the one we obtain with the help of the Laplace transforms

$$e^{-t\sqrt{V}} = \frac{t}{2\sqrt{\pi}} \int_0^\infty \frac{d\xi}{\xi\sqrt{\xi}} e^{-\frac{t^2}{4\xi}-\xi V}, \; t > 0. \tag{13}$$

The solution has the following integral form, provided it converges:

$$F(x,t) = e^{-\frac{t}{2}\hat{\varepsilon}(x)} \frac{t}{4\sqrt{\pi}} \int_0^\infty \frac{d\xi}{\xi\sqrt{\xi}} e^{-\frac{t^2}{16\xi}} e^{-\xi\hat{\varepsilon}^2(x)} e^{-4\xi\hat{D}(x)} f(x). \tag{14}$$

The ability to perform analytical integration in (14) depends on the explicit form of the operators $\hat{\varepsilon}$, $\hat{D}$ and of the initial function $f(x)$; the numerical calculation can be done though. Accounting for the explicit form of the operators $\hat{D} = (\alpha - v^2)\partial_x^2 - \varepsilon v \partial_x + \delta v \partial_x^3 + \kappa$ and $\hat{\varepsilon} = \varepsilon + 2v\partial_x - \delta\partial_x^2$ in Equation (7), we obtain from (14) the following integral:

$$F(x,t) = e^{-\frac{t}{2}(\varepsilon+2v\partial_x)} \frac{t}{4\sqrt{\pi}} \int_0^\infty \frac{d\xi}{\xi\sqrt{\xi}} e^{-\frac{t^2}{16\xi}-\xi(\varepsilon^2+4\kappa)} e^{(\frac{t}{2}\delta+2\xi(\varepsilon\delta-2\alpha))\partial_x^2} e^{-\xi\delta^2\partial_x^4} f(x), \tag{15}$$

which benefits from the use of the operational identity for $\hat{p} = \sqrt{a}\hat{D}$, $\hat{D} = \partial^2$ (see [59,77]):

$$e^{\hat{p}^2} = \frac{1}{\sqrt{\pi}} \int_{-\infty}^\infty \exp\left(-\xi^2 + 2\xi\hat{p}\right) d\xi, \tag{16}$$

and yields

$$e^{a\hat{D}^2} f(x) = \frac{1}{\sqrt{\pi}} \int_{-\infty}^\infty \exp(-\xi^2 + 2\xi\sqrt{a}\hat{D}) f(x) d\xi. \tag{17}$$

Applying formula (17) to the fourth-order exponential differential operator $e^{-\xi\delta^2\partial_x^4} f(x) = \int_{-\infty}^\infty e^{-\zeta^2+2i\zeta\delta\sqrt{\xi}\partial_x^2} f(x) d\zeta/\sqrt{\pi}$, we get the heat operator [59] $\hat{S} = e^{\nu\partial_x^2}$; collecting the second-order derivative terms in the exponential, we get $\int_{-\infty}^\infty e^{-\zeta^2+((t\delta/2)-4\xi\alpha+2\xi\varepsilon\delta+2i\zeta\delta\sqrt{\xi})\partial_x^2} f(x) d\zeta$. This yields the following particular bounded solution to GK-type heat equation with substantial derivative (7):

$$F(x,t) = \frac{e^{-\frac{t}{2}\varepsilon t}}{4\pi} \hat{\Theta} \int_0^\infty \frac{d\xi}{\xi\sqrt{\xi}} e^{-\frac{t^2}{16\xi}-\xi(\varepsilon^2+4\kappa)} \int_{-\infty}^\infty e^{-\zeta^2} \hat{S} f(x) d\zeta, \tag{18}$$

which involves the shift operator $\hat{\Theta} = e^{y\partial_x}$, where $y = -vt$, and the heat operator $\hat{S} = e^{\eta\partial_x^2}$, where $\eta = 2i\zeta\sqrt{\xi}\delta - \xi 2(2\alpha + \varepsilon\delta) + t\delta/2$. The shift operator produces the translation along x $e^{y(\partial_x+\alpha)} f(x) = e^{y\alpha} f(x+y)$ and the heat operator produces Gauss transforms (17) and operational relations [59], so that the solution with account for the motion contains the translation $x - vt$ of the stationary solution.

Let us consider the initial polynomial function $f(x) = \sum_k x^k$. The particular solution (18) to GK-type equation with substantial derivative (7) arises upon the application of the operational rule

$$\hat{S} e^{\gamma x} x^k = e^{\eta\partial_x^2} x^k e^{\gamma x} = e^{\gamma x + \gamma^2 \nu} H_k(x + 2\gamma\nu, \nu), \; k \in \text{Integers}, \; \gamma \in \text{Reals}, \tag{19}$$

which is easy to prove [57,58]. For $\gamma = 0$ we immediately obtain the following integral:

$$F(x,t)|_{f(x)=\sum_k x^k} = \sum_k \frac{te^{-\frac{t}{2}\varepsilon}}{4\pi} \int_0^\infty \frac{d\xi}{\xi\sqrt{\xi}} e^{-\frac{t^2}{16\xi}-\xi(\varepsilon^2+4\kappa)} \int_{-\infty}^{\infty} e^{-\zeta^2} H_k(x-vt,\eta) d\zeta, \tag{20}$$

where

$$\eta = 2\xi\varepsilon\delta - 4\xi\alpha + t\delta/2 + i\zeta 2\sqrt{\xi}\delta. \tag{21}$$

The above integral can be taken in elementary functions if we account for the explicit form of the Hermite polynomials:

$$H_k(x-vt,\eta) = k! \sum_{r=0}^{[k/2]} \frac{(x-vt)^{k-2r}\eta^r}{(k-2r)!r!} = (-i)^k \eta^{\frac{k}{2}} H_k\left(\frac{i(x-vt)}{2\sqrt{\eta}}\right). \tag{22}$$

For example, for $f(x) = x^2$ the solution reads as follows:

$$F(x,t)|_{f(x)=x^2} = e^{-\frac{t}{2}(\sqrt{V}+\varepsilon)} \left((x-vt)^2 + t\delta + \frac{t}{\sqrt{V}}(\delta\varepsilon - 2\alpha)\right), \; V = \varepsilon^2 + 4\kappa^2. \tag{23}$$

More general case of the exponential-polynomial function $f(x) = x^k e^{\gamma x}$ is cumbersome and we omit proper expressions for conciseness. However, we have performed the integration explicitly in Wolfram Mathematica program and below we give the example of the solution for $F(x,0) = x^2 e^{-x}$:

$$\begin{aligned}
&F(x,t)|_{f(x)=x^2e^{-x}} = e^{-x-\frac{t}{2}(\sqrt{r}-2v-\delta+\varepsilon)}\left(x^2 + \tfrac{t}{r^{3/2}}(a+bx) + \tfrac{t^2}{r^{3/2}}c\right),\\
&r = 4\alpha + \delta^2 - 2\delta\varepsilon + \varepsilon^2 + 4\kappa,\\
&a = \sqrt{r}\delta\left(4\alpha + (\varepsilon-\delta)^2\right) + \delta^2(-6\alpha - \delta^2 - 3\varepsilon(\varepsilon-\delta)) +\\
&\varepsilon^2(-2\alpha + \delta\varepsilon) - 4\kappa(2\alpha + \sqrt{r}\delta - 3\delta^2 + \delta\varepsilon),\\
&b = 2r(2\alpha + \delta^2 - \sqrt{r}v - \delta(\sqrt{r}+\varepsilon)),\\
&c = r^{3/2}v^2 - 2rvd + 2p + \delta^2\Delta\\
&d = 2\alpha + \delta(-\sqrt{r} + \delta - \varepsilon),\\
&p = 2\alpha^2(\sqrt{r} - 4\delta) - 2\alpha\delta(3\delta^2 + \sqrt{r}\varepsilon + \varepsilon^2 - 2\delta(\sqrt{r}+2\varepsilon) + 4\kappa),\\
&\Delta = \sqrt{r}(\varepsilon-\delta)^2 + (\varepsilon-\delta)^3 + 2\kappa(\sqrt{r} + 2(\varepsilon-\delta)).
\end{aligned} \tag{24}$$

Evidently, all the values in the above expressions are dimensionless as well as the coordinate x and the time t.

In the harmonic ansatz the evolution of the initial function $f(x) = \exp(inx)$ can be easily obtained from (18) as follows:

$$F(x,t)|_{f(x)=e^{inx}} = \exp\left(in(x-vt) - \frac{t}{2}\left(\bar{\varepsilon} + \sqrt{\bar{\varepsilon}^2 + 4(\kappa^2 - \alpha n^2)}\right)\right), \; \bar{\varepsilon} = \varepsilon + n^2\delta. \tag{25}$$

Other solutions and their detailed study will be performed in forthcoming publications. All the above solutions to equations with substantial derivative contain the shift, $x - vt$, respectively to the solution in a stationary media, which depends on x. Further examples can be easily considered with the help of the operational approach. In the following chapter we will explore in details the particular case of the harmonic solution. Evidently, it can be easily generalized for any periodic solution, expandable in Fourier series. Similarly, the Fourier integral transforms technique apply.

4. Exact Periodic Solutions to GK-Type Equation with Substantial Derivative

An exact harmonic solution $\propto e^{inx}$ for the inhomogeneous system of PDE (9) and (10) can be obtained straight from the operational solution (18) (see (25)), or by separating the variables:

$T(x,t) = X(x)y(t)$. GK-type equation with substantial derivative (7) in the harmonic ansatz, $T(x,t) = e^{inx}y(t) \equiv F(x,t)$ reduces to the following ordinary differential equation for $y(t)$:

$$y''(t) + \varepsilon_{eff}y'(t) + \lambda_{eff}y(t) = 0, \tag{26}$$

where $\varepsilon_{eff} = \bar{\varepsilon} + 2vin$, $\bar{\varepsilon} = \varepsilon + n^2\delta$, and $\lambda_{eff} = (\alpha - v^2)n^2 + ivn\bar{\varepsilon} - \kappa$. The function $F(x,t) \equiv T(x,t)$ is introduced here for clarity of notations. The exact solution to GK equation with substantial derivative (7) then easily follows from the solution to (26):

$$\begin{gathered} F(x,t) = e^{inx}y(t),\ y(t) = e^{-\frac{t}{2}\varepsilon_{eff}}\left(C_1e^{-\frac{t}{2}\sqrt{U}} + C_2e^{\frac{t}{2}\sqrt{U}}\right), \\ U = \varepsilon_{eff}^2 - 4\lambda_{eff} = \bar{\varepsilon}^2 + 4\left(\kappa - \alpha n^2\right), \end{gathered} \tag{27}$$

where the constants C_1, C_2 are determined either from the initial or boundary conditions. According to the theory of separation of variables, the whole solution has to satisfy initial or boundary conditions. Evidently, telegrapher's equation with substantial derivative in the harmonic ansatz similarly reads

$$y''(t) + \tilde{\varepsilon}y'(t) + \tilde{\lambda}y(t) = 0, \tag{28}$$

and differs from (26) by the substitutions $\varepsilon_{eff} \to \tilde{\varepsilon} = \varepsilon + 2vin$, $\bar{\varepsilon} \to \varepsilon$, $\lambda_{eff} \to \tilde{\lambda} = (\alpha - v^2)n^2 + ivn\varepsilon - \kappa$. It's solution $y(t)$ is evidently (27), where $\delta = 0$.

Consider, for example, the Cauchy initial conditions

$$F(x,0) = Ae^{inx},\ \partial_t F(x,0) = Be^{inx}. \tag{29}$$

In the stationary case, $v = 0$, the Cauchy conditions (37), where initially $\frac{\partial\theta(x,t)}{\partial t} = 0$, have the meaning of zero heat flux q if $B = 0$ for both ballistic and diffusive components of quasi-temperature [30]. If $v \neq 0$, then the dimensionless equation $\theta'_t + v\theta'_x + q = 0$ describes the one-dimensional energy balance. The initially zero heat flux q then corresponds to $B = -invA$.

Then, GK-type Equation (7) with substantial derivative in the harmonic ansatz has the solution (27), satisfying ODE (26), where the coefficients C_1, C_2 are determined from the Cauchy conditions (29): $C_1 + C_2 = A$ and $-C_1\left(\varepsilon_{eff} + \sqrt{U}\right) - C_2\left(\varepsilon_{eff} - \sqrt{U}\right) = 2B$, and read as follows:

$$C_1 = \frac{A}{2} - \frac{B + A\varepsilon_{eff}/2}{\sqrt{U}},\ C_2 = \frac{A}{2} + \frac{B + A\varepsilon_{eff}/2}{\sqrt{U}}. \tag{30}$$

TE with substantial derivative (8) has the harmonic solution, whose time-dependent part $y(t)$ satisfies ODE (28):

$$F(x,t) = e^{inx}y(t),\ y(t) = e^{-\frac{t}{2}\tilde{\varepsilon}}\left(B_1e^{-\frac{t}{2}\sqrt{u}} + B_2e^{\frac{t}{2}\sqrt{u}}\right),\ u = \tilde{\varepsilon}^2 - 4\tilde{\lambda} = \varepsilon^2 + 4\left(\kappa - \alpha n^2\right), \tag{31}$$

where the coefficients B_1, B_2 are obtained from the Cauchy initial conditions (29):

$$B_1 = \frac{A}{2} - \frac{B + A\tilde{\varepsilon}/2}{\sqrt{u}},\ B_2 = \frac{A}{2} + \frac{B + A\tilde{\varepsilon}/2}{\sqrt{u}}. \tag{32}$$

Now, let us consider the Dirichlet boundary conditions in the moments of time $t = 0$ and $t = T$:

$$F(x,0) = Ae^{inx},\ F(x,T) = Ge^{inx}. \tag{33}$$

The harmonic solutions (27) and (31) satisfy respectively GK Equation (7) and telegrapher's Equation (8) with substantial derivative with the coefficients, determined by Equation (33):

$$C_1 = -e^{\frac{T}{2}\varepsilon_{eff}} \frac{Ae^{-\frac{T}{2}(\varepsilon_{eff}-\sqrt{U})} - G}{e^{-\frac{T}{2}\sqrt{U}} - e^{\frac{T}{2}\sqrt{U}}},\ C_2 = A - C_1, \tag{34}$$

and

$$B_1 = -e^{\frac{T}{2}\tilde{\varepsilon}} \frac{Ae^{-\frac{T}{2}(\tilde{\varepsilon}-\sqrt{u})} - G}{e^{-\frac{T}{2}\sqrt{u}} - e^{\frac{T}{2}\sqrt{u}}},\ B_2 = A - B_1. \tag{35}$$

Similarly, we can consider evolution of any function, expandable in Fourier series.

Note, that for some values of the coefficients α and δ the quantities U and u can assume negative values. Albeit it is not very obvious in the form of the harmonic solution (27) to GK-type equation with substantial derivative (26) and the solution (31) to TE with substantial derivative (28), these solutions remains real at any moment of time for a real initial function, because of the complex exponential is compensated by proper complex parts in the coefficients $C_{1,2}$ and $B_{1,2}$.

5. Exact Periodic Solutions to Ballistic Heat Transport in Thin Films

Ballistic heat transport in thin films is described by the system of PDE (9) and (10). In order to obtain the exact solution to this problem we reduced GK-type Equation (9) in the harmonic ansatz to (26) and solved it (see (27)); the result naturally involves the particular solution (25). This solution now constitutes the r.h.s. of Equation (10) for θ_d. Consider the Cauchy initial problem for θ_b:

$$\theta_b(x,0) = Ae^{inx},\ \partial_t\theta_b(x,0) = Be^{inx}, \tag{36}$$

and the Cauchy conditions for θ_d:

$$\theta_d(x,0) = Ve^{inx},\ \partial_t\theta_d(x,0) = We^{inx},\ V, W = \text{const}. \tag{37}$$

The quasi-temperature component $\theta_b(x,t)$ is in fact given by Equation (27):

$$\theta_b(x,t) = e^{inx}\left(C_1 e^{-\frac{t}{2}E_1} + C_2 e^{-\frac{t}{2}E_2}\right),\ E_1 = \left(\varepsilon_{eff} + \sqrt{U}\right),\ E_2 = \left(\varepsilon_{eff} - \sqrt{U}\right), \tag{38}$$

where $\varepsilon_{eff} = \bar{\varepsilon} + 2ivn$, $U = \bar{\varepsilon}^2 + 4(\kappa - \alpha_b n^2)$, $\bar{\varepsilon} = \varepsilon_b + n^2\delta$, $\alpha_b = 10Kn_b^2/3$, $\varepsilon_b = 2$, $\delta = 3Kn_b^2$, $\kappa_b = -1$, v is the speed of the media; for the Cauchy initial conditions (36) (see (29) with $F \equiv \theta_b$) the coefficients $C_{1,2}$ are given by Equation (30). The solution (38) for $\theta_b(x,t)$ contributes to the r.h.s. of the telegrapher's Equation (6) for the component $\theta_d(x,t)$. The exponential differential operators $e^{\hat{D}(x)}$ do not bring new harmonics to the initial content, $\theta_d(x,t) = \Theta_d(t)e^{inx}$, and Equation (6) reduces to the following inhomogeneous ODE for $\Theta_d(t)$:

$$\left(\frac{d^2}{dt^2} + \tilde{\varepsilon}_d\frac{d}{dt} + \tilde{\lambda}_d\right)\Theta_d(t) = +Pe^{-\frac{t}{2}E_1} + Qe^{-\frac{t}{2}E_2}, \tag{39}$$

where

$$\begin{gathered}\tilde{\varepsilon}_d = \varepsilon_d + i2vn,\ \tilde{\lambda}_d = \left(\alpha_d - v^2\right)n^2 + ivn\varepsilon_d,\ \varepsilon_d = \frac{Kn_b^2}{Kn_d^2},\ \alpha_d = \frac{Kn_b^4}{3Kn_d^2}, \\ P = C_1\left(\varepsilon_d - \frac{\bar{\varepsilon} + \sqrt{U}}{2} + ivn\right),\ Q = C_2\left(\varepsilon_d - \frac{\bar{\varepsilon} - \sqrt{U}}{2} + ivn\right),\end{gathered} \tag{40}$$

where $C_{1,2}$ are given by Equation (30). The above Equation (39) for $\Theta_d(t)$ possesses analytical solution, which, in turn, yields the evolving in time harmonic solution for the diffusive quasi-temperature component $\theta_d(x,t)$ in the following form:

$$\theta_d(x,t) = e^{inx}\left(4\left(\frac{P}{S}e^{-\frac{t}{2}E_1} + \frac{Q}{L}e^{-\frac{t}{2}E_2}\right) + D_1 e^{-\frac{t}{2}(\widetilde{\varepsilon}_d+R)} + D_2 e^{-\frac{t}{2}(\widetilde{\varepsilon}_d-R)}\right), \tag{41}$$

$$\begin{aligned} R &= \sqrt{\widetilde{\varepsilon}_d^2 - 4\widetilde{\lambda}_d} = \sqrt{\varepsilon_d^2 - 4n^2\alpha_d}, \\ S &= E_1^2 + 4\widetilde{\lambda}_d - 2E_1\widetilde{\varepsilon}_d = \left(\overline{\varepsilon} - \varepsilon_d + \sqrt{U}\right)^2 + 4\alpha_d n^2 - \varepsilon_d^2, \\ L &= E_2^2 + 4\widetilde{\lambda}_d - 2E_2\widetilde{\varepsilon}_d = \left(\overline{\varepsilon} - \varepsilon_d - \sqrt{U}\right)^2 + 4\alpha_d n^2 - \varepsilon_d^2, \end{aligned} \tag{42}$$

where the constants $D_{1,2}$ are determined from the Cauchy initial conditions (37) for $\theta_d(x,t)$:

$$D_1 = \frac{V}{2}\left(1 - \frac{\widetilde{\varepsilon}_d}{R}\right) - \frac{1}{R}\left(W + \frac{2P}{S}\left(\overline{\varepsilon} - \varepsilon_d + \sqrt{U} + R\right) + \frac{2Q}{L}\left(\overline{\varepsilon} - \varepsilon_d - \sqrt{U} + R\right)\right), \tag{43}$$

$$D_2 = \frac{V}{2}\left(1 + \frac{\widetilde{\varepsilon}_d}{R}\right) + \frac{1}{R}\left(W + \frac{2P}{S}\left(\overline{\varepsilon} - \varepsilon_d + \sqrt{U} - R\right) + \frac{2Q}{L}\left(\overline{\varepsilon} - \varepsilon_d - \sqrt{U} - R\right)\right). \tag{44}$$

The exact analytical harmonic solutions for the $\theta_b(x,t)$ is given by (38) and (30), and for $\theta_d(x,t)$, it is given by (41)–(44); their sum represents the complete solution for the ballistic heat transfer in thin films [30] with the Cauchy initial conditions (36), (37), valid for arbitrary values of the Knudsen numbers Kn_d, Kn_b. Note, that a real initial distribution of quasi-temperature $\theta(x,0)$ evolves in real domain even for imaginary exponentials in the solutions due to complex coefficients C_1, C_2 and D_1, D_2.

Now let us consider the Dirichlet conditions at the boundaries

$$\theta_b(x,0) = Ae^{inx},\ \theta_b(x,T) = Ge^{inx},\ \theta_d(x,0) = Ve^{inx},\ \theta_d(x,T) = We^{inx}, \tag{45}$$

where $A, G, V, W = \text{const}$ (see also Equation (33)). The evolution of $\theta_b(x,t)$ is given by Equation (38), where $C_{1,2}$ are set by Equation (34). The evolution of the counterpart $\theta_d(x,t)$ is governed by Equations (39) and (40). The r.h.s. of Equation (39) is set by Equation (38) for the ballistic component $\theta_b(x,t)$, where $C_{1,2}$ are given by (34). The explicit solution $\theta_d(x,t)$ to inhomogeneous DE (39) is possible, although it is much more cumbersome than in the Cauchy case:

$$\theta_d(x,t) = e^{inx}\left(\frac{b_2}{w_1}e^{-E_2 t} + \frac{b_1}{w_2}e^{-E_1 t} + \frac{q_1}{w_1 w_2}e^{-\frac{1}{2}t(\varepsilon_d+R)} - \frac{q_2}{w_1 w_2}e^{-\frac{1}{2}t(\varepsilon_d-R)}\right), \tag{46}$$

where

$$\begin{aligned} & w_1 = E_2^2 - E_2\widetilde{\varepsilon}_d + \widetilde{\lambda}_d,\ w_2 = E_1^2 - E_1\widetilde{\varepsilon}_d + \widetilde{\lambda}_d, \\ & q_1 = \frac{1}{e^{RT}-1}\left(e^{-E_2T}w_1 b_1 d_2 + e^{-E_2T}b_2 d_2 w_2 + d_1 w_1 w_2\right), \\ & q_2 = \frac{1}{e^{RT}-1}\left(e^{-E_2T}\left(\widetilde{\lambda}_d - w_2\right)(b_2 d_6 - w_1 d_4) + e^{-E_1T}\left(w_1\left(\widetilde{\lambda}_d d_3 - b_1 d_5\right) + \widetilde{\lambda}_d b_2 d_7\right)\right), \\ & b_1 = C_1\left(\widetilde{\varepsilon}_d - \frac{\varepsilon_{eff} + \sqrt{U}}{2} + ivn\right),\ b_2 = C_2\left(\widetilde{\varepsilon}_d - \frac{\varepsilon_{eff} - \sqrt{U}}{2} + ivn\right),\ d_1 = Ve^{RT} - We^{\frac{T}{2}(\widetilde{\varepsilon}_d+R)}, \\ & d_2 = e^{\frac{T}{2}(\widetilde{\varepsilon}_d+R)} - e^{T(E_1+R)},\ d_3 = Ve^{E_1T} - We^{(2E_1+\widetilde{\varepsilon}_d+R)\frac{T}{2}},\ d_4 = Ve^{E_2T} - We^{(2E_2+\widetilde{\varepsilon}_d+R)\frac{T}{2}}, \\ & d_5 = e^{E_1T} - e^{(\widetilde{\varepsilon}_d+R)\frac{T}{2}},\ d_6 = e^{E_2T} - e^{(\widetilde{\varepsilon}_d+R)\frac{T}{2}},\ d_7 = -e^{E_1T} + e^{(\widetilde{\varepsilon}_d+R+2(E_1-E_2))\frac{T}{2}}. \end{aligned} \tag{47}$$

The solution for $\theta_d(x,t)$ in the form (46), (47) explicitly involves the Dirichlet conditions and together with the solution for $\theta_b(x,t)$, given in the Dirichlet case by (38) and (34), they describe the evolution of the complete harmonic quasi-temperature $\theta = \theta_d + \theta_b$.

Let us now consider some examples of the harmonic solutions to GK-type heat equation with substantial derivative. Consider, for example, the Cauchy problem with initial conditions $\theta_b(x,0) = \theta_d(x,0) = e^{ix}$, $\partial_t\theta_b(x,0) = \partial_t\theta_d(x,0) = 0$. Let's first consider $Kn_b = Kn_d = 1$. Proper solutions for $\theta_b(x,t)$, $\theta_d(x,t)$ and $\theta(x,t)$ in the stationary media are presented in Figure 1.

It can be seen in the top left plot in Figure 1 as the ballistic component $\theta_b(x,t)$ rapidly relaxes to the stationary state. The diffusive component $\theta_d(x,t)$ behaves similarly, but the relaxation process takes much more time. The behavior of ballistic and diffusive components in the domain is more distinct for $Kn < 1$. The complete quasi-temperature $\theta(x,t)$ (see bottom plot in Figure 1) follows the behavior of both the ballistic $\theta_b(x,t)$ and diffusive $\theta_d(x,t)$ constituents. However, the latter, being the solution of Equation (39), is strongly influenced by the solution (38) for the ballistic component $\theta_b(x,t)$, which sets the inhomogeneous part in Equation (39). Thus, the evolution of the quasi-temperature $\theta(x,t)$ in thin films has two speeds: the ballistic constituent $\theta_b(x,t)$ develops rapidly and sets the r.h.s. in Equation (39) for the diffusive constituent $\theta_d(x,t)$. The evolution of $\theta_d(x,t)$ is much slower: after the first instances it proceeds under the influence of already relaxed true ballistic constituent $\theta_b(x,t)$.

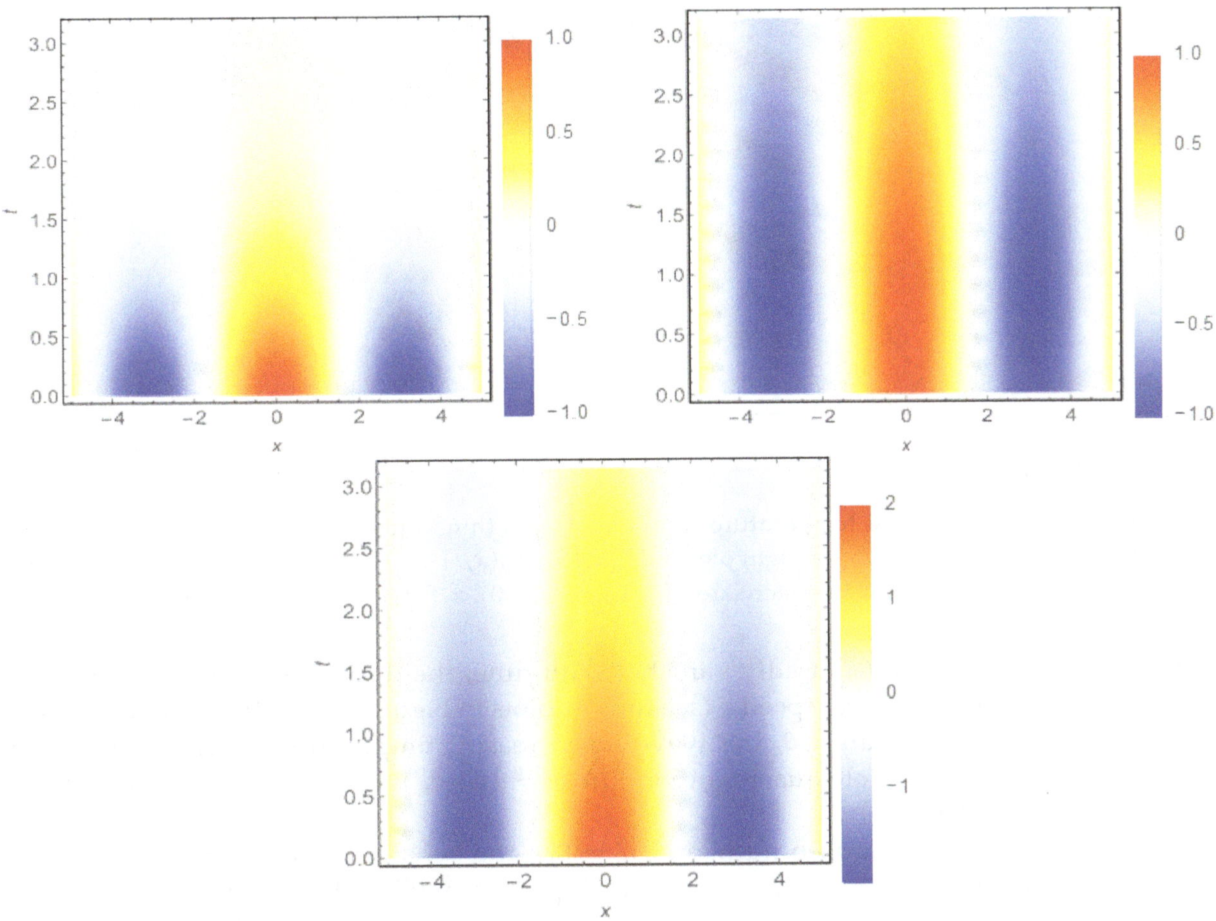

Figure 1. Evolution of quasi-temperature components in a thin stationary film with $Kn_b = Kn_d = 1$ for $v = 0$: $\theta_b(x,t)$—top left plot, $\theta_d(x,t)$—top right plot and $\theta(x,t) = \theta_b(x,t) + \theta_d(x,t)$—bottom plot. The Cauchy conditions for PDEs (9) and (10) system are $\theta_b(x,0) = \theta_d(x,0) = e^{inx}$, $\partial_t\theta_b(x,0) = \partial_t\theta_d(x,0) = 0$.

In the case when the rate of change of temperature for a given position in a field depends both on the instantaneous rate of change of temperature at that location $(\partial/\partial t)$ as well as on the rate at which the temperature is convected to that location by the fluid motion, the behaviors of the solutions change significantly. The solutions for $Kn_d = Kn_b = 1$ and $v = 10$, are presented in Figure 2, for $v = -10$ in Figure 3.

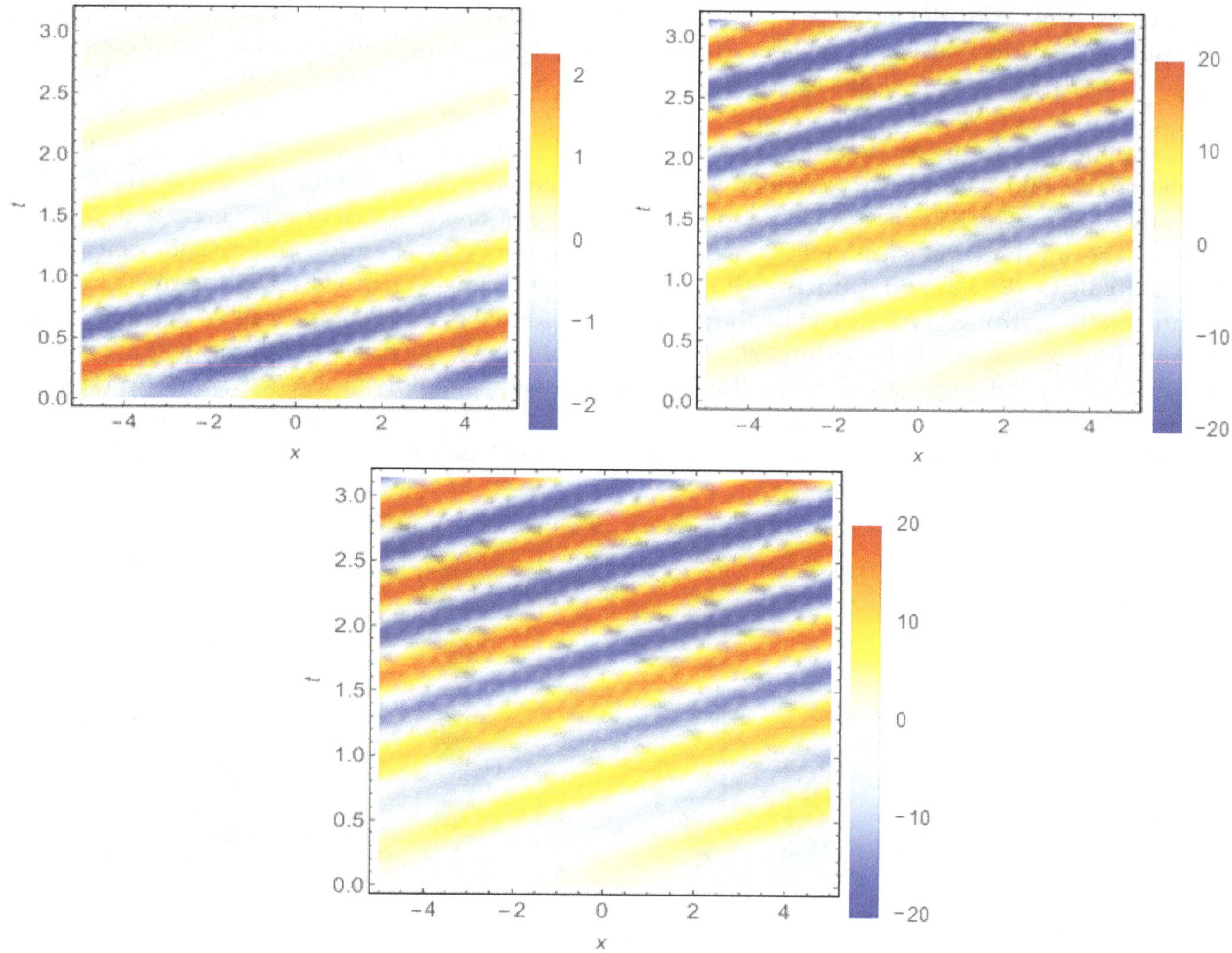

Figure 2. Evolution of quasi-temperature components in a thin film with $Kn_b = Kn_d = 1$ for $v = +10$: $\theta_b(x,t)$—top left plot, $\theta_d(x,t)$—top right plot and $\theta(x,t) = \theta_b(x,t) + \theta_d(x,t)$—bottom plot. The Cauchy conditions for PDEs (9) and (10) system are $\theta_b(x,0) = \theta_d(x,0) = e^{inx}$, $\partial_t\theta_b(x,0) = \partial_t\theta_d(x,0) = 0$.

The ballistic constituent θ_b rapidly vanishes with time; the direction of the colorful waves in the plot depends on the sign of the speed v (see top left plots in Figures 2 and 3). Diffusive counterpart of the complete quasi-temperature θ_d has waves of increasing amplitude as seen in top right plots in Figures 2 and 3. The complete quasi-temperature $\theta = \theta_d + \theta_b$ has obvious non-Fourier behavior, which largely follows that of θ_d.

For smaller values of Knudsen number, $Kn = 0.1$, the behavior of the harmonic solutions to ballistic heat propagation in thin films, obeying the Eqs. system (9) and (10) with Cauchy conditions (36) (37), has less distinctive wave interference with slower fade of the solution for θ_b and weaker amplitude increase speed for θ_d; proper solutions for $v = +10$ are presented in Figure 4.

It is important to note, that the whole system is symmetric under the speed inversion, which becomes obvious with account for formulae (38)–(44).

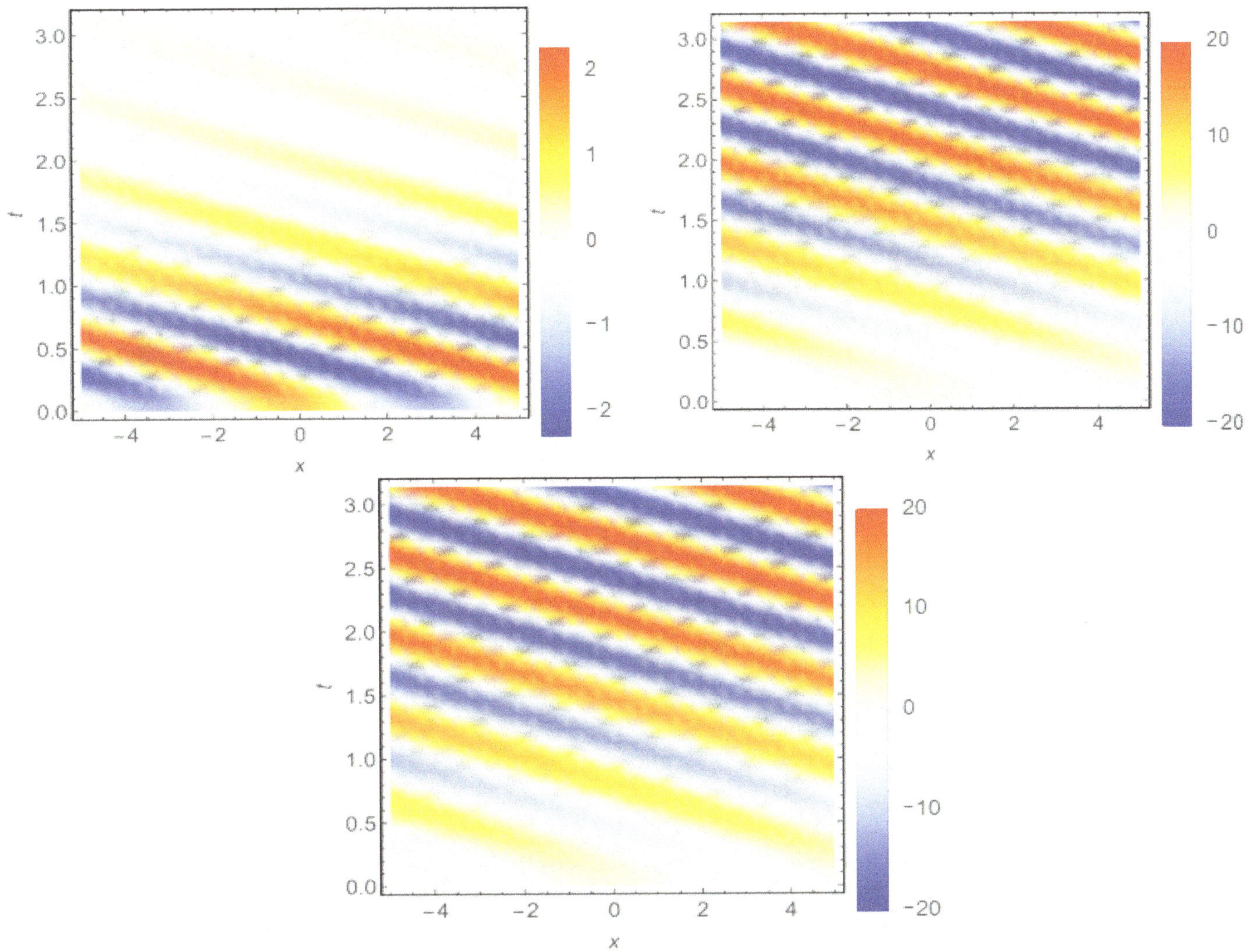

Figure 3. Evolution of quasi-temperature components in a thin film with $Kn_b = Kn_d = 1$ for $v = -10$: $\theta_b(x,t)$—top left plot, $\theta_d(x,t)$—top right plot and $\theta(x,t) = \theta_b(x,t) + \theta_d(x,t)$—bottom plot. The Cauchy conditions for PDEs (9) and (10) system are $\theta_b(x,0) = \theta_d(x,0) = e^{inx}$, $\partial_t\theta_b(x,0) = \partial_t\theta_d(x,0) = 0$.

Note, that for $Kn_b = Kn_d = 0.1$ $\theta_b(x,t)$ and $\theta_d(x,t)$ constituents behave differently (see Figure 4). The diffusive constituent slowly grows as the ballistic constituent shows gradual amplitude decrease, as shown in Figure 5. Moreover, for $v = 10$ the complete quasi-temperature $\theta(x,t) = \theta_b(x,t) + \theta_d(x,t)$ monotonously increases, following the behavior of the diffusive constituent as shown in Figure 5. The maximum principle, established though for parabolic equations, is violated for the diffusive component and complete quasi-temperature with Cauchy conditions for non-Fourier ballistic heat transport in thin films. In all of the cases, the evolution of $\theta_b(x,t)$ occurs faster than that of $\theta_d(x,t)$; the two-speed heat propagation process can be seen in every set of plots in Figures 1–5.

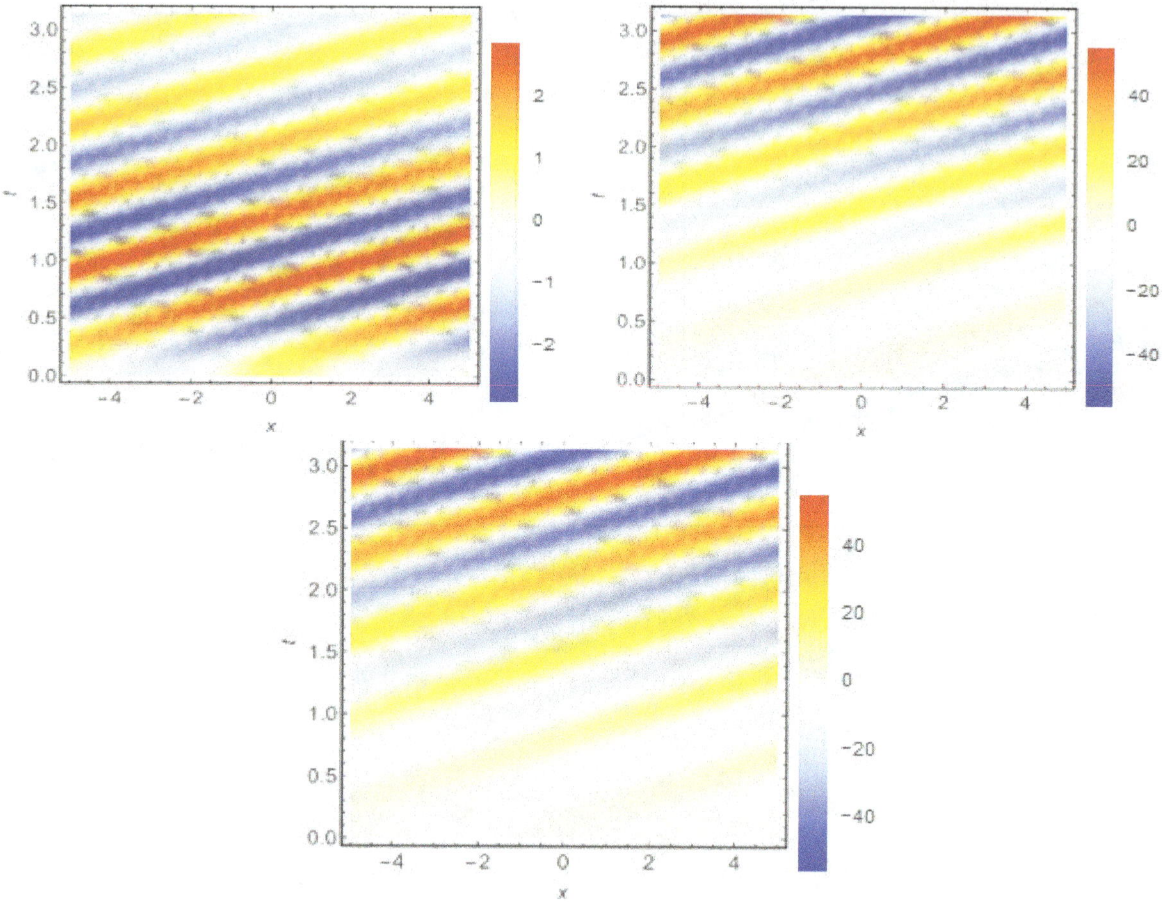

Figure 4. Evolution of quasi-temperature components in a thin film with $Kn_b = Kn_d = 0.1$ for $v = +10$: $\theta_b(x,t)$—top left plot, $\theta_d(x,t)$—top right plot and $\theta(x,t) = \theta_b(x,t) + \theta_d(x,t)$—bottom plot. The Cauchy conditions for PDEs (9) and (10) system are $\theta_b(x,0) = \theta_d(x,0) = e^{inx}$, $\partial_t\theta_b(x,0) = \partial_t\theta_d(x,0) = 0$.

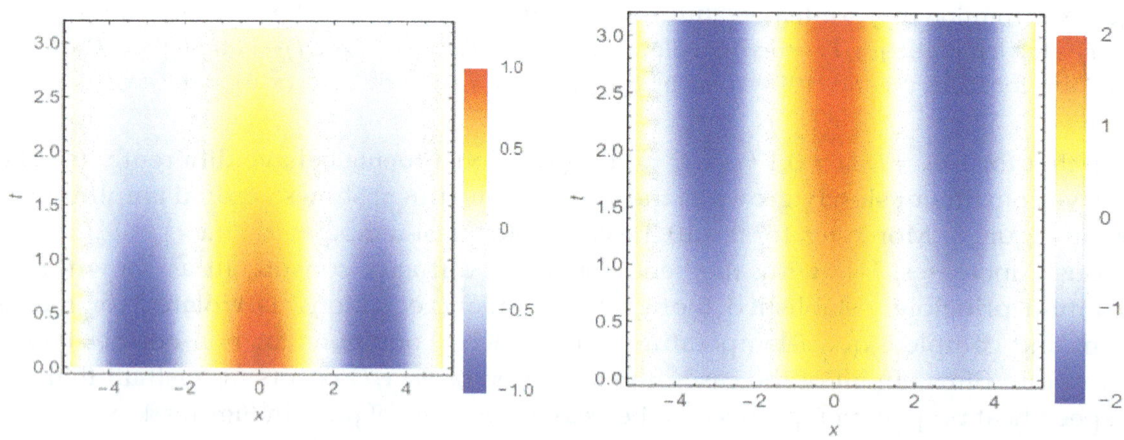

Figure 5. Evolution of quasi-temperature components in a stationary thin film with $Kn_b = Kn_d = 0.1$ for $v = 0$: $\theta_b(x,t)$—left plot, $\theta_d(x,t)$—right plot. The Cauchy conditions for PDEs (9) and (10) system are $\theta_b(x,0) = \theta_d(x,0) = e^{inx}$, $\partial_t\theta_b(x,0) = \partial_t\theta_d(x,0) = 0$.

The exploration of the solutions for Dirichlet conditions will be done elsewhere.

It should be noted that the one-dimensional energy balance equation yields zero value for the initial heat flux, if $B = -invA$. Above we have obtained the solutions, which allow nonzero initial conditions on the first time derivative; we have investigated some examples while writing this paper.

For example, for the case of $Kn_b = Kn_d = 1, v = 10$ the maximum principle is violated for both zero and nonzero initial conditions imposed on the first time-derivative of the ballistic component; however, there was no significant qualitative difference in the behaviors of the diffusive component in this case.

Explicit study for the solution in thin films with zero initial heat flux and non-zero velocity v will be done in forthcoming publications.

6. Conclusions

In the present work we have studied and analytically solved one-dimensional heat transport equations of Cattaneo- and Guyer–Krumhansl-type with substantial derivative. Exponential differential operators and the operational method were used to obtain integral forms of exact particular solutions to heat transport equations. The media speed is accounted for by the exponential differential operator $e^{-vt\partial_x}$, which produces a shift $x - vt$. Together with the heat operator $\hat{S}$ they transform the initial temperature profile and determine its evolution. Their action on the initial polynomial $\sum_n x^n$ yields sums of the Hermite polynomials $H_n(x,y)$; the action on the monomial-exponential function yields more complicated sums of Hermite polynomials. Proper exact operational solutions were obtained in the sums of converging integrals of elementary functions. Several examples were considered, the solutions for $F(x,0) = x^2$, $F(x,0) = x^2e^{-x}$ and $F(x,0) = e^{inx}$ were given. The solutions exactly satisfy the heat transport equations, which has been proven by direct substitution.

The equations with substantial derivative of Guyer-Krumhansl-type and of telegrapher's-type are demonstrated to have similar structure in the harmonic ansatz. Their exact harmonic solutions have been obtained by both operational method and the separation of variables. The latter reduces these PDEs to ODEs. We have shown that the harmonic solutions may have local and global extremums in the domain due to the interference of the Cattaneo heat waves; proper second-order time-derivative term also insures finite speed of the temperature perturbation propagation. We have shown that in a GK-type equation with dominant ballistic term δ, the Cattaneo heat waves are damped.

We have obtained the exact analytical solution to the ballistic heat transport in thin films by solving the system of inhomogeneous PDEs [30] in the harmonic ansatz; both Cauchy and Dirichlet conditions were explored. We found that true ballistic constituent of quasi-temperature, θ_b, evolves much faster than the diffusive counterpart of quasi-temperature, θ_d. We have studied the evolution of the complete quasi-temperature $\theta(x,t) = \theta_b(x,t) + \theta_d(x,t)$ under the ballistic conditions; the latter apply when the phonon mean free path l is comparable with the system scale L. The quasi-temperature θ first senses fast evolving true ballistic part θ_b. The contribution of θ_b sets the r.h.s. in the inhomogeneous DE for the component θ_d. After relatively fast relaxation of θ_b, further evolution of the complete quasi-temperature θ follows the diffusive constituent θ_d. Thus the initial fast-developing true ballistic component θ_b determines the evolution of the complete quasi-temperature of the system θ by imposing distribution for the diffusive quasi-temperature θ_d at the beginning of the process.

The obtained harmonic solutions for the Cauchy problem (9), (10), (36), (37) were applied for the study of the temperature patterns in thin films for the Knudsen number values $Kn = 0.1$ and $Kn = 1$. In the stationary case $v = 0$, the true ballistic component θ_b monotonously relaxes (see Figures 1 and 5); its diffusive counterpart θ_d behavior shows noticeable dependence on the value of the Knudsen numbers: for $Kn = 1$ it slowly relaxes (see Figure 5) and for $Kn = 0.1$ it droningly grows (see Figure 1). The effect of the speed of the observer on the perceived heat transport in thin films is distinguished for the pure ballistic and diffusive quasi-temperature constituents. In the presence of the motion, the extremum of the ballistic component θ_b occurs inside the domain after the initial moment $t = 0$ (see Figures 2–4). The diffusive quasi-temperature component monotonously grows in the domain both in the weak ballistic (see Figure 4) and strong ballistic cases (see Figures 2 and 3). This behavior contrasts common Fourier heat diffusion, for which the maximum principle holds. It is also noticeable that the system shows symmetric behavior while the value of the speed is reversed, which does not contradict the Second law. The Knudsen number variation from 0.1 to 1 in the presence of the speed $v = 10$, are practically not sensed by the system. Thus the speed of the observer v influences the registered temperature rather than the ballistic heat transport contribution.

Author Contributions: K.Z. has made the major work on the paper. N.G. has made a contribution in the Section 3. D.O. has made the contribution in Sections 4 and 5.

Acknowledgments: The authors are grateful to Anatolii Borisov, Andrey Lobanov for the discussions of the heat equations; to Laszlo I. Kiss (UQAC, Department of Applied Sciences, Quebec University) and Yasar Demirel (Department of Chemical and Biological Engineering, University of Nebraska, Lincoln) for discussions of the heat evolution patterns.

References

1. Fourier, J.P.J. *The Analytical Theory of Heat*; Cambridge University Press: London, UK, 1878.
2. Ackerman, C.C.; Guyer, R.A. Temperature pulses in dielectric solids. *Ann. Phys.* **1968**, *50*, 128–185. [CrossRef]
3. Ackerman, C.C.; Overton, W.C. Second sound in solid helium-3. *Phys. Rev. Lett.* **1969**, *22*, 764. [CrossRef]
4. McNelly, T.F.; Rogers, S.J.; Channin, D.J.; Rollefson, R.; Goubau, W.M.; Schmidt, G.E.; Krumhansl, J.A.; Pohl, R.O. Heat pulses in NaF: Onset of second sound. *Phys. Rev. Lett.* **1970**, *24*, 100. [CrossRef]
5. Narayanamurti, V.; Dynes, R.D. Observation of second sound in Bismuth. *Phys. Rev. Lett.* **1972**, *26*, 1461–1465. [CrossRef]
6. Peshkov, V. Second sound in Helium II. *J. Phys. (Moscow)* **1944**, *8*, 381.
7. Cattaneo, C. Sur une forme de l'equation de la chaleur eliminant le paradoxe d'une propagation instantanee. *C. R. l'Acad. Sci. Paris* **1958**, *247*, 431–433.
8. Terman, F.E. *Radio Engineers' Handbook*, 1st ed.; McGraw-Hill: New York, NY, USA, 1943.
9. Both, S.; Czél, B.; Fülöp, T.; Ván, P.; Verhás, J. Deviation from the Fourier law in room-temperature heat pulse experiments. *J. Non-Equilib. Thermodyn.* **2016**, *41*, 41–48. [CrossRef]
10. Kovács, R.; Ván, P. Models of ballistic propagation of heat at low temperatures. *Int. J. Thermophys.* **2016**, *37*, 95. [CrossRef]
11. Ván, P.; Berezovski, A.; Fülöp, T.; Gróf, G.; Kovács, R.; Lovas, Á.; Verhás, J. Guyer-Krumhansl-type heat conduction at room temperature. *EPL* **2017**, *118*, 50005. [CrossRef]
12. Cahill, D.G. Thermal conductivity measurement from 30 to 750 K: The 3ω method. *Rev. Sci. Instrum.* **1990**, *61*, 802–808. [CrossRef]
13. Tang, D.W.; Araki, N. Non-Fourier heat conduction behaviour in finite mediums under pulse surface heating. *Mater. Sci. Eng. A* **2000**, *292*, 173–178. [CrossRef]
14. Kaminski, W. Hyperbolic heat conduction equations for materials with a nonhomogeneous inner structure. *J. Heat Transf.* **1990**, *112*, 555–560. [CrossRef]
15. Mitra, K.; Kumar, S.; Vedavarz, A.; Moallemi, M.K. Experimental evidence of hyperbolic heat conduction in processed meat. *J. Heat Transf.* **1995**, *117*, 568–573. [CrossRef]
16. Herwig, H.; Beckert, K. Fourier versus non-Fourier heat conduction in materials with a nonhomogeneous inner structure. *J. Heat Transf.* **2000**, *122*, 363–365. [CrossRef]
17. Roetzel, W.; Putra, N.; Das, S.K. Experiment and analysis for non-Fourier conduction in materials with non-homogeneous inner structure. *Int. J. Therm. Sci.* **2003**, *42*, 541–552. [CrossRef]
18. Sazhin, S.S. Modelling of fuel droplet heating and evaporation: Recent results and unsolved problems. *Fuel* **2017**, *196*, 69–101. [CrossRef]
19. Scott, E.P.; Tilahun, M.; Vick, B. The question of thermal waves in heterogeneous and biological materials. *J. Biomech. Eng.* **2009**, *131*, 074518. [CrossRef] [PubMed]
20. Ricciu, R.; Besalduch, L.A.; Galatioto, A.; Ciulla, G. Thermal characterization of insulating materials. *Renew. Sustain. Energy Rev.* **2018**, *82*, 1765–1773. [CrossRef]
21. Porrà, J.M.; Masoliver, J.; Weiss, G.H. When the telegrapher's equation furnishes a better approximation to the transport equation than the diffusion approximation. *Phys. Rev. E* **1997**, *55*, 7771–7774. [CrossRef]
22. Shiomi, J.; Maruyama, S. Non-Fourier heat conduction in a single-walled carbon nanotube: Classical molecular dynamics simulations. *Phys. Rev. B* **2006**, *73*, 205420. [CrossRef]

23. Baringhaus, J.; Ruan, M.; Edler, F.; Tejeda, A.; Sicot, M.; Taleb-Ibrahimi, A.; Li, A.P.; Jiang, Z.; Conrad, E.H.; Berger, C.; et al. Exceptional ballistic transport in epitaxial graphene nanoribbons. *Nature* **2014**, *506*, 349–354. [CrossRef] [PubMed]
24. Hochbaum, A.I.; Chen, R.; Delgado, R.D.; Liang, W.; Garnett, E.C.; Najarian, M.; Majumdar, A.; Yang, P. Enhanced thermoelectric performance of rough silicon nanowires. *Nature (London)* **2008**, *451*, 163–167. [CrossRef] [PubMed]
25. Boukai, A.I.; Bunimovich, Y.; Tahir-Kheli, J.; Yu, J.-K.; Goddard, W.A.; Heath, J.R. Silicon nanowires as efficient thermoelectric materials. *Nature (London)* **2008**, *451*, 168. [CrossRef] [PubMed]
26. Paddock, C.A.; Eesley, G.L. Transient thermoreflectance from thin metal films. *J. Appl. Phys.* **1986**, *60*, 285–290. [CrossRef]
27. Maldovan, M. Transition between ballistic and diffusive heat transport regimes in silicon materials. *Appl. Phys. Lett.* **2012**, *101*, 113110. [CrossRef]
28. Guyer, R.A.; Krumhansl, J.A. Solution of the linearized phonon Boltzmann equation. *Phys. Rev.* **1966**, *148*, 766–778. [CrossRef]
29. Guyer, R.A.; Krumhansl, J.A. Thermal conductivity, second sound and phonon hydrodynamic phenomena in non-metallic crystals. *Phys. Rev.* **1966**, *148*, 778–788. [CrossRef]
30. Lebon, G.; Machrafi, H.; Gremela, M.; Dubois, C. An extended thermodynamic model of transient heat conduction at sub-continuum scales. *Proc. R. Soc. A* **2011**, *467*, 3241–3256. [CrossRef]
31. Müller, I.; Ruggeri, T. *Rational Extended Thermodynamics*; Springer: Berlin/Heidelberg, Germany, 1998.
32. Rogolino, P.; Kovács, R.; Ván, P.; Cimmelli, V.A. Generalized heat-transport equations: Parabolic and hyperbolic models. *Contin. Mech. Thermodyn.* **2018**. [CrossRef]
33. Moosaie, A. Non-Fourier heat conduction in a finite medium with insulated boundaries and arbitrary initial conditions. *Int. Commun. Heat Mass Transf.* **2008**, *35*, 103–111. [CrossRef]
34. Ahmadikia1, H.; Rismanian, M. Analytical solution of non-Fourier heat conduction problem on a fin under periodic boundary conditions. *J. Mech. Sci. Technol.* **2011**, *25*, 2919–2926. [CrossRef]
35. Yen, C.C.; Wu, C.Y. Modelling hyperbolic heat conduction in a finite medium with periodic thermal disturbance and surface radiation. *Appl. Math. Model.* **2003**, *27*, 397–408. [CrossRef]
36. Lewandowska, M. Hyperbolic heat conduction in the semi-infinite body with a time-dependent laser heat source. *Heat Mass Transf.* **2001**, *37*, 333–342. [CrossRef]
37. Lewandowska, M.; Malinowski, L. An analytical solution of the hyperbolic heat conduction equation for the case of a finite medium symmetrically heated on both sides. *Int. Commun. Heat Mass Transf.* **2006**, *33*, 61–69. [CrossRef]
38. Saedodin, S.; Torabi, M. Analytical solution of non-Fourier heat conduction in cylindrical coordinates. *Int. Rev. Mech. Eng.* **2009**, *3*, 726–732.
39. Challamel, N.; Grazide, C.; Picandet, V.; Perrot, A.; Zhang, Y. A nonlocal Fourier's law and its application to the heat conduction of one-dimensional and two-dimensional thermal lattices. *C. R. Mec.* **2016**, *344*, 388–401. [CrossRef]
40. Saedodin, S.; Torabi, M. Algebraically explicit analytical solution of three-dimensional hyperbolic heat conduction equation. *Adv. Theor. Appl. Mech.* **2010**, *3*, 369–383.
41. Chen, G. Ballistic-diffusive heat-conduction equations. *Phys. Rev. Lett.* **2001**, *86*, 2297–2300. [CrossRef] [PubMed]
42. Hsiao, T.K.; Chang, H.K.; Liou, S.C.; Chu, M.W.; Lee, S.C.; Chang, C.W. Observation of room-temperature ballistic thermal conduction persisting over 8.3 mm in SiGe nanowires. *Nat. Nanotechnol.* **2013**, *8*, 534–538. [CrossRef] [PubMed]
43. Zhang, Y.; Ye, W. Modified ballistic–diffusive equations for transient non-continuum heat conduction. *Int. J. Heat Mass Transf.* **2015**, *83*, 51–63. [CrossRef]
44. Kovacs, R.; Van, P. Generalized heat conduction in heat pulse experiments. *Int. J. Heat Mass Transf.* **2015**, *83*, 613–620. [CrossRef]
45. KZhukovsky, V. Operational method of solution of linear non-integer ordinary and partial differential equations. *SpringerPlus* **2016**, *5*, 119. [CrossRef] [PubMed]
46. Zhukovsky, K. Operational approach and solutions of hyperbolic heat conduction equations. *Axioms* **2016**, *5*, 28. [CrossRef]

47. Zhukovsky, K.V.; Srivastava, H.M. Analytical solutions for heat diffusion beyond Fourier law. *Appl. Math. Comput.* **2017**, *293*, 423–437. [CrossRef]
48. Zhukovsky, K.V. Violation of the maximum principle and negative solutions with pulse propagation in Guyer–Krumhansl model. *Int. J. Heat Mass Transf.* **2016**, *98*, 523–529. [CrossRef]
49. Zhukovsky, K.V. Exact solution of Guyer–Krumhansl type heat equation by operational method. *Int. J. Heat Mass Transf.* **2016**, *96*, 132–144. [CrossRef]
50. Zhukovsky, K. Exact negative solutions for Guyer–Krumhansl type equation and the violation of the maximum principle. *Entropy* **2017**, *19*, 440. [CrossRef]
51. Zhukovsky, K. Operational solution for some types of second order differential equations and for relevant physical problems. *J. Math. Anal. Appl.* **2017**, *446*, 628–647. [CrossRef]
52. Zhukovsky, K.V. A method of inverse differential operators using ortogonal polynomials and special functions for solving some types of differential equations and physical problems. *Mosc. Univ. Phys. Bull.* **2015**, *70*, 93–100. [CrossRef]
53. Zhukovsky, K. Solution of some types of differential equations: Operational calculus and inverse differential operators. *Sci. World J.* **2014**, *2014*. [CrossRef] [PubMed]
54. Zhukovsky, K.V. Solving evolutionary-type differential equations and physical problems using the operator method. *Theor. Math. Phys.* **2017**, *190*, 52–68. [CrossRef]
55. Dattoli, G.; Srivastava, H.M.; Zhukovsky, K.V. Operational methods and Differential Equations with applications to initial-value problems. *Appl. Math. Comput.* **2007**, *184*, 979–1001. [CrossRef]
56. Zhukovsky, K.V. Operational solution of differential equations with derivatives of non-integer order, Black–Scholes type and heat conduction. *Mosc. Univ. Phys. Bull.* **2016**, *71*, 237–244. [CrossRef]
57. Dattoli, G.; Srivastava, H.M.; Zhukovsky, K.V. Orthogonality properties of the Hermite and related polynomials. *J. Comput. Appl. Math.* **2005**, *182*, 165–172. [CrossRef]
58. Dattoli, G.; Srivastava, H.M.; Zhukovsky, K.V. A new family of integral transforms and their applications. *Integral Transforms Spec. Funct.* **2006**, *17*, 31–37. [CrossRef]
59. Srivastava, H.M.; Manocha, H.L. *A Treatise on Generating Functions*; Halsted Press (Ellis Horwood Limited): Chichester, UK; John Wiley and Sons: New York, NY, USA; Chichester, UK; Brisbane, Australia; Toronto, ON, Canada, 1984.
60. Boucetta, A.; Ghodbane, H.; Ayad, M.Y.; Bahri, M. A review on the performance and modelling of proton exchange membrane fuel cells. *AIP Conf. Proc.* **2016**, *1758*, 030019.
61. Arato, E.; Pinna, M.; Mazzoccoli, M.; Bosio, B. Gas-phase mass-transfer resistances at polymeric electrolyte membrane fuel cells electrodes: Theoretical analysis on the effectiveness of interdigitated and serpentine flow arrangements. *Energies* **2016**, *9*, 229. [CrossRef]
62. Maidhily, M.; Rajalakshmi, N.; Dhathathreyan, K.S. Electrochemical impedance spectroscopy as a diagnostic tool for the evaluation of flow field geometry in polymer electrolyte membrane fuel cells. *Renew. Energy* **2013**, *51*, 79–84. [CrossRef]
63. St-Pierre, J. Hydrogen mass transport in fuel cell gas diffusion electrodes. *Fuel Cells* **2011**, *11*, 263–273. [CrossRef]
64. Zhukovsky, K.; Pozio, A. Maximum current limitations of the PEM fuel cell with serpentine gas supply channels. *J. Power Sources* **2004**, *130*, 95–105. [CrossRef]
65. Zhukovsky, K.V. Three Dimensional model of gas transport in a porous diffuser of a polymer electrolyte fuel cell. *AIChE J.* **2003**, *49*, 3029–3036. [CrossRef]
66. Zhukovsky, K. Modeling of the Current Limitations of PEFC. *AIChE J.* **2006**, *52*, 2356–2366. [CrossRef]
67. Weber, A.Z.; Newman, J. Modeling transport in polymer-electrolyte fuel cells. *Chem. Rev.* **2004**, *104*, 4679–4726. [CrossRef] [PubMed]
68. Kawase, M.; Sato, K.; Mitsui, R.; Asonuma, H.; Kageyama, M.; Yamaguchi, K.; Inoue, G. Electrochemical reaction engineering of polymer electrolyte fuel cell. *AIChE J.* **2017**, *63*, 249–256. [CrossRef]
69. Rogié, B.; Monier-Vinard, E.; Nguyen, M.-N.; Bissuel, V.; Laraqi, N. Practical analytical modeling of 3D multi-layer Printed Wired Board with buried volumetric heating sources. *Int. J. Therm. Sci.* **2018**, *129*, 404–415. [CrossRef]
70. Calvo-Schwarzwälder, M.; Hennessy, M.G.; Torres, P.; Myers, T.G.; Alvarez, F.X. A slip-based model for the size-dependent effective thermal conductivity of nanowires. *Int. Commun. Heat Mass Transf.* **2018**, *91*, 57–63. [CrossRef]
71. Zhukovskij, K.V. Gas flow in long microchannels. *Vestn. Mosk. Univ. Ser. 3 Fiz. Astron.* **2001**, *3*, 49–54.

72. Zhukovsky, K.V. A harmonic solution for the hyperbolic heat conduction equation and its relationship to the Guyer–Krumhansl Equation. *Mosc. Univ. Phys. Bull.* **2018**, *73*, 45–52. [CrossRef]
73. Zhukovsky, K. Exact harmonic solution to ballistic type heat propagation in thin films and wires. *Int. J. Heat Mass Transf.* **2018**, *120*, 944–955. [CrossRef]
74. Zhukovsky, K.; Oskolkov, D. Exact harmonic solutions to Guyer–Krumhansl-type equation and application to heat transport in thin films. *Contin. Mech. Thermodyn.* **2018**. [CrossRef]
75. Ali, Y.M.; Zhang, L.C. Relativistic heat conduction. *Int. J. Heat Mass Transf.* **2005**, *48*, 2397–2406. [CrossRef]
76. Al-Khairy, R.T.; Al-Ofey, Z.M. Analytical solution of the hyperbolic heat conduction equation for moving semi-infinite medium under the effect of time-dependent laser heat source. *J. Appl. Math.* **2009**, *2009*. [CrossRef]
77. Wolf, K.B. *Integral Transforms in Science and Engineering*; New York Plenum Press: New York, NY, USA, 1979.

15

New Bell–Sheffer Polynomial Sets

Pierpaolo Natalini [1,*] and Paolo Emilio Ricci [2]

1 Dipartimento di Matematica e Fisica, Università degli Studi Roma Tre, Largo San Leonardo Murialdo, 1, 00146 Roma, Italy
2 Sezione di Matematica, International Telematic University UniNettuno, Corso Vittorio Emanuele II, 39, 00186 Roma, Italy; paoloemilioricci@gmail.com
* Correspondence: natalini@mat.uniroma3.it

Abstract: In recent papers, new sets of Sheffer and Brenke polynomials based on higher order Bell numbers, and several integer sequences related to them, have been studied. The method used in previous articles, and even in the present one, traces back to preceding results by Dattoli and Ben Cheikh on the monomiality principle, showing the possibility to derive explicitly the main properties of Sheffer polynomial families starting from the basic elements of their generating functions. The introduction of iterated exponential and logarithmic functions allows to construct new sets of Bell–Sheffer polynomials which exhibit an iterative character of the obtained shift operators and differential equations. In this context, it is possible, for every integer r, to define polynomials of higher type, which are linked to the higher order Bell-exponential and logarithmic numbers introduced in preceding papers. Connections with integer sequences appearing in Combinatorial analysis are also mentioned. Naturally, the considered technique can also be used in similar frameworks, where the iteration of exponential and logarithmic functions appear.

Keywords: Sheffer polynomials; generating functions; monomiality principle; shift operators; combinatorial analysis

1. Introduction

In recent articles [1,2], new sets of Sheffer [3] and Brenke [4] polynomials, based on higher order Bell numbers [2,5–7], have been studied. Furthermore, several integer sequences associated [8] with the considered polynomials sets both of exponential [9,10] and logarithmic type have been introduced [1].

It is worth noting that exponential and logarithmic polynomials have been recently studied in the multidimensional case [11–13].

In this article, new sets of Bell–Sheffer polynomials are considered and some particular cases are analyzed.

It is worth noting that the Sheffer A-type 0 polynomial sets have been also approached with elementary methods of linear algebra (see, e.g., [14–16] and the references therein).

Connection with umbral calculus has been recently emphasized (see, e.g., [17,18] and the references therein).

2. Sheffer Polynomials

The Sheffer polynomials $\{s_n(x)\}$ are introduced [3] by means of the exponential generating function [19] of the type:

$$A(t)\exp(xH(t)) = \sum_{n=0}^{\infty} s_n(x)\,\frac{t^n}{n!}\,, \tag{1}$$

where

$$A(t) = \sum_{n=0}^{\infty} a_n \frac{t^n}{n!}, \qquad (a_0 \neq 0),$$
$$H(t) = \sum_{n=0}^{\infty} h_n \frac{t^n}{n!}, \qquad (h_0 = 0). \tag{2}$$

According to a different characterization (see [20], p. 18), the same polynomial sequence can be defined by means of the pair $(g(t), f(t))$, where $g(t)$ is an invertible series and $f(t)$ is a delta series:

$$g(t) = \sum_{n=0}^{\infty} g_n \frac{t^n}{n!}, \qquad (g_0 \neq 0),$$
$$f(t) = \sum_{n=0}^{\infty} f_n \frac{t^n}{n!}, \qquad (f_0 = 0,\ f_1 \neq 0). \tag{3}$$

Denoting by $f^{-1}(t)$ the compositional inverse of $f(t)$ (i.e., such that $f\left(f^{-1}(t)\right) = f^{-1}\left(f(t)\right) = t$), the exponential generating function of the sequence $\{s_n(x)\}$ is given by

$$\frac{1}{g[f^{-1}(t)]} \exp\left(x f^{-1}(t)\right) = \sum_{n=0}^{\infty} s_n(x) \frac{t^n}{n!}, \tag{4}$$

so that

$$A(t) = \frac{1}{g[f^{-1}(t)]}, \qquad H(t) = f^{-1}(t). \tag{5}$$

When $g(t) \equiv 1$, the Sheffer sequence corresponding to the pair $(1, f(t))$ is called the associated Sheffer sequence $\{\sigma_n(x)\}$ for $f(t)$, and its exponential generating function is given by

$$\exp\left(x f^{-1}(t)\right) = \sum_{n=0}^{\infty} \sigma_n(x) \frac{t^n}{n!}. \tag{6}$$

A list of known Sheffer polynomial sequences and their associated ones can be found in [21].

3. New Bell–Sheffer Polynomial Sets

We introduce, for shortness, the following compact notation.
Put, by definition:

$E_0(t) := \exp(t) - 1,$
$E_1(t) := E_0(E_0(t)) = \exp(\exp(t) - 1) - 1$
....
$E_r(t) := E_0(E_{r-1}(t)) = \exp(\ldots \exp(\exp(t) - 1) - 1) \cdots - 1 \qquad [(r+1) - \text{ times } \exp],$

and in a similar way:

$\Lambda_0(t) := \log(t+1)$
$\Lambda_1(t) := \Lambda_0(\Lambda_0(t)) = \log(\log(t+1) + 1)$
....
$\Lambda_r(t) := \Lambda_0(\Lambda_{r-1}(t)) = \log\left(\log\left(\ldots\left(\log(t+1) + 1\right)\ldots\right) + 1\right), \qquad [(r+1) - \text{ times } \log].$

Remark 1. *Note that, for all integers* $r, k, h,$

$$E_r(\Lambda_r(t)) = t\,, \qquad \Lambda_r(E_r(t)) = t\,,$$

$$(\text{if } k > h) \qquad E_k(\Lambda_h(t)) = E_{k-h-1}(t)\,, \quad E_h(\Lambda_k(t)) = \Lambda_{k-h-1}(t)\,,$$

$$(\text{if } k > h) \qquad \Lambda_k(E_h(t)) = \Lambda_{k-h-1}(t)\,, \quad \Lambda_h(E_k(t)) = E_{k-h-1}(t)\,,$$

$$e^{E_r(t)} = E_{r+1}(t) + 1\,, \qquad e^{\Lambda_r(t)} = \Lambda_{r-1}(t) + 1\,.$$

Remark 2. *Note that the coefficients of the Taylor expansion of* $E_1(t)$ *are given by the Bell numbers* $b_n = b_n^{[1]}$

$$E_1(t) = \sum_{n=1}^{\infty} b_n^{[1]} \frac{t^n}{n!}\,,$$

and, in general, the coefficients of the Taylor expansion of $E_r(t)$ *are given by the higher order Bell numbers* $b_n^{[r]}$

$$E_r(t) = \sum_{n=1}^{\infty} b_n^{[r]} \frac{t^n}{n!}\,.$$

The higher order Bell numbers, also known as higher order exponential numbers, have been considered in [5,7,22], and used in [2] in the framework of Brenke and Sheffer polynomials.

Remark 3. *Note that the coefficients of the Taylor expansion of* $\Lambda_0(t)$ *are given by the logarithmic numbers* $l_n^{[1]} = (-1)^{n-1}(n-1)!$

$$\Lambda_0(t) = \sum_{n=1}^{\infty} l_n^{[1]} \frac{t^n}{n!} = \sum_{n=1}^{\infty} (-1)^{n-1}(n-1)!\,\frac{t^n}{n!}\,,$$

and, in general, the coefficients of the Taylor expansion of $\Lambda_{r-1}(t)$ *are given by the higher order logarithmic numbers* $l_n^{[r]}$

$$\Lambda_{r-1}(t) = \sum_{n=1}^{\infty} l_n^{[r]} \frac{t^n}{n!}\,.$$

The higher order logarithmic numbers, which are the counterpart of the higher order Bell (exponential) numbers, have been considered in [1], *and used there in the framework of new sets of Sheffer polynomials.*

3.1. The Polynomials $\mathcal{E}_k^{(1)}(x)$

Therefore, we consider the Sheffer polynomials, defined through their generating function, by putting

$$A(t) = E_1(t) + 1 = e^{E_0(t)}, \qquad H(t) = E_0(t)\,,$$
$$G(t,x) = \exp\left[(1+x)E_0(x)\right] = \sum_{k=0}^{\infty} \mathcal{E}_k^{(1)}(x)\,\frac{t^k}{k!}\,. \tag{7}$$

3.2. Recurrence Relation for the $\mathcal{E}_k^{(1)}(x)$

Theorem 1. *For any $k \geq 0$, the polynomials $\mathcal{E}_k^{(1)}(x)$ satisfy the following recurrence relation:*

$$\mathcal{E}_{k+1}^{(1)}(x) = \sum_{h=0}^{k} \binom{k}{h} (1+x)\, \mathcal{E}_h^{(1)}(x) . \tag{8}$$

Proof. Differentiating $G(t,x)$ with respect to t, we have

$$\frac{\partial G(t,x)}{\partial t} = G(t,x)\, e^t\, (1+x) , \tag{9}$$

and therefore

$$\sum_{k=0}^{\infty} (1+x)\, \mathcal{E}_k^{(1)}(x) \frac{t^k}{k!} \sum_{k=0}^{\infty} \frac{t^k}{k!} = \sum_{k=0}^{\infty} \mathcal{E}_{k+1}^{(1)}(x) \frac{t^k}{k!} ,$$

i.e.,

$$\sum_{k=0}^{\infty} \sum_{h=0}^{k} \binom{k}{h} (1+x)\, \mathcal{E}_h^{(1)}(x) \frac{t^k}{k!} = \sum_{k=0}^{\infty} \mathcal{E}_{k+1}^{(1)}(x) \frac{t^k}{k!}$$

so that the recurrence relation (8) follows. □

3.3. Generating Function's PDEs

Theorem 2. *The generating function* $(7)_2$ *satisfies the homogeneous linear PDEs:*

$$\left(1-e^{-t}\right) \frac{\partial G(t,x)}{\partial t} = (1+x) \frac{\partial G(t,x)}{\partial x} , \tag{10}$$

$$\frac{\partial G(t,x)}{\partial t} = \frac{\partial G(t,x)}{\partial x} + G(t,x)(1+xe^t) , \tag{11}$$

$$\frac{\partial G(t,x)}{\partial t} = (1+x) \left[\frac{\partial G(t,x)}{\partial x} + G(t,x)\right] . \tag{12}$$

Proof. Differentiating $G(t,x)$ with respect to x, we have

$$\frac{\partial G(t,x)}{\partial x} = G(t,x)\, (e^t - 1) . \tag{13}$$

By taking the ratio between the members of Equations (9) and (13), we find Equation (10). The other ones easily follows by elementary algebraic manipulations. □

3.4. Shift Operators

We recall that a polynomial set $\{p_n(x)\}$ is called quasi-monomial if and only if there exist two operators $\hat{P}$ and $\hat{M}$ such that

$$\hat{P}\,(p_n(x)) = n p_{n-1}(x) , \qquad \hat{M}\,(p_n(x)) = p_{n+1}(x) , \qquad (n = 1,2,\dots). \tag{14}$$

$\hat{P}$ is called the *derivative* operator and $\hat{M}$ the *multiplication* operator, as they act in the same way of classical operators on monomials.

This definition traces back to a paper by Steffensen [23], recently improved by Dattoli [24] and widely used in several applications.

Ben Cheikh [25] proved that every polynomial set is quasi-monomial under the action of suitable derivative and multiplication operators. In particular, in the same article (Corollary 3.2), the following result is proved.

Theorem 3. *Let $(p_n(x))$ denote a Boas–Buck polynomial set, i.e., a set defined by the generating function*

$$A(t)\psi(xH(t)) = \sum_{n=0}^{\infty} p_n(x)\frac{t^n}{n!}, \tag{15}$$

where

$$\begin{aligned} A(t) &= \sum_{n=0}^{\infty} a_n t^n, \qquad (a_0 \neq 0), \\ \psi(t) &= \sum_{n=0}^{\infty} \gamma_n t^n, \qquad (\gamma_n \neq 0 \quad \forall n), \end{aligned} \tag{16}$$

with $\psi(t)$ not a polynomial, and lastly

$$H(t) = \sum_{n=0}^{\infty} h_n\, t^{n+1}, \qquad (h_0 \neq 0). \tag{17}$$

Let $\sigma \in \Lambda^{(-)}$ the lowering operator defined by

$$\sigma(1) = 0, \qquad \sigma(x^n) = \frac{\gamma_{n-1}}{\gamma_n}\, x^{n-1}, \qquad (n = 1, 2, \dots). \tag{18}$$

Put

$$\sigma^{-1}(x^n) = \frac{\gamma_{n+1}}{\gamma_n}\, x^{n+1} \quad (n = 0, 1, 2, \dots). \tag{19}$$

Denoting, as before, by $f(t)$ the compositional inverse of $H(t)$, the Boas–Buck polynomial set $\{p_n(x)\}$ is quasi-monomial under the action of the operators

$$\hat{P} = f(\sigma), \qquad \hat{M} = \frac{A'[f(\sigma)]}{A[f(\sigma)]} + xD_x H'[f(\sigma)]\sigma^{-1}, \tag{20}$$

where prime denotes the ordinary derivatives with respect to t.

Note that, in our case, we are dealing with a Sheffer polynomial set, so that since we have $\psi(t) = e^t$, the operator σ defined by Equation (16) simply reduces to the derivative operator D_x. Furthermore, we have:

$$\begin{aligned} f(t) &= H^{-1}(t) = \Lambda_0(t), \\ \frac{A'(t)}{A(t)} &= e^t, \qquad H'(t) = e^t, \end{aligned}$$

and, consequently,

$$f(\sigma) = \Lambda_0(D_x)\,, \qquad \frac{A'[\Lambda_0(D_x)]}{A[\Lambda_0(D_x)]} = D_x + 1\,,$$

$$H'[f(\sigma)] = H'[\Lambda_0(D_x)] = D_x + 1\,.$$

Theorem 4. *The Bell–Sheffer polynomials $\{\mathcal{E}_k^{(1)}(x)\}$ are quasi-monomial under the action of the operators*

$$\hat{P} = \Lambda_0(D_x) = \sum_{k=0}^{\infty} (-1)^{k+1} \frac{D_x^k}{k}\,, \qquad \hat{M} = (1+x)(D_x+1)\,. \tag{21}$$

3.5. Differential Equation for the $\mathcal{E}_k^{(1)}(x)$

According to the results of monomiality principle [24], the quasi-monomial polynomials $\{p_n(x)\}$ satisfy the differential equation

$$\hat{M}\hat{P}\, p_n(x) = n\, p_n(x)\,. \tag{22}$$

In the present case, recalling Equation (22), we have

Theorem 5. *The Bell–Sheffer polynomials $\{\mathcal{E}_k^{(1)}(x)\}$ satisfy the differential equation*

$$(1+x) \sum_{k=1}^{n} (-1)^{k+1} \left[\frac{D_x^{k+1} + D_x^k}{k!}\right] \mathcal{E}_n^{(1)}(x) = n\, \mathcal{E}_n^{(1)}(x)\,. \tag{23}$$

Proof. Equation (22), by using Equation (21), becomes

$$\hat{M}\hat{P}\, \mathcal{E}_n^{(1)}(x) = (1+x)\,(D_x+1)\,\Lambda_0(D_x)\, \mathcal{E}_n^{(1)}(x) =$$

$$= (1+x)\,(D_x+1) \sum_{k=1}^{n} (-1)^{k+1} \frac{D_x^k}{k!}\, \mathcal{E}_n^{(1)}(x) = n\, \mathcal{E}_n^{(1)}(x)\,,$$

i.e.,

$$(1+x) \sum_{k=1}^{\infty} (-1)^{k+1} \left[\frac{D_x^{k+1} + D_x^k}{k!}\right] \mathcal{E}_n^{(1)}(x) = n\, \mathcal{E}_n^{(1)}(x)\,,$$

and, furthermore, for any fixed n, the last series expansion reduces to a finite sum, with upper limit $n-1$, when it is applied to a polynomial of degree n because the last not vanishing term (for $k = n-1$) contains the derivative of order n. □

3.6. First Few Values of the $\mathcal{E}_k^{(1)}(x)$

Here, we show the first few values for the Bell–Sheffer polynomials $\mathcal{E}_k^{(1)}(x)$, defined by the generating function (7)$_2$

$$\begin{aligned}
&\mathcal{E}_0^{(1)}(x) = 1,\\
&\mathcal{E}_1^{(1)}(x) = x+1,\\
&\mathcal{E}_2^{(1)}(x) = x^2+3x+2,\\
&\mathcal{E}_3^{(1)}(x) = x^3+6x^2+10x+5,\\
&\mathcal{E}_4^{(1)}(x) = x^4+10x^3+31x^2+37x+15,\\
&\mathcal{E}_5^{(1)}(x) = x^5+15x^4+75x^3+160x^2+151x+52.
\end{aligned}$$

Further values can be easily achieved by using Wolfram Alpha© (2009, Wolfram Research, Champaign, IL, USA).

Remark 4. *Note that the numerical values $\mathcal{E}_k^{(1)}(0)$ of the considered Bell–Sheffer polynomials*

$$(1,1,2,5,15,52,203,877,\dots)$$

appears in the Encyclopedia of Integer Sequences [8] under A000110: Bell or exponential numbers: number of ways to partition a set of n labeled elements.

The same sequence also appears under A164864, A164863, A276723, A276724, A276725, A276726, A287278, A287279, A287280.

4. Iterated Bell–Sheffer Polynomial Sets

Here, we iterate the procedure introduced in Section 3, by considering the Sheffer polynomial sets defined by putting

$$\begin{aligned}
&A(t) = E_2(t)+1 = e^{E_1(t)}, \qquad H(t) = E_1(t),\\
&G(t,x) = \exp\left[(1+x)E_1(x)\right] = \sum_{k=0}^{\infty} \mathcal{E}_k^{(2)}(x)\,\frac{t^k}{k!}\,.
\end{aligned} \tag{24}$$

We find:

$$f(t) = H^{-1}(t) = \Lambda_1(t)\,,$$

$$\frac{A'(t)}{A(t)} = H'(t) = \left[E_1(t)+1\right]e^t = \left[E_1(t)+1\right]\left[E_0(t)+1\right],$$

and, consequently,

$$f(\sigma) = \Lambda_1(D_x)\,, \qquad H'[f(\sigma)] = H'[\Lambda_1(D_x)] = [D_x+1]\,[\Lambda_0(D_x)+1]\,,$$

$$\frac{A'[\Lambda_1(D_x)]}{A[\Lambda_1(D_x)]} = \left[E_1(\Lambda_1(D_x))+1\right]e^{\Lambda_1(D_x)} = (D_x+1)\,[\Lambda_0(D_x)+1]\,.$$

Theorem 6. *The Bell–Sheffer polynomials $\{\mathcal{E}_k^{(2)}(x)\}$ are quasi-monomial under the action of the operators*

$$\begin{aligned}
&\hat{P} = \Lambda_1(D_x)\,,\\
&\hat{M} = (1+x)\,(D_x+1)\,[\Lambda_0(D_x)+1]\,.
\end{aligned} \tag{25}$$

4.1. Differential Equation for the $\mathcal{E}_k^{(2)}(x)$

According to the results of monomiality principle [24,26], the quasi-monomial polynomials $\{p_n(x)\}$ satisfy the differential equation

$$\hat{M}\hat{P}\, p_n(x) = n\, p_n(x)\,. \tag{26}$$

In the present case, recalling Equation (19), we have

Theorem 7. *The Bell–Sheffer polynomials* $\{\mathcal{E}_k^{(2)}(x)\}$ *satisfy the differential equation*

$$(1+x)\,(D_x+1)\,\left[\Lambda_0(D_x)+1\right]\,\Lambda_1(D_x)\,\mathcal{E}_n^{(2)}(x) = n\,\mathcal{E}_n^{(2)}(x)\,. \tag{27}$$

4.2. First Few Values of the $\mathcal{E}_k^{(2)}(x)$

Here, we show the first few values for the Bell–Sheffer polynomials $\mathcal{E}_k^{(2)}(x)$, defined by the generating function $(7)_2$

$$\begin{aligned}
&\mathcal{E}_0^{(2)}(x) = 1,\\
&\mathcal{E}_1^{(2)}(x) = x+1,\\
&\mathcal{E}_2^{(2)}(x) = x^2+4x+3,\\
&\mathcal{E}_3^{(2)}(x) = x^3+9x^2+20x+12,\\
&\mathcal{E}_4^{(2)}(x)) = x^4+16x^3+74x^2+119x+60,\\
&\mathcal{E}_5^{(2)}(x) = x^5+25x^4+200x^3+635x^2+817x+358.
\end{aligned}$$

Further values can be easily achieved by using Wolfram Alpha©.

Remark 5. *Note that the numerical values* $\mathcal{E}_k^{(2)}(0)$ *of the considered Bell–Sheffer polynomials*

$$(1,1,3,12,60,358,2471,19302,\dots)$$

appear in the Encyclopedia of Integer Sequences under A000258: McLaurin coefficients of the function $\mathrm{E}_2(x)$.

5. The General Case

In general, by putting

$$\begin{aligned}
&A(t) = \mathrm{E}_r(t)+1 = e^{\mathrm{E}_{r-1}(t)}, \qquad H(t) = \mathrm{E}_{r-1}(t)\,,\\
&G(t,x) = \exp\left[(1+x)E_{r-1}(t)\right] = \sum_{k=0}^{\infty} \mathcal{E}_k^{(r)}(x)\,\frac{t^k}{k!}\,,
\end{aligned} \tag{28}$$

we find:

$$f(t) = H^{-1}(t) = \Lambda_{r-1}(t)\,,$$

$$\frac{A'(t)}{A(t)} = H'(t) = \prod_{\ell=1}^{r-1}\left[\mathrm{E}_\ell(t)+1\right]e^t = \prod_{\ell=0}^{r-1}\left[\mathrm{E}_\ell(t)+1\right],$$

and, consequently,

$$f(\sigma) = \Lambda_{r-1}(D_x), \qquad \frac{A'[\Lambda_{r-1}(D_x)]}{A[\Lambda_{r-1}(D_x)]} = \prod_{\ell=0}^{r-1} [E_\ell(\Lambda_{r-1}(D_x)) + 1],$$

$$H'[f(\sigma)] = H'[\Lambda_{r-1}(D_x)] = \prod_{\ell=0}^{r-1} [E_\ell(\Lambda_{r-1}(D_x)) + 1].$$

Recalling Remark 3.1, we find

$$E_\ell(\Lambda_{r-1}(D_x)) = \Lambda_{r-\ell-2}(D_x),$$

$$\prod_{\ell=0}^{r-1} [E_\ell(\Lambda_{r-1}(D_x)) + 1] = (D_x + 1) \prod_{k=0}^{r-2} [\Lambda_k(D_x) + 1],$$

so that we have the theorem:

Theorem 8. *The Bell–Sheffer polynomials* $\{\mathcal{E}_k^{(r)}(x)\}$ *are quasi-monomial under the action of the operators*

$$\hat{P} = \Lambda_{r-1}(D_x),$$
$$\hat{M} = (1+x)\ (D_x + 1) \prod_{k=0}^{r-2} [\Lambda_k(D_x) + 1]. \tag{29}$$

Differential Equation for the $\mathcal{E}_k^{(r)}(x)$

According to the results of monomiality principle [24], the quasi-monomial polynomials $\{p_n(x)\}$ satisfy the differential equation

$$\hat{M}\hat{P}\, p_n(x) = n\, p_n(x). \tag{30}$$

In the present case, recalling Equation (19), we have

Theorem 9. *The Bell–Sheffer polynomials* $\{\mathcal{E}_k^{(r)}(x)\}$ *satisfy the differential equation*

$$(1+x)\ (D_x+1) \prod_{k=0}^{r-2} [\Lambda_k(D_x) + 1]\, \Lambda_{r-1}(D_x)\, \mathcal{E}_n^{(r)}(x) = n\, \mathcal{E}_n^{(r)}(x). \tag{31}$$

6. Conclusions

By introducing iterated exponential and logarithmic functions, we have shown how to construct new sets of Bell-Sheffer polynomials which exhibit an iterative character. We have found their main properties by using the monomiality property and a general result by Y. Ben Cheikh which gives explicitly the derivative and multiplication operators for polynomials of Sheffer type. The tools we used are internal to Sheffer's polynomial theory and do not use external techniques. In our opinion the demonstrated properties (in particular the differential equations for polynomials of higher order) could hardly be achieved by other methods.

Author Contributions: The authors declare to have both contributed to the final version of this article.

References

1. Bretti, G.; Natalini, P.; Ricci, P.E. A new set of Sheffer-Bell polynomials and logarithmic numbers. *Georgian Math. J.* **2018**, in print.
2. Ricci, P.E.; Natalini, P.; Bretti, G. Sheffer and Brenke polynomials associated with generalized Bell numbers. *Jnanabha Vijnana Parishad India* **2017**, *47*, 337–352.
3. Sheffer, I.M. Some properties of polynomials sets of zero type. *Duke Math. J.* **1939**, *5*, 590–622. [CrossRef]
4. Brenke, W.C. On generating functions of polynomial systems. *Am. Math. Mon.* **1945**, *52*, 297–301. [CrossRef]
5. Natalini, P.; Ricci, P.E. Remarks on Bell and higher order Bell polynomials and numbers. *Cogent Math.* **2016**, *3*, 1–15. [CrossRef]
6. Bernardini, A.; Natalini, P.; Ricci, P.E. Multi-dimensional Bell polynomials of higher order. *Comput. Math. Appl.* **2005**, *50*, 1697–1708. [CrossRef]
7. Natalini, P.; Ricci, P.E. An extension of the Bell polynomials. *Comput. Math. Appl.* **2004**, *47*, 719–725. [CrossRef]
8. Sloane, N.J.A. The On-Line Encyclopedia of Integer Sequences. 2016. Available online: http://oeis.org (accessed on 24 May 2018).
9. Bell, E.T. Exponential polynomials. *Ann. Math.* **1934**, *35*, 258–277. [CrossRef]
10. Bell, E.T. The Iterated Exponential Integers. *Ann. Math.* **1938**, *39*, 539–557. [CrossRef]
11. Qi, F.; Niu, D.W.; Lim, D.; Guo, B.N. Some Properties and an Application of Multivariate Exponential Polynomials. HAL Archives. 2018. Available online: https://hal.archives-ouvertes.fr/hal-01745173 (accessed on 13 September 2018).
12. Qi, F. Integral representations for multivariate logarithmic potentials. *J. Comput. Appl. Math.* **2018**, *336*, 54–62. [CrossRef]
13. Qi, F. On multivariate logarithmic polynomials and their properties. *Indag. Math.* **2018**, *336*. [CrossRef]
14. Costabile, F.A.; Longo, E. A new recurrence relation and related determinantal form for Binomial type polynomial sequences. *Mediterr. J. Math.* **2016**, *13*, 4001–4017. [CrossRef]
15. Yang, Y.; Youn, H. Appell polynomial sequences: a linear algebraic approach. *JP J. Algebra Number Theory Appl.* **2009**, *13*, 65–98.
16. Yang, S.L. Recurrence relations for Sheffer sequences. *Linear Algebra Appl.* **2012**, *437*, 2986–2996. [CrossRef]
17. Kim, T.; Kim, D.S. On λ-Bell polynomials associated with umbral calculus. *Russ. J. Math. Phys.* **2017**, *24*, 69–78. [CrossRef]
18. Kim, T.; Kim, D.S.; Jang, G.-W. Some formulas of ordered Bell numbers and polynomials arising from umbral calculus. *Proc. Jangjeon Math. Soc.* **2017**, *20*, 659–670.
19. Srivastava, H.M.; Manocha, H.L. *A Treatise on Generating Functions*; Ellis Horwood: Chichester, UK, 1984.
20. Roman, S.M. *The Umbral Calculus*; Academic Press: New York, NY, USA, 1984.
21. Boas, R.P.; Buck, R.C. *Polynomial Expansions of Analytic Functions*; Springer: Berlin/Heidelberg, Germany; Gottingen, Germany; New York, NY, USA, 1958.
22. Natalini, P.; Ricci, P.E. Higher order Bell polynomials and the relevant integer sequences. *Appl. Anal. Discret. Math.* **2017**, *11*, 327–339. [CrossRef]
23. Steffensen, J.F. The poweroid, an extension of the mathematical notion of power. *Acta Math.* **1941**, *73*, 333–366. [CrossRef]
24. Dattoli, G. Hermite-Bessel and Laguerre-Bessel functions: A by-product of the monomiality principle. In *Advanced Special Functions and Applications, Proceedings of the Melfi School on Advanced Topics in Mathematics and Physics, Melfi, Italy, 9–12 May 1999*; Cocolicchio, D., Dattoli, G., Srivastava, H.M., Eds.; Aracne Editrice: Rome, Italy, 2000; pp. 147–164.
25. Cheikh, Y.B. Some results on quasi-monomiality. *Appl. Math. Comput.* **2003**, *141*, 63–76.
26. Dattoli, G.; Ricci, P.E.; Srivastava, H.M. (Eds.) Advanced Special Functions and Related Topics in Probability and in Differential Equations. In *Applied Mathematics and Computation, Proceedings of the Melfi School on Advanced Topics in Mathematics and Physics, Melfi, Italy, 24–29 June 2001*; Aracne Editrice: Rome, Italy, 2000; Volume 141, pp. 1–230.

16

A Short Note on Integral Transformations and Conversion Formulas for Sequence Generating Functions

Maxie D. Schmidt

School of Mathematics, Georgia Institute of Technology, Atlanta, GA 30318, USA; maxieds@gmail.com or mschmidt34@gatech.edu

Abstract: The purpose of this note is to provide an expository introduction to some more curious integral formulas and transformations involving generating functions. We seek to generalize these results and integral representations which effectively provide a mechanism for converting between a sequence's ordinary and exponential generating function (OGF and EGF, respectively) and vice versa. The Laplace transform provides an integral formula for the EGF-to-OGF transformation, where the reverse OGF-to-EGF operation requires more careful integration techniques. We prove two variants of the OGF-to-EGF transformation integrals from the Hankel loop contour for the reciprocal gamma function and from Fourier series expansions of integral representations for the Hadamard product of two generating functions, respectively. We also suggest several generalizations of these integral formulas and provide new examples along the way.

Keywords: generating function; series transformation; gamma function; Hankel contour

---

1. Introduction

1.1. Definitions

Given a sequence $\{f_n\}_{n\geq 0}$, we adopt the notation for the respective ordinary generating function (OGF), $F(z)$, and exponential generating function (EGF), $\widehat{F}(z)$, of the sequence in some formal indeterminate parameter $z \in \mathbb{C}$:

$$F(z) = \sum_{n\geq 0} f_n z^n \qquad (1)$$
$$\widehat{F}(z) = \sum_{n\geq 0} \frac{f_n}{n!} z^n.$$

Notice that we can always construct these functions over any sequence $\{f_n\}_{n\in\mathbb{N}}$ and formally perform operations on these functions within the ring of formal power series in z without any considerations on the constraints imposed by the convergence of the underlying series as a complex function of z. If we assume that the respective series for $F(z)$ or $\widehat{F}(z)$ is analytic, or converges absolutely, for all $z \in \mathbb{C}$ with $0 < |z| < \sigma_f$, then we can apply complex function theory to these sequence generating functions and treat them as analytic functions of z on this region.

We can precisely define the form of an integral transformation (in one variable) as [1] (§ 1.4)

$$\mathcal{I}[f(x)](k) := \int_a^b \mathcal{K}(x,k) f(x) dx, \qquad (2)$$

for $-\infty \leq b < a \leq +\infty$ and where the function $\mathcal{K} : \mathbb{R} \times \mathbb{C} \to \mathbb{C}$ is called the kernel of the transformation. When the function f which we operate on in the formula given by the last equation corresponds to an OGF or EGF of a sequence with which we are concerned in applications, we consider integrals of the form in (2) to be so-called generating function transformations. Such generating function transformations are employed to transform the ordinary power series of the target generating function for one sequence into the form of a generating function which enumerates another sequence we are interested in studying.

Generating function transformations form a useful combinatorial and analytic method (depending on perspective) which can be combined and employed to study new sequences of many forms. Our focus in this article is to motivate the constructions of generating function transformations as meaningful and indispensable tools in enumerative combinatorics, combinatorial number theory, and in the theory of partitions, among other fields where such applications live. The particular modus operandi within this article shows the evolution of integral transforms for the reciprocal gamma function, and its multi-factorial integer sequence special cases, as a motivating method for enumerating several types of special sequences and series which we will consider in the next sections.

The references [2–4] provide a much broader sense of the applications of generating function techniques in general to those readers who are not familiar with this topic as a means for sequence enumeration. A comprehensive array of analytic and experimental techniques in the theory of integral transformations is also treated in the references [1,5]. We focus on only a comparatively few concrete examples of integral and sequence transformations in the next subsections with hopes to motivate our primary results proved in this article from this perspective. We hope that the discussion of these techniques in this short note provide motivation and useful applications to readers in a broader range of mathematical areas.

1.2. *From Hobby To Short Note: OGF-to-EGF Conversion Formulas*

A time consuming hobby that the author assumes from time to time is rediscovering old and unusual identities in mathematics textbooks– particularly in the areas of combinatorics and discrete mathematics. Favorite books to search include Comtet's *Advanced Combinatorics* and the exercises and their solutions found in *Concrete Mathematics* by Graham, Knuth and Patashnik. One curious and interesting conversion operation discussed in the exercises to Chapter 7 of the latter book involves a pair of integral formulas for converting an arbitrary sequence OGF into its EGF and vice versa provided the resulting integral is suitably convergent. The exercise listed in Concrete Mathematics suggests the second form of the operation. Namely, that of converting a sequence EGF into its OGF.

In this direction, we have an easy conversion integral for converting from the EGF of a sequence $\{f_n\}_{n\geq 0}$, denoted by $\widehat{F}(z)$, and its corresponding OGF, denoted by $F(z)$, given by the Laplace–Borel transform [6] (§ B.14):

$$\mathcal{L}[\widehat{F}](z) = F(z) = \int_0^\infty \widehat{F}(tz)e^{-t}dt.$$

Other integral formulas for conversions between specified generating function "types" can be constructed similarly as well (see Section 1.3). The key facets in constructing these semi-standard, or at least known, conversion integrals is in applying a termwise series operation which generates a factor, or reciprocal factor, of the gamma function $\Gamma(z+1)$ when $z \in \mathbb{N}$. The corresponding "reversion" operation of converting from a sequence's OGF to its EGF requires a more careful treatment of the properties of the reciprocal gamma function, $1/\Gamma(z+1)$, and the construction of integral formulas which generate it for $z \in \mathbb{N}$ involving the Hankel loop contour described in Section 2.

That being said, Graham, Knuth and Patshnik already suggest a curious "known" integral formula for performing this corresponding OGF-to-EGF conversion operation of the following form [3] (p. 566):

$$\widehat{F}(z) = \frac{1}{2\pi} \int_{-\pi}^{\pi} F\left(ze^{-\imath t}\right) e^{e^{\imath t}} dt. \tag{3}$$

The statement of this result is given without proof in the identity-full appendix section of the textbook. When first (re)-discovered many years back, the author assumed that the motivation for this integral transformation must correspond to the non-zero paths of a complex contour integral for the reciprocal gamma function. For many years the precise formulation of a proof of this termwise integral formula and its generalization to enumerating terms of reciprocal generalized multi-factorial functions, such as $1/(2n-1)!!$, remained a mystery and curiosity of periodic interest to the author. In the summer of 2017, the author finally decided to formally inquire about the proof and possible generalizations in an online mathematics forum. The question went unanswered for over a year until by chance the author stumbled onto a Fourier series identity which finally motivated a rigorous proof of the formula in (3). This note explains this proof and derives another integral formula for this operation of OGF-to-EGF inversion based on the Hankel loop contour. The preparation of this article is intended to be expository in nature in the hope of inspiring the creativity of more researchers towards developing related integral transformations of sequence generating functions.

1.3. Examples: Integral transformations of a Sequence Generating Function

Integral transformations are a powerful and convenient formal and analytic tool which are used to study sequences and their properties. Moreover, they are easy to parse and apply in many contexts with only basic knowledge of infinitesimal calculus making them easy-to-understand operations which we can apply to sequence generating functions. The author is an enthusiast for particularly pretty or interesting integral representations (cf. [5,7]) and has taken a special research interest in finding integral formulas of the ordinary generating function of sequence which transform the series into another generating function enumerating a modified special sequence.

One notable example of such an integral transformation given in [8] (§ 2) allows us to construct generalized polylogarithm-like and Dirichlet-like series over any prescribed sequence in the following forms for integers $r \geq 1$:

$$\sum_{n\geq 0} \frac{f_n}{(n+1)^r} z^n = \frac{(-1)^{r-1}}{(r-1)!} \int_0^1 \log^{r-1}(t) F(tz) dt \tag{4}$$
$$= \frac{1}{r!} \int_0^\infty t^{r-1} e^{-t} F\left(e^{-t} z\right) dt.$$

Another source of generating function transformation identities correspond to the bilateral series given by Lindelöf in [9] (§ 2) of the form

$$\sum_{n=-\infty}^{\infty} f(n) z^n = -\frac{1}{2\pi \imath} \oint_\gamma \pi \cot(\pi w) f(w) z^w dw, \tag{5}$$

where γ is any closed contour in $\mathbb{C}$ which contains all of the singular points of f in its interior. In this note, we will focus on integral formulas for generating function transformations of an arbitrary sequence, $\{f_n\}_{n\geq 0}$.

Additional series transformations involving a sequence generating function into the form of $\sum_{n\geq 0} f_n z^n / g(n)^s$ for $\operatorname{Re}(s) > 1$ and non-zero sequences $\{g(n)\}_{n\geq 0}$ are proved in [10,11]. Note that the harmonic-number-related coefficients implicit to these series transformations satisfy summation formulas which are readily expressed by Nörlund-Rice contour integral formulas as well. The author has proved in [12] so-called square series transformations providing that

$$\sum_{n\geq 0} f_n q^{n^2} z^n = \frac{1}{\sqrt{2\pi}} \int_0^\infty \left[\sum_{b=\pm 1} F\left(e^{bt\sqrt{2\operatorname{Log}(q)}}\right)\right] e^{-t^2/2} dt, \ |q|, |qz| < 1. \tag{6}$$

Applications of these square series integral representations include many new integral formulas for theta functions and classical q-series identities such as the Jacobi triple product and the partition function generating function, $(q;q)_\infty^{-1}$, expanded by Euler's pentagonal number theorem.

There are more general Meinardus methods for computing asymptotics of the coefficients of classes of partition number generating functions of the form [13]

$$\sum_{n\geq 0} p_n(b)z^n := \prod_{k\geq 1}\left(1-z^k\right)^{-b_k}, \tag{7}$$

where $p_n(b)$ denotes the number of weighted partitions of n corresponding to the parameter weights b_k for $k \geq 1$. Generating functions enumerating partition function sequences of this type are related to a known Euler transform of a sequence $\{a_n\}_{n\geq 1}$ given by [14]

$$1+\sum_{n\geq 1} b_n z^n := \prod_{j\geq 1}\frac{1}{(1-z^j)^{a_j}} \implies \log\left(1+B(z)\right) = \sum_{k\geq 1}\frac{A(z^k)}{k}, \tag{8}$$

where $A(z) := \sum_n a_n z^n$ and $B(z) := \sum_n b_n z^n$ are the respective OGFs of the component sequences. In this case the right-hand-side generating function in the last equation is generated succinctly by a q-integral for the q-beta function of the form [15]

$$\frac{1}{1-q}\int_0^1 f(x)d(z,x) = \sum_{i\geq 0} f(z^i)z^i,$$

where inputting the modified generating function, $\widetilde{A}_z(t) := A(t)\log(z)/(t\log t)$ for fixed z, into this integral formula generates the second to last series result.

1.4. Results Proved in This Note

In this short note we provide proofs of known integral formulas providing an ordinary-to-exponential generating function operation. We prove the following theorem using the Hankel loop contour for the reciprocal gamma function in Section 2.

Theorem 1 (OGF-to-EGF Integral Formula I). *For any real $c > 0$, provided that $F(z)$ is analytic for $0 < |z| \leq c$, we have that*

$$\widehat{F}(z) = \sum_{n\geq 0} f_n z^n \int_{-\infty}^{\infty} \frac{e^{c+\imath t}}{(c+\imath t)^{n+1}}dt = \int_{-\infty}^{\infty}\frac{e^{c+\imath t}}{(c+\imath t)}F\left(\frac{z}{c+\imath t}\right)dt.$$

We also give a rigorous proof of the next integral formula relating $F(z)$ and $\widehat{F}(z)$.

Theorem 2 (OGF-to-EGF Integral Formula II). *If $F(z)$ is analytic for $0 < |z| < \sigma_f$, we have that* (3) *holds. Namely, we have that*

$$\widehat{F}(z) = \frac{1}{2\pi}\int_{-\pi}^{\pi} F\left(ze^{\imath t}\right)e^{e^{\imath t}}dt.$$

The proof of Theorem 2 is given in Section 3.

2. Integral Representations of the Reciprocal Gamma Function

Since $\Gamma(z)$ is a meromorphic function of z with poles at the non-positive integers, it follows that the reciprocal gamma function, $1/\Gamma(z)$, is an entire function (of order one) with zeros at $z = 0, -1, -2, \ldots$ [16] (§ 5.1). Indeed, as $|z| \to \infty$ at a constant $|\arg(z)| < \pi$, we can expand

$$\log\left[\frac{1}{\Gamma(z)}\right] \sim -z\log z + z + \frac{1}{2}\log\left(\frac{z}{2\pi}\right) - \frac{1}{12z} + \frac{1}{360z^3} - \frac{1}{1260z^5}, \tag{9}$$

which can be computed via the infinite products

$$\frac{1}{\Gamma(z)} = z\prod_{n\geq 1}\frac{\left(1+\frac{z}{n}\right)}{\left(1+\frac{1}{n}\right)^z} = ze^{\gamma z}\prod_{n\geq 1}\left(1+\frac{z}{n}\right)e^{-z/n},$$

where $\gamma \approx 0.577216$ is Euler's gamma constant. Classically, Karl Weierstrass called the function $1/\Gamma(z)$ the "factorielle" function, and used its representation to prove his famous Weierstrass factorization theorem in complex analysis [17] (§ 2).

For $z \in \mathbb{C}$ such that $\mathrm{Re}(z) > 0$ we have a known series expansion for the reciprocal gamma function given by

$$\frac{1}{\Gamma(z)} = \sum_{k=1}^{\infty} a_k z^k = z + \gamma z^2 + \left(\frac{\gamma^2}{2} - \frac{\pi^2}{12}\right)z^3 + \left(\frac{\gamma^3}{6} - \frac{\gamma\pi^2}{12} + \frac{\zeta(3)}{3}\right)z^4 + \cdots. \tag{10}$$

The coefficients a_k in this expansion satisfy many known recurrence relations and expansions by the Riemann zeta function. In [18] an exact integral formula for these coefficients is given by

$$a_n = \frac{(-1)^n}{\pi \cdot n!}\int_0^\infty e^{-t}\,\mathrm{Im}\left\{(\log t - \imath\pi)^n\right\}dt.$$

This integral formula is obtained in the reference using Euler's reflection formula for the gamma function given by

$$\frac{1}{\Gamma(z)} = \frac{\sin(\pi z)}{\pi}\Gamma(1-z),$$

and then applying a standard known real integral to express the gamma function on the right-hand-side of the previous equation. Equivalently, the reflection formula can be stated as

$$\frac{1}{\Gamma(1+z)\Gamma(1-z)} = \frac{\sin(\pi z)}{\pi z}.$$

2.1. The Hankel Loop Contour for the Reciprocal Gamma Function

We seek an exact integral representation for the reciprocal gamma function, not just an integral formula defining the coefficients of its Taylor series expansion about zero in this case. To find such a formula we must use the Hankel loop contour $H_{\delta,\varepsilon}$ shown in Figure 1 and consider the contributions of each component section of the contour in the limiting cases for increasingly small $\delta, \varepsilon \to 0$. We prove Theorem 1 using the next lemma derived from this contour below.

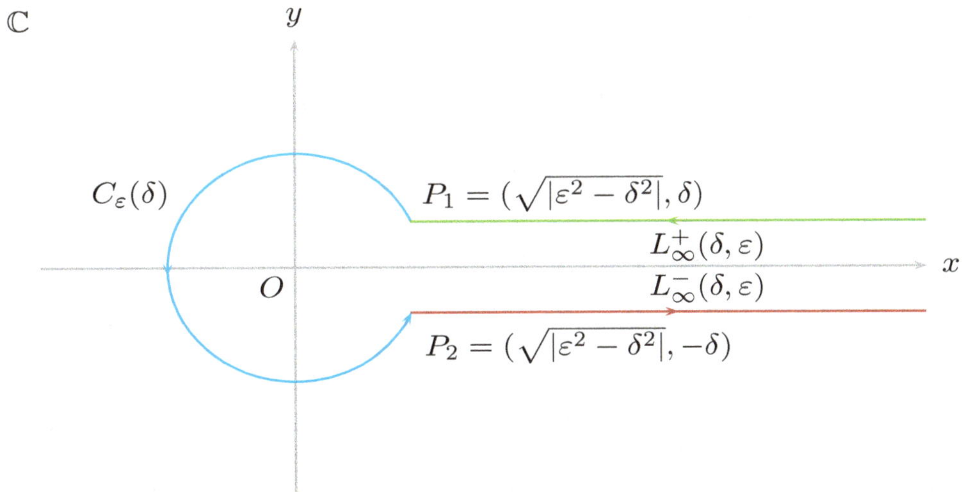

Figure 1. The Hankel loop contour providing an integral representation of the reciprocal gamma function when $\mathrm{Re}(z) > 0$. This contour starts positively from the right, traverses the horizontal line $L_\infty^+(\delta,\varepsilon)$ at distance $+\delta$ from the x-axis from $+\infty \to \sqrt{|\varepsilon^2-\delta^2|}$, then enters the semi-circular loop about the origin of radius ε denoted by $C_\varepsilon(\delta)$ at the point P_1, and then at the point $P_2 = (\sqrt{|\varepsilon^2-\delta^2|}, -\delta)$ traverses the last horizontal line $L_\infty^-(\delta,\varepsilon)$ back to infinity parallel to the x-axis.

Lemma 1. *For any real $c > 0$ and $z \in \mathbb{C}$ such that* $\mathrm{Re}(z) > 0$,

$$\frac{1}{\Gamma(z)} = \frac{1}{2\pi} \int_{-\infty}^{\infty} (c+\imath t)^{-z} e^{c+\imath t} dt. \tag{11}$$

Proof. Working from the figure, we have that [16] (§5.9)

$$\frac{1}{\Gamma(z)} = \lim_{d,\varepsilon\to 0} \frac{1}{2\pi\imath} \oint_{H_{H_{\delta,\varepsilon}}} (-t)^{-z} e^{-t} dt \tag{12a}$$

$$= \lim_{d,\varepsilon\to 0} \frac{1}{2\pi\imath} \left[\int_{C_\varepsilon(\delta)} + \int_{L_\infty^+(\delta,\varepsilon)} + \int_{L_\infty^-(\delta,\varepsilon)}\right] \left(e^{-\imath\pi z} t^{-z} e^{-t}\right) dt. \tag{12b}$$

We will first approach the contribution of the section of the contour given by C_ε which is a path enclosing the origin along the circle of radius ε centered at $(0,0)$. This portion of the contour is oriented in the positive direction and begins at the point $P_1 := (\sqrt{|\varepsilon^2-\delta^2|}, \delta)$ and ends at the point $P_2 := (\sqrt{|\varepsilon^2-\delta^2|}, -\delta)$. By parameterizing t along this circle, we obtain the real integral giving

$$I_C := \lim_{d,\varepsilon\to 0} \int_{\sin^{-1}\left(-\frac{\delta}{\varepsilon}\right)}^{\sin^{-1}\left(\frac{\delta}{\varepsilon}\right)} \imath\varepsilon^2 e^{-\imath\pi z} e^{-2izt} e^{-e^{2it}} dt = 0, \tag{12c}$$

since $\sin^{-1}\left(\frac{\delta}{\varepsilon}\right) = \frac{\delta}{\varepsilon} + \frac{\delta^3}{6\varepsilon^3} + O\left(\frac{\delta^5}{\varepsilon^5}\right) \to 0$ as δ, ε independently tend to zero. Now we can easily parameterize each of the sections of the contour on the horizontal lines each at distance δ from the x-axis. In particular, let's define our integrand in the complex parameters z, w as $f_\Gamma(z,w) := e^{-\imath\pi z} w^{-z} e^{-w}$. Then we consider the limiting cases of the following parameterizations of the two line segments $\{(s, \pm\delta) : s \in [\sqrt{|\varepsilon^2-\delta^2|}, T]\}$ on $L_\infty^+(\delta,\varepsilon)$ and $L_\infty^-(\delta,\varepsilon)$, respectively, by evaluating the limit of $\delta, \varepsilon \to 0$ and then letting T tend to $+\infty$:

$$z_\pm(\delta,\varepsilon;t) := \sqrt{|\varepsilon^2-\delta^2|} \pm \imath\delta + t\left(T - \sqrt{|\varepsilon^2-\delta^2|}\right) \tag{12d}$$

$$z'_\pm(\delta,\varepsilon;t) = T - \sqrt{|\varepsilon^2-\delta^2|}, \quad \text{for } t \in [0,1]. \tag{12e}$$

When we take the first small-order limits we obtain

$$\lim_{\delta,\varepsilon\to 0}\int_0^1 f_\Gamma\left(z_\pm(\delta,\varepsilon;t)\right)\cdot z'_\pm(\delta,\varepsilon;t)dt = \frac{1}{2\pi\imath}\int_0^T e^{-\imath\pi z}s^{-z}e^{-s}ds, \tag{12f}$$

which by substitution provides us with the symmetric bounds of integration given by

$$\lim_{T\to\infty}\frac{1}{2\pi\imath}\int_0^T e^{-\imath\pi z}s^{-z}e^{-s}ds = \pm\int_{\mp\infty}^0 s^{-z}e^{s}ds. \tag{12g}$$

We then finally arrive at the stated known integral formula for the reciprocal gamma function which holds for any fixed real $c > 0$. □

Proof of Theorem 1. Since we are initially motivated by finding a general conversion integral from a sequence OGF into its EGF, we notice that we require an application of (11) termwise to the Taylor series expansions of our prescribed generating function by setting $z = n+1$. For example, if we assume that our sequence OGF at hand is well enough behaved when its argument satisfies $0 < \mathrm{Re}(z) < c$ for some fixed choice of the real $c > 0$ in the integral formula from above, we can sum the integrand of (11) termwise to obtain

$$\widehat{F}(z) = \sum_{n\geq 0} f_n z^n \int_{-\infty}^{\infty}\frac{e^{c+\imath t}}{(c+\imath t)^{n+1}}dt = \int_{-\infty}^{\infty}\frac{e^{c+\imath t}}{(c+\imath t)}F\left(\frac{z}{c+\imath t}\right)dt. \quad \square$$

2.2. Examples: Applications of the Integral Formula on the Real Line

We can perform the same "trick" of the generating function trades to sum a "doubly exponential" sequence generating function when we replace the sequence OGF by its EGF in the previous equation:

$$\sum_{n\geq 0}\frac{f_n z^n}{(n!)^2} = \int_{-\infty}^{\infty}\frac{e^{c+\imath t}}{(c+\imath t)}\widehat{F}\left(\frac{z}{c+\imath t}\right)dt. \tag{13}$$

Perhaps at first glance this iterated integral formula is somewhat unsatisfying since we have really just repeated the procedure for constructing the first integral twice, but in fact there are notable special case applications which we can derive from this method of summation which provide new integral representations for otherwise hard-to-sum hypergeometric series.

For example, if we take the geometric series sequence case where $f_n \equiv 1$ for all $n \geq 0$, then we can arrive at a new integral formula for the doubly exponential series expansion of the incomplete Bessel function, $I_0(2\sqrt{z}) = \sum_{n\geq 0} z^n/(n!)^2$ [3] (§ 5.5). In particular, we easily obtain that

$$I_0(2\sqrt{z}) = \int_{-\infty}^{\infty}\frac{e^{c+\imath t}}{c+\imath t}\exp\left(\frac{z}{c+\imath t}\right)dt. \tag{14}$$

There is an integral representation for this function which is simpler to evaluate in the general case given in (3). We elaborate more on this identity, its proof, and the corresponding series involving Stirling numbers which it implies in the next section.

3. An Integral Formula from Fourier Analysis

One curious identity that the author has come across relating the OGF of a sequence to its EGF is found in the appendices of the Concrete Mathematics reference [3] (p. 566). It states (3) without proof, again providing that

$$\widehat{F}(z) = \frac{1}{2\pi}\int_{-\pi}^{\pi} F\left(ze^{-\imath t}\right)e^{e^{\imath t}}dt.$$

Finding a precise method of verifying this unproven identity is the initial motivation for this note. Given the discussion and lead up to an integral for the reciprocal gamma function taken over the real line via the Hankel loop contour in the last section, the author initially assumed—and asked with no replies in online math forums—that this computationally correct integral representation must correspond to the non-zero components of some complex contour integral. It turns out that this formula follows from the basic theory and constructions of Fourier analysis.

Proof of Theorem 2. Given a sequence, $\{f_n\}_{n\geq 0}$, its (mostly convergent) Fourier series is given by $f(x) = \sum_{n\geq 0} f_n e^{\imath\pi n}$. The terms of this sequence are then generated by this Fourier series according to the standard integral formula [1,19]

$$f_m = \frac{1}{2\pi}\int_{-\pi}^{\pi} f(x)e^{-\imath m x}dx,$$

for natural numbers $m \geq 0$. If we can assume that the Fourier series, $f(x)$, or equivalently the OGF, $F(e^{\imath x})$, is absolutely convergent for all $x \in [-\pi, \pi]$ then we can sum over the integral formula in the previous equation to obtain the first key component to this proof:

$$\sum_{m\geq 0}\frac{f_m z^m}{m!} = \frac{1}{2\pi}\int_{-\pi}^{\pi} F(e^{\imath x})e^{ze^{-\imath x}}dx.$$

The change of variables $e^{\imath x} = z\cdot e^{-\imath t}$ for fixed z shows that this formula is equivalent to the integral formula in (3) directly by a change of variables. Also, by expanding the integrand in powers of $e^{\pm\imath x}$ where

$$\int_{-\pi}^{\pi} e^{\imath(n-k)x}dx = 2\pi\cdot\delta_{n,k},$$

it is apparent that these two formulas in fact generate the same power series representation for $\widehat{F}(z)$. □

Alternate Proof of Theorem 2. Another satisfyingly less analytical and more formally motivated explanation for this behavior can be given by considering known integral formulas for the Hadamard product of two series given in terms of the orthogonal set $\{e^{\imath kx}\}_{k=-\infty}^{\infty}$ for x on the symmetric interval $[-\pi, \pi]$ [2] (§ 1.12(V); Ex. 1.30, p. 85) [4] (*cf.* §6.3). This perspective on the formulations of these two series allows us to swap the series variables $ze^{\pm\imath x} \mapsto e^{\pm\imath x}$ from the input of one function in the product to another and similarly in the reverse direction. Thus we can effectively pick and choose where we would like to position the generating function parameter z in each component of the integrand—whether it be situated more naturally as an argument to F as in (3), or whether we choose to keep it nested in the corresponding multiplier function as in the previous equation. We shall see other examples of these integral formula variants in the next remark and following examples. □

Remark 1 (Generalizations of series expansions from Fourier series). *This technique of using a convergent Fourier series and the corresponding integral operation for extracting its coefficients can be generalized to generate many other series variants. For example, there are many zeta function and polylogarithm-related series which are summed by modifying a polylogarithmic series of the form expanded in Section 1.2 by the reciprocal of the central binomial coefficients,* $\binom{2n}{n}$. *In particular, in the exponential-series-based generating function cases we have that*

$$\sum_{n\geq 0}\frac{f_n z^n}{n!\cdot\binom{2n}{n}} = \frac{2}{\pi}\int_{-\pi}^{\pi} F(e^{-\imath x})\frac{\left[\sqrt{4-ze^{\imath x}}+\sqrt{ze^{\imath x}}\sin^{-1}\left(\frac{\sqrt{ze^{\imath x}}}{2}\right)\right]}{(4-ze^{\imath x})^{3/2}}dx \tag{15}$$

$$= \frac{2}{\pi}\int_{-\pi}^{\pi} F(ze^{-\imath x})\frac{\left[\sqrt{4-e^{\imath x}}+\sqrt{e^{\imath x}}\sin^{-1}\left(\frac{\sqrt{e^{\imath x}}}{2}\right)\right]}{(4-e^{\imath x})^{3/2}}dx,$$

and in the geometric-series-based OGF cases we recover the exponential error function by

$$\sum_{n\geq 0}\frac{f_n z^n}{\binom{2n}{n}} = \frac{1}{4\pi}\int_{-\pi}^{\pi} F(e^{-\imath x})\left[2+e^{\frac{ze^{\imath x}}{4}}\sqrt{\pi z e^{\imath x}}\,\mathrm{erf}\left(\frac{\sqrt{ze^{\imath x}}}{2}\right)\right]dx \tag{16}$$
$$= \frac{1}{4\pi}\int_{-\pi}^{\pi} F(ze^{-\imath x})\left[2+e^{\frac{e^{\imath x}}{4}}\sqrt{\pi e^{\imath x}}\,\mathrm{erf}\left(\frac{\sqrt{e^{\imath x}}}{2}\right)\right]dx.$$

There are many other possibilities for constructing integral transformations for modified generating function types. All one needs to do is be creative and consult a detailed reference of compendia such as [5,7].

Examples: Generalizations and Solutions to a Long-Standing Forum Post

The primary goal of the first post [20] mentioned in the introduction was to eventually generalize the integral formula in (3) to enumerate the modified EGF sequences of the form

$$\widehat{F}_{a,b}(z) := \sum_{n\geq 0}\frac{f_n z^n}{\Gamma(an+b+1)},$$

for integers $a \geq 1$ and $b \geq 0$, or over factors of the generalized integer multifactorials defined in [21] as

$$\ddot{F}_{a,d}(z) := \sum_{n\geq 0}\frac{f_n z^n}{(an+d)!_{(a)}}.$$

In the spirit of our realization that the integral representation in (3) is derived from a Fourier series coefficient formula, we may similarly complete our initial goal to sum the second forms of these series in the special cases where $(a,b) = (2,0),(2,1)$. In particular, we can sum these cases of the modified EGFs defined above in closed-form as explicit integral formulas in the forms

$$\ddot{F}_{2,0}(z) = \frac{1}{2\pi}\int_{-\pi}^{\pi} F\left(ze^{-\imath t}\right)e^{\frac{1}{2}e^{\imath t}}\,dt \tag{17}$$
$$\ddot{F}_{2,1}(z) = \frac{1}{2\pi}\int_{-\pi}^{\pi} F\left(ze^{-\imath t}\right)e^{\frac{1}{2}\left[e^{\imath t}-\imath t\right]}\,\mathrm{erf}\left(\sqrt{\frac{e^{\imath t}}{2}}\right)dt.$$

The modified exponential series of the first type identified above are primarily summed in closed-form using expansions of the Mittag–Leffler functions, $E_{a,b}(z) := \sum_{n\geq 0} z^n/\Gamma(an+b)$, and powers of primitive a^{th} roots of unity [16] (§ 10.46). For example, let's take $(a,b) := (3,0)$ and observe that

$$E_{3,0}(t) = \sum_{m\geq 0}\frac{t^m}{\Gamma(3m+1)} = \frac{e^{t^{1/3}}}{3} + \frac{2e^{-t^{1/3}/2}}{3}\cos\left(\frac{\sqrt{3}t^{1/3}}{2}\right). \tag{18}$$

Then we arrive at a corresponding explicit integral representation for the modified EGF of any sequence of the form

$$\widehat{F}_{3,0}(z) = \frac{1}{2\pi}\int_{-\pi}^{\pi} F(ze^{-\imath t})E_{3,0}\left(e^{\imath t}\right)dt.$$

4. Concluding Remarks

We have proved two key new forms of integral representations for the reciprocal gamma function on the real line. By composition and the uniform convergence of power series for functions defined on some disc $|z| < \sigma_f$, these results effectively provide us with OGF-to-EGF conversion formulas between the generating functions for some $F(z)$. These integral formulas for OGF-to-EGF conversion can be applied termwise, or in analytic estimates of the asymptotic growth of the coefficients in the power series expansions of the functions defined by the corresponding integral transformation.

We have provided several examples of motivating cases of our so-termed generating function transformations by integral-based methods in Section 1.3. The broader applications of these transformation methods to other fields and phrasings of problems is certainly possible given a suitable context waiting for a new method from which to be approached. We hope that readers come away from this article with a new understanding of how useful and sometimes indispensable integral transformation methods are in sequence analysis.

References

1. Debnath, L.; Bhatta, D. *Integral Transforms And Their Applications*, 3rd ed.; CRC Press: Boca Raton, FL, USA, 2015.
2. Comtet, L. *Advanced Combinatorics: The Art of Finite and Infinite Expansions*; D. Reidel Publishing Company: Dordrecht, The Netherlands, 1974.
3. Graham, R.L.; Knuth, D.E.; Patashnik, O. *Concrete Mathematics: A Foundation for Computer Science*; Addison-Wesley: Boston, MA, USA, 1994.
4. Stanley, R.P. *Enumerative Combinatorics, Volume 2*; Cambridge University Press: Cambridge, UK, 2005.
5. Boros, G.; Moll, V.H. *Irresistible Integrals: Symbolics, Analysis and Experiements in the Evaluation of Integrals*; Cambridge University Press: Cambridge, UK, 2006.
6. Flajolet, P.; Sedgewick, R. *Analytic Combinatorics*; Cambridge University Press: Cambridge, UK, 2009.
7. Gradshetyn, I.S.; Rhyzhik, I.M. *Table of Integrals, Series, and Products*; Academic Press: Cambridge, MA, USA, 1994.
8. Borwein, D.; Borwein, J.M.; Girgensohn, R. Explicit evaluation of Euler sums. *Proc. Edinb. Math. Soc.* **1995**, *38*, 277–294. [CrossRef]
9. Milgram, M.S. Integral and Series Representations of Riemann's Zeta Function, Dirichlet's eta Function and A Medley of Related Results. *arXiv* **2012**, arXiv:1208.3429.
10. Schmidt, M.D. Zeta Series Generating Function Transformations Related to Generalized Stirling Numbers and Partial Sums of the Hurwitz Zeta Function. *Online J. Anal. Comb.* **2018**, *13*, 1–31.
11. Schmidt, M.D. Zeta series generating function transformztions related to polylogarithm functions and the *k*–order harmonic numbers. *Online J. Anal. Comb.* **2017**, *12*, 1–22.
12. Schmidt, M.D. Square series generating function transformations. *Inequalities Spec. Funct.* **2017**, arxiv:1609.02803.
13. Granovskya, B.L.; Starkb, D.; Erlihson, M. Meinardus' theorem on weighted partitions: Extensions and a probabilistic proof. *Adv. Appl. Math.* **2008**, *41*, 307–328. [CrossRef]
14. Bernstein, M.; Sloane, N.J.A. Some Cannonical Sequences of Integers. *arXiv* **2002**, arxiv:math/0205301.
15. Andrews, G.E. *q-Series: Their Development and Application in Analysis, Number Theory, Combinatorics, Physics, and Computer Algebra*; American Mathematical Society: Providence, RI, USA, 1986.
16. Olver, F.W.J.; Lozier, D.W.; Boisvert, R.F.; Clark, C.W. (Eds.) *NIST Handbook of Mathematical Functions*; Cambridge University Press: Cambridge, UK, 2010.
17. Remmert, R. *Classical Topics in Complex Function Theory*; Springer: Berlin/Heidelberg, Germany, 1991.
18. Fekih-Ahmed, L. On the power series expansions of the reciprocal gamma function. *arXiv* **2017**, arxiv:abs/1407.5983.
19. Tolstov, G.P. *Fourier Series*; Dover Publications: Mineola, NY, USA, 2014.
20. An Integral Formula For The Reciprocal Gamma Function. Available online: https://math.stackexchange.com/questions/2274972/an-integral-formula-for-the-reciprocal-gamma-function (accessed on 22 May 2018).
21. Schmidt, M.D. Generalized *j*–factorial functions, polynomials, and applications. *J. Integer Seq.* **2010**, *13*, 1–54.

17

Solutions to Abel's Integral Equations in Distributions

Chenkuan Li *, Thomas Humphries and Hunter Plowman

Department of Mathematics and Computer Science, Brandon University, Brandon, MB R7A 6A9, Canada; humphrte65@brandonu.ca (T.H.); plowmahh10@brandonu.ca (H.P.)

* Correspondence: lic@brandonu.ca;

Abstract: The goal of this paper is to study fractional calculus of distributions, the generalized Abel's integral equations, as well as fractional differential equations in the distributional space $\mathcal{D}'(R^+)$ based on inverse convolutional operators and Babenko's approach. Furthermore, we provide interesting applications of Abel's integral equations in viscoelastic systems, as well as solving other integral equations, such as $\int_\theta^{\pi/2} \frac{y(\varphi)}{\cos^\beta \varphi(\cos\theta - \cos\varphi)^\alpha} d\varphi = f(\theta)$, and $\int_0^\infty x^{1/2} g(x) y(x+t)dx = f(t)$.

Keywords: distribution; fractional calculus; Mittag–Leffler function; Abel's integral equation; convolution

---

1. Introduction

Fractional modeling is an emergent tool which uses fractional differential and integral equations to describe non-local dynamic processes associated with complex systems [1–8]. Integral and fractional differential equations arise in numerous physical problems [9–12], in the fields of chemistry, biology, electronics, noncommutative quantum field theories [13], and quantum mechanics [14]. Mathematical models of systems and processes in the mentioned areas of engineering [15] and scientific disciplines involve integrals of unknown functions and derivatives of fractional order. As far as we know, fractional calculus provides an excellent tool to construct certain electro-chemical problems and characterizes long-term behaviors [16,17], allometric scaling laws, hereditary properties of various materials and so on [18]. This is the main advantage of fractional differential equations, in comparison with classical integer-order models in practice. Recently, Srivastava et al. presented the model under-actuated mechanical system with fractional order derivative [19]. Many initial and boundary value problems associated with ordinary (or partial) differential equations, can be converted into Volterra integral equations [1,20]. The Volterra's population growth model, biological species living together, and the heat change can all be characterized by integral equations. For example, Gorenflo and Mainardi [21] provided applications of Abel's integral equations, of the first and second kind, in solving the partial differential equation which describes the problem of the heating (or cooling) of a semi-infinite rod by influx (or efflux) of heat across the boundary into (or from) it's interior. In 1985, Hatcher [22] worked on a nonlinear Hilbert problem of a power type, solved in closed form by representing a sectionally holomorphic function by means of an integral with power kernel, and transformed the problem to one of solving a generalized Abel's integral equation. The development of integral equations has led to the construction of many real world problems, such as mathematical physics models [23,24], scattering in quantum mechanics and water waves. There have been lots of techniques, such as numerical analysis and integral transforms [25–27], thus far to studying fractional differential and integral equations, including Abel's equations, with many applications [1,20,28–42].

Kilbas et al. [43] presented a solution in a closed form of multi-dimensional integral equations of the first kind with the Gauss hypergeometric function in the kernel over special pyramidal domains.

Raina et al. [44] later on investigated the solvability of the one-dimensional Abel-type hypergeometric integral equation, given by

$$\frac{(x-a)^{-\alpha}}{\Gamma(\gamma)}\int_a^x (x-t)^{\gamma-1}F\left(\alpha,\beta;\gamma;\frac{x-t}{x-a}\right)\phi(t)dt = f(x)$$

where $x > a$ with $\alpha, \beta \in R$ and $0 < \gamma < 1$, as well as the multidimensional Abel-type hypergeometric integral equation over a pyramidal domain in R^n. The generalized fractional integral and differential operators are introduced and their properties are investigated systematically based on the results obtained.

Srivastava and Buschman [20] presented the comprehensive theory and numerous applications of the integral equations of convolution type, and of certain classes of integro-differential and non-linear integral equations, including Abel's integral equations, in the classical sense.

We start with the necessary concepts and definitions of fractional calculus of distributions in $\mathcal{D}'(R^+)$ based on the generalized convolution in the Schwartz space. Using inverse convolutional operators and Babenko's approach, we study and solve several Abel's integral (for all $\alpha \in R$) and fractional differential equations, as convergent series or in terms of the Mittag–Leffler functions. Many of the results derived can not be archived in the classical sense including numerical analysis methods, or by the Laplace transform. Applications are presented at the end in viscoelastic systems, and for solving other types of integral equations which can be converted into Abel's ones.

2. Abel's Integral Equations in Distribution

In order to study Abel's integral and fractional differential equations distributionally, we briefly introduce the following basic concepts in distribution. Let $\mathcal{D}(R)$ be the Schwartz space (testing function space) [45] of infinitely differentiable functions with compact support in R, and $\mathcal{D}'(R)$ the (dual) space of distributions defined on $\mathcal{D}(R)$. A sequence $\phi_1, \phi_2, \cdots, \phi_n, \cdots$ goes to zero in $\mathcal{D}(R)$ if and only if these functions vanish outside a certain fixed bounded set, and converge to zero uniformly together with their derivatives of any order. We further assume that $\mathcal{D}'(R^+)$ is the subspace of $\mathcal{D}'(R)$ with support contained in R^+.

The functional $\delta^{(n)}(x-x_0)$ for $x_0 \in R$ is defined as

$$(\delta^{(n)}(x-x_0), \phi(x)) = (-1)^n\phi^{(n)}(x_0)$$

where $\phi \in \mathcal{D}(R)$. Clearly, $\delta^{(n)}(x-x_0)$ is a linear and continuous functional on $\mathcal{D}(R)$, and hence $\delta^{(n)}(x-x_0) \in \mathcal{D}'(R)$.

Define

$$f(x) = \begin{cases} \sin x & \text{if } 0 < x < 1, \\ 0 & \text{otherwise.} \end{cases}$$

Then, $f(x)$ is a locally integrable function on R (clearly not continuous) and

$$(f(x), \phi(x)) = \int_0^1 \sin x\, \phi(x)dx \quad \text{for} \quad \phi \in \mathcal{D}(R), \tag{1}$$

defines a regular distribution $f(x) \in \mathcal{D}'(R^+)$.

Let $f \in \mathcal{D}'(R)$. The distributional derivative of f, denoted by f' or df/dx, is defined as

$$(f', \phi) = -(f, \phi')$$

for $\phi \in \mathcal{D}(R)$.

Assume f is a distribution in $\mathcal{D}'(R)$ and g is a function in $C^\infty(R)$. Then the product fg is well defined by

$$(fg, \phi) = (f, g\phi)$$

for all functions $\phi \in \mathcal{D}(R)$ as $g\phi \in \mathcal{D}(R)$.

Clearly, $f' \in \mathcal{D}'(R)$ and every distribution has a derivative.

It can be shown that the ordinary rules of differentiation also apply to distributions. For instance, the derivative of a sum, is the sum of the derivatives, and a constant can be commuted with the derivative operator.

It follows from [45] that $\Phi_\lambda = \dfrac{x_+^{\lambda-1}}{\Gamma(\lambda)} \in \mathcal{D}'(R^+)$ is an entire function of λ on the complex plane, and

$$\left.\frac{x_+^{\lambda-1}}{\Gamma(\lambda)}\right|_{\lambda=-n} = \delta^{(n)}(x), \quad \text{for} \quad n = 0, 1, 2, \cdots. \tag{2}$$

Clearly, the Laplace transform of Φ_λ is given by

$$\mathcal{L}\{\Phi_\lambda(x)\} = \int_0^\infty e^{-sx}\Phi_\lambda(x)dx = \frac{1}{s^\lambda}, \quad \mathrm{Re}\lambda > 0, \quad \mathrm{Re}s > 0$$

which plays an important role in solving integral equations [46].

For the functional $\Phi_\lambda = \dfrac{x_+^{\lambda-1}}{\Gamma(\lambda)}$, the (distributional) derivative formula is simpler than that for x_+^λ. In fact,

$$\frac{d}{dx}\Phi_\lambda = \frac{d}{dx}\frac{x_+^{\lambda-1}}{\Gamma(\lambda)} = \frac{(\lambda-1)x_+^{\lambda-2}}{\Gamma(\lambda)} = \frac{x_+^{\lambda-2}}{\Gamma(\lambda-1)} = \Phi_{\lambda-1}. \tag{3}$$

The convolution of certain pairs of distributions is usually defined as follows, see Gel'fand and Shilov [45] for example.

Definition 1. *Let f and g be distributions in $\mathcal{D}'(R)$ satisfying either of the following conditions:*

(a) either f or g has bounded support (set of all essential points), or
(b) the supports of f and g are bounded on the same side.

Then the convolution $f * g$ is defined by the equation

$$((f * g)(x), \phi(x)) = (g(x), (f(y), \phi(x+y)))$$

for $\phi \in \mathcal{D}(R)$.

The classical definition of the convolution is as follows:

Definition 2. *If f and g are locally integrable functions, then the convolution $f * g$ is defined by*

$$(f * g)(x) = \int_{-\infty}^{\infty} f(t)g(x-t)dt = \int_{-\infty}^{\infty} f(x-t)g(t)dt$$

for all x for which the integrals exist.

Note that if f and g are locally integrable functions satisfying either of the conditions in (a) or (b) in Definition 1, then Definition 1 is in agreement with Definition 2. It also follows that if the convolution $f * g$ exists by Definitions 1 or 2, then the following equations hold:

$$f * g = g * f \tag{4}$$

$$(f * g)' = f * g' = f' * g \tag{5}$$

where all the derivatives above are in the distributional sense.

Let λ and μ be arbitrary complex numbers. Then it is easy to show

$$\Phi_\lambda * \Phi_\mu = \Phi_{\lambda+\mu} \tag{6}$$

by Equation (3), without any help of analytic continuation mentioned in all current books.

Let λ be an arbitrary complex number and $g(x)$ be the distribution concentrated on $x \geq 0$. We define the primitive of order λ of g as convolution in the distributional sense

$$g_\lambda(x) = g(x) * \frac{x_+^{\lambda-1}}{\Gamma(\lambda)} = g(x) * \Phi_\lambda. \tag{7}$$

Note that the convolution on the right-hand side is well defined since supports of g and Φ_λ are bounded on the same side.

Thus Equation (7) with various λ will not only give the fractional derivatives, but also the fractional integrals of $g(x) \in \mathcal{D}'(R^+)$ when $\lambda \notin Z$, and it reduces to integer-order derivatives or integrals when $\lambda \in Z$. We shall define the convolution

$$g_{-\lambda} = g(x) * \Phi_{-\lambda}$$

as the fractional derivative of the distribution $g(x)$ with order λ, writing it as

$$g_{-\lambda} = \frac{d^\lambda}{dx^\lambda} g$$

for $\text{Re}\lambda \geq 0$. Similarly, $\frac{d^\lambda}{dx^\lambda} g$ is interpreted as the fractional integral if $\text{Re}\lambda < 0$.

In 1996, Matignon [47] also studied fractional derivatives in the distributional sense using the kernel distribution $\Phi_{-\lambda}$, and defined the fractional derivative of order λ of a continuous (in the normal sense) causal (zero for $t < 0$) function g, as $g_{-\lambda} = g(x) * \Phi_{-\lambda}$, and further obtained a relation between the distributional derivative and the classical one for a smooth function. Mainardi [42] extended Matignon's work and formally defined the fractional derivative of order $\lambda > 0$ of a causal function (not necessarily continuous) as

$$\frac{d^\lambda}{dx^\lambda} g(x) = \Phi_{-\lambda} * g = \frac{1}{\Gamma(-\lambda)} \int_0^x \frac{g(\zeta)d\zeta}{(x-\zeta)^{1+\lambda}}, \quad \lambda \in R^+.$$

The limit case $\lambda = 0$ is defined as

$$\frac{d^0}{dx^0} g(x) = \Phi_0 * g = \delta * g = g.$$

In addition, Podlubny [46] investigated fractional calculus of generalized functions by the distributional convolution and derived the following identities of fractional derivatives and integrals

$$\frac{d\lambda}{dx^\lambda}\theta(x-a) = \frac{(x-a)_+^{-\lambda}}{\Gamma(-\lambda+1)},$$

$$\frac{d\lambda}{dx^\lambda}\delta^{(k)}(x-a) = \frac{(x-a)_+^{-k-\lambda-1}}{\Gamma(-k-\lambda)}$$

where $\lambda \in C$ and k is a nonnegative integer.

The following theorem can be obtained from [41] with a minor change in the proof.

Theorem 1. *Let $G(x)$ be a given distribution and y be an unknown distribution in $\mathcal{D}'(R^+)$. Then the generalized Abel's integral equation of the first kind*

$$G(x) = \frac{1}{\Gamma(\alpha)} \int_0^x (x-\zeta)^{\alpha-1} y(\zeta) d\zeta \tag{8}$$

has the solution

$$y(x) = G(x) * \Phi_{-\alpha}$$

where α is any real number in R. In particular, if $G(x) = \theta(x-x_0)g(x)$, where $g(x)$ is an infinitely differentiable function on $[0,\infty]$ and $x_0 \geq 0$, then we have four different cases depending on the value of α.

(i) *If $m < \alpha < m+1$ for $m = 0, 1, \ldots,$ then*

$$\begin{aligned} y(x) &= \frac{d^{m+1}}{dx^{m+1}} \theta(x-x_0)g(x) * \frac{x_+^{-\alpha+m}}{\Gamma(-\alpha+m+1)} \\ &= g(x_0)\frac{(x-x_0)_+^{-\alpha}}{\Gamma(-\alpha+1)} + \cdots + g^{(m)}(x_0)\frac{(x-x_0)_+^{-\alpha+m}}{\Gamma(-\alpha+m+1)} \\ &\quad + \frac{1}{\Gamma(-\alpha+m+1)} \int_{x_0}^x g^{(m+1)}(\zeta)(x-\zeta)^{-\alpha+m} d\zeta \end{aligned}$$

for $x \geq x_0$.

(ii) *If $\alpha = 1, 2, \ldots,$ then*

$$y(x) = g(x_0)\delta^{(\alpha-1)}(x-x_0) + \cdots + g^{(\alpha-1)}(x_0)\delta(x-x_0) + \theta(x-x_0)g^{(\alpha)}(x).$$

(iii) *If $\alpha = 0$, then $y(x) = \theta(x-x_0)g(x)$.*

(iv) *If $\alpha < 0$, then for $x \geq x_0$*

$$y(x) = \frac{1}{\Gamma(-\alpha)} \int_{x_0}^x g(\zeta)(x-\zeta)^{-\alpha-1} d\zeta$$

which is well defined.

Example 1. *Let k be a nonnegative integer, and $n \in N$, $s \in R$ with $s \neq -1, -2, \cdots$. Then, the integral equation*

$$x_+^s + \delta^{(k)}(x) = \int_0^x y(\tau)(x-\tau)^{n/2-1} d\tau \tag{9}$$

has the solution in the space $\mathcal{D}'(R^+)$

$$y(x) = \frac{2^{(n-1)/2}}{(n-2)!!\sqrt{\pi}} \left(\Gamma(s+1)\Phi_{-n/2+s+1}(x) + \Phi_{-n/2-k}(x)\right).$$

Proof. Equation (9) is equivalent to

$$x_+^s + \delta^{(k)}(x) = \frac{\Gamma(n/2)}{\Gamma(n/2)} \int_0^x y(\tau)(x-\tau)^{n/2-1} d\tau$$

which gives

$$x_+^s + \delta^{(k)}(x) = \Gamma(n/2)\left(y(x) * \Phi_{n/2}(x)\right).$$

Theorem 1 implies

$$y(x) = \frac{1}{\Gamma(n/2)} \left(\Phi_{-n/2}(x) * (x_+^s + \delta^{(k)}(x))\right)$$

which simplifies to

$$\begin{aligned}
y(x) &= \frac{1}{\Gamma(n/2)}((\Phi_{-n/2}(x) * x_+^s) + (\Phi_{-n/2}(x) * \delta^{(k)}(x))) \\
&= \frac{1}{\Gamma(n/2)}((\Phi_{-n/2} * \frac{\Gamma(s+1)}{\Gamma(s+1)} x_+^s) + (\Phi_{-n/2}(x) * \Phi_{-k}(x))) \\
&= \frac{1}{\Gamma(n/2)}((\Phi_{-n/2} * \Gamma(s+1)\Phi_{s+1}(x)) + \Phi_{-n/2-k}(x)) \\
&= \frac{1}{\Gamma(n/2)}((\Gamma(s+1)\Phi_{-n/2+s+1}) + \Phi_{-n/2-k}(x)) .
\end{aligned}$$

Using the formula

$$\Gamma(n/2) = \frac{(n-2)!!\sqrt{\pi}}{2^{(n-1)/2}},$$

we infer that

$$y(x) = \frac{2^{(n-1)/2}}{(n-2)!!\sqrt{\pi}}(\Gamma(s+1)\Phi_{-n/2+s+1}(x) + \Phi_{-n/2-k}(x)) .$$

This completes the proof of Example 1. □

In particular, the integral equation

$$x_+^{-3/2} + \delta'(x) = \int_0^x y(\tau)(x-\tau)^{-1/2} d\tau \tag{10}$$

has the solution in the space $\mathcal{D}'(R^+)$

$$y(x) = -2\delta'(x) + \frac{1}{\sqrt{\pi}}\Phi_{-3/2}(x)$$

using

$$\Gamma(-1/2) = -2\sqrt{\pi}.$$

Remark 1. *We must mention that Equation (10) cannot be solved by the Laplace transform since the distribution* $x_+^{-3/2}$ *is not locally integrable and its Laplace transform does not exist.*

Similarly, the integral equation

$$x_+ + \delta(x) = \int_0^x y(\tau)(x-\tau) d\tau$$

has the solution in the space $\mathcal{D}'(R^+)$

$$y(x) = \delta(x) + \delta''(x)$$

by Equation (2).

Example 2. *Let* $s, \alpha \in R$, *and* $\alpha \neq 0, -1, -2 \cdots$. *Then, the integral equation*

$$\sin x_+ + \Phi_s(x) = \int_0^x y(\tau)(x-\tau)^{\alpha-1} d\tau \tag{11}$$

has the solution in the space $\mathcal{D}'(R^+)$

$$y(x) = \frac{1}{\Gamma(\alpha)}\left(\sum_{k=0}^{\infty}(-1)^k\Phi_{2k-\alpha+2}(x) + \Phi_{s-\alpha}(x)\right),$$

where

$$\sin x_+ = \begin{cases} \sin x & \text{if } x \geq 0, \\ 0 & \text{otherwise.} \end{cases}$$

Proof. Equation (11) can be written as

$$\sin x_+ + \Phi_s(x) = \frac{\Gamma(\alpha)}{\Gamma(\alpha)} \int_0^x y(\tau)(x-\tau)^{\alpha-1} d\tau$$

which is equal to

$$\sin x_+ + \Phi_s(x) = \Gamma(\alpha)(\Phi_\alpha * y)(x).$$

Applying Theorem 1, we get

$$y(x) = \frac{1}{\Gamma(\alpha)}(\Phi_{-\alpha}(x) * (\sin x_+ + \Phi_s(x)))$$

which distributes to

$$y(x) = \frac{1}{\Gamma(\alpha)}(\Phi_{-\alpha}(x) * \sin x_+ + \Phi_{-\alpha}(x) * \Phi_s(x)).$$

The Taylor expansion

$$\sin x_+ = \sum_{k=0}^{\infty} \frac{(-1)^k x_+^{2k+1}}{(2k+1)!} = \sum_{k=0}^{\infty} (-1)^k \Phi_{2k+2}(x)$$

gives

$$y(x) = \frac{1}{\Gamma(\alpha)} \left(\sum_{k=0}^{\infty} (-1)^k \Phi_{2k-\alpha+2}(x) + \Phi_{s-\alpha}(x) \right).$$

We note that the series

$$\sum_{k=0}^{\infty} (-1)^k \Phi_{2k-\alpha+2}$$

is absolutely convergent by the ratio test. Indeed,

$$\begin{aligned} \lim_{k\to\infty} \frac{\Phi_{2(k+1)-\alpha+2}(x)}{\Phi_{2k-\alpha+2}(x)} &= \lim_{k\to\infty} \frac{\dfrac{x_+^{2k-\alpha+3}}{\Gamma(2k-\alpha+4)}}{\dfrac{x_+^{2k-\alpha+1}}{\Gamma(2k-\alpha+2)}} \\ &= \lim_{k\to\infty} \frac{x_+^2}{(2k-\alpha+3)(2k-\alpha+2)} = 0. \end{aligned}$$

This completes the proof of Example 2. □

Remark 2. *We should point out that the series* $\sum_{k=0}^{\infty} (-1)^k \Phi_{2k-\alpha+2}(x)$ *is the sum of singular and regular distributions. Indeed, let* j *be the largest non-negative integer, such that* $2j - \alpha + 2 \leq 0$. *Then*

$$\sum_{k=0}^{\infty} (-1)^k \Phi_{2k-\alpha+2}(x) = \sum_{k=0}^{j} (-1)^k \Phi_{2k-\alpha+2}(x) + \sum_{k=j+1}^{\infty} (-1)^k \Phi_{2k-\alpha+2}(x)$$

where the term

$$\sum_{k=0}^{j}(-1)^k\Phi_{2k-\alpha+2}(x)$$

is a singular distribution, while

$$\sum_{k=j+1}^{\infty}(-1)^k\Phi_{2k-\alpha+2}(x)$$

is regular.

In the special case of $\alpha < 2$, we get

$$\begin{aligned} y(x) &= \frac{1}{\Gamma(\alpha)}\left(\sum_{k=0}^{\infty}\frac{(-1)^k x_+^{2k-\alpha+1}}{\Gamma(2k-\alpha+2)} + \Phi_{s-\alpha}(x)\right) \\ &= \frac{1}{\Gamma(\alpha)}\left(x_+^{-\alpha+1}\sum_{k=0}^{\infty}\frac{(-x_+^2)^k}{\Gamma(2k-\alpha+2)} + \Phi_{s-\alpha}(x)\right) \\ &= \frac{1}{\Gamma(\alpha)}\left(x_+^{-\alpha+1}E_{2,-\alpha+2}(-x_+^2) + \Phi_{s-\alpha}(x)\right). \end{aligned}$$

Note that the function $x_+^{-\alpha+1}$ is locally integrable on R.
In particular, we have for $\alpha = 1$ that

$$y(x) = \theta(x)\cosh(ix) + \Phi_{s-1}(x) = \theta(x)\cos x + \Phi_{s-1}(x)$$

using

$$E_{2,1}(z) = \cosh\sqrt{z}.$$

This also can be derived directly from Equation (11). In fact, it becomes for $\alpha = 1$

$$\sin x_+ + \Phi_s(x) = \int_0^x y(\tau)d\tau$$

which claims that

$$y(x) = \frac{d}{dx}(\theta(x)\sin x + \Phi_s(x)) = \theta(x)\cos x + \Phi_{s-1}(x)$$

by noting that

$$\frac{d}{dx}\theta(x)\sin x = \delta(x)\sin x + \theta(x)\cos x = \theta(x)\cos x.$$

We further note that Equation (11) becomes

$$\sin x_+ + \delta^{(-s)}(x) = \int_0^x y(\tau)(x-\tau)^{\alpha-1}d\tau.$$

for $s = 0, -1, -2, \cdots$.

Similarly, the integral equation

$$\cos x_+ + e_+^x = \int_0^x y(\tau)(x-\tau)^{\alpha-1}d\tau \qquad (12)$$

has the solution in the space $\mathcal{D}'(R^+)$

$$y(x) = \frac{1}{\Gamma(\alpha)}\left(\sum_{k=0}^{\infty}(-1)^k\Phi_{2k-\alpha+1}(x) + \sum_{k=0}^{\infty}\Phi_{k-\alpha+1}(x)\right),$$

where $\alpha \neq 0, -1, \cdots$ and

$$e_+^x = \begin{cases} e^x & \text{if } x \geq 0, \\ 0 & \text{otherwise.} \end{cases}$$

Indeed, we apply Theorem 1 to get

$$y(x) = \frac{1}{\Gamma(\alpha)}(\Phi_{-\alpha}(x) * (\cos x_+ + e_+^x)).$$

Applying the following Taylor's expansions

$$\cos x_+ = \sum_{k=0}^{\infty} \frac{(-1)^k}{(2k)!} x_+^{2k} = \sum_{k=0}^{\infty} (-1)^k \Phi_{2k+1}(x),$$
$$e_+^x = \sum_{k=0}^{\infty} \frac{x_+^k}{k!} = \sum_{k=0}^{\infty} \Phi_{k+1}(x)$$

we arrive at

$$y(x) = \frac{1}{\Gamma(\alpha)} \left(\sum_{k=0}^{\infty} (-1)^k \Phi_{2k+1-\alpha}(x) + \sum_{k=0}^{\infty} \Phi_{k+1-\alpha}(x) \right).$$

This completes the proof by noting that both $\sum_{k=0}^{\infty} (-1)^k \Phi_{2i-\alpha+1}(x)$ and $\sum_{k=0}^{\infty} \Phi_{k-\alpha+1}(x)$ are absolutely convergent by the ratio test.

In the case of $\alpha < 1$ we get

$$y(x) = \frac{x_+^{-\alpha}}{\Gamma(\alpha)}(E_{2,-\alpha+1}(-x_+^2) + E_{1,-\alpha+1}(x_+)).$$

In particular when $\alpha = 1/2$

$$f(x) = \frac{x_+^{-1/2}}{\sqrt{\pi}} \left(E_{2,1/2}(-x_+^2) + E_{1,1/2}(x_+) \right).$$

We shall extend the techniques used by Yu. I. Babenko in his book [48], for solving various types of fractional differential and integral equations in the classical sense, to generalized functions. The method itself is close to the Laplace transform method in the ordinary sense, but it can be used in more cases [46], such as solving integral or fractional differential equations with distributions whose Laplace transforms do not exist in the classical sense as indicated below. Clearly, it is always necessary to show convergence of the series obtained as solutions. In [46], Podlubny also provided interesting applications to solving certain partial differential equations for heat and mass transfer by Babenko's method.

To illustrate Babenko's approach in detail, we solve the following Abel's integral equation of the second kind in the space $\mathcal{D}'(R^+)$

$$\Phi_{-1/2} = y(x) + \int_0^x (x-\zeta)^\alpha y(\zeta) d\zeta$$

for $\alpha > 0$. Note that the Laplace transform does not work for this equation, since the Laplace transform of $\Phi_{-1/2}$ does not exist. However, this equation can be converted into

$$\Phi_{-1/2} = y(x) + \frac{\Gamma(\alpha+1)}{\Gamma(\alpha+1)} \int_0^x (x-\zeta)^\alpha y(\zeta) d\zeta = (\delta + \Gamma(\alpha+1)\Phi_{\alpha+1}) * y(x)$$

This implies by Babenko's method that

$$
\begin{aligned}
y(x) &= \left(\delta + \Gamma(\alpha+1)\Phi_{\alpha+1}\right)^{-1} * \Phi_{-1/2} \\
&= \sum_{n=0}^{\infty}(-1)^n \Gamma^n(\alpha+1)\Phi_{\alpha+1}^n * \Phi_{-1/2} \\
&= \sum_{n=0}^{\infty}(-1)^n \Gamma^n(\alpha+1)\Phi_{(\alpha+1)n} * \Phi_{-1/2} \\
&= \sum_{n=0}^{\infty}(-1)^n \Gamma^n(\alpha+1)\Phi_{(\alpha+1)n-1/2} \\
&= \Phi_{-1/2} + \sum_{n=1}^{\infty}(-1)^n \Gamma^n(\alpha+1)\Phi_{(\alpha+1)n-1/2} \\
&= \Phi_{-1/2} + \sum_{n=0}^{\infty}(-1)^{n+1} \Gamma^{n+1}(\alpha+1)\Phi_{(\alpha+1)n+\alpha+1/2} \\
&= \Phi_{-1/2} - \Gamma(\alpha+1)x_+^{\alpha-1}E_{\alpha+1,\alpha+1/2}\left(-\Gamma(\alpha+1)x_+^{\alpha+1}\right)
\end{aligned}
$$

using

$$\Phi_{\alpha+1}^n = \Phi_{(\alpha+1)n}.$$

Example 3. *Let $\alpha > \beta \geq 0$, and $\gamma \geq 0$. Then the fractional differential equation*

$$ay^{(\alpha)}(x) + by^{(\beta)}(x) = c\Phi_\gamma(x) + x_+^m \tag{13}$$

has the solution in the space $\mathcal{D}'(R^+)$

$$y(x) = \frac{cx_+^{\alpha+\gamma-1}}{a}E_{\alpha-\beta,\alpha+\gamma}\left(-\frac{b}{a}x_+^{\alpha-\beta}\right) + \frac{m!\,x_+^{\alpha+m}}{a}E_{\alpha-\beta,\alpha+m+1}\left(-\frac{b}{a}x_+^{\alpha-\beta}\right)$$

where $m = 0, 1, 2...$, and $a, b, c \in R$ with $a \neq 0$.

Proof. We see that Equation (13) is equivalent to

$$ay(x) * \Phi_{-\alpha}(x) + by(x) * \Phi_{-\beta}(x) = c\Phi_\gamma(x) + x_+^m.$$

Applying Φ_α to both sides, we get

$$
\begin{aligned}
ay(x) + by(x) * \Phi_{\alpha-\beta}(x) &= ay(x) * \left(\delta(x) + \frac{b}{a}\Phi_{\alpha-\beta}(x)\right) \\
&= c\Phi_{\alpha+\gamma}(x) + x_+^m * \Phi_\alpha(x).
\end{aligned}
$$

This implies, by Babenko's approach

$$
\begin{aligned}
y(x) &= \frac{1}{a}\left(\delta(x) + \frac{b}{a}\Phi_{\alpha-\beta}(x)\right)^{-1}\left(c\Phi_{\alpha+\gamma}(x) + x_+^m * \Phi_\alpha(x)\right) \\
&= \frac{1}{a}\sum_{k=0}^{\infty}(-1)^k\left(\frac{b}{a}\Phi_{\alpha-\beta}(x)\right)^k * \left(c\Phi_{\alpha+\gamma}(x) + x_+^m * \Phi_\alpha(x)\right) \\
&= \frac{c}{a}\sum_{k=0}^{\infty}\frac{(-b)^k}{a^k}\Phi_{k(\alpha-\beta)+\alpha+\gamma}(x) + \frac{m!}{a}\sum_{k=0}^{\infty}\frac{(-b)^k}{a^k}\Phi_{k(\alpha-\beta)+\alpha+m+1}(x) \\
&= \frac{c}{a}\sum_{k=0}^{\infty}\frac{(-b)^k x_+^{k(\alpha-\beta)+\alpha+\gamma-1}}{a^k\Gamma(k(\alpha-\beta)+\alpha+\gamma)} + \frac{m!}{a}\sum_{k=0}^{\infty}\frac{(-b)^k x_+^{k(\alpha-\beta)+\alpha+m}}{a^k\Gamma(k(\alpha-\beta)+\alpha+m+1)} \\
&= \frac{cx_+^{\alpha+\gamma-1}}{a}E_{\alpha-\beta,\alpha+\gamma}\left(-\frac{b}{a}x_+^{\alpha-\beta}\right) + \frac{m!\,x_+^{\alpha+m}}{a}E_{\alpha-\beta,\alpha+m+1}\left(-\frac{b}{a}x_+^{\alpha-\beta}\right).
\end{aligned}
$$

This completes the proof of Example 3. □

In particular, the ordinary differential equation

$$ay'(x) + by(x) = c\Phi_2(x) + x_+,$$

has the solution in the space $\mathcal{D}'(R^+)$

$$y(x) = \frac{(c+1)x_+^2}{a} E_{1,3}\left(-\frac{bx_+}{a}\right)$$

where

$$E_{1,3}(z) = \frac{e^z - z - 1}{z^2}.$$

On the other hand, we derive that the fractional differential equation

$$ay'(x) + by^{(1/2)}(x) = c\delta(x) + x_+^m, \quad a \neq 0$$

has the solution in the space $\mathcal{D}'(R^+)$

$$y(x) = \frac{c\theta(x)}{a} e^{\frac{b^2}{a^2}x_+} \operatorname{erfc}\left(\frac{b}{a}\sqrt{x_+}\right) + \frac{x_+^{1+m} m!}{a} E_{0.5,2+m}\left(-\frac{b}{a}\sqrt{x_+}\right).$$

by using

$$E_{0.5,1}(z) = e^{z^2} \operatorname{erfc}(-z),$$

where erfc is the complement to the error function (erf),

$$\operatorname{erfc}(z) = \frac{2}{\sqrt{\pi}} \int_z^\infty e^{-u^2} du = 1 - \operatorname{erf}(z), \quad z \in C.$$

Clearly, this example can also be solved using the Laplace transform. Applying the Laplace transform to the equation

$$ay^{(\alpha)}(x) + by^{(\beta)}(x) = c\Phi_\gamma(x) + x_+^m,$$

we come to

$$y^*(s) = \frac{cs^{m+1} + m!s^\gamma}{as^{\alpha+\gamma+m+1} + bs^{\beta+\gamma+m+1}} = \left(\frac{c}{a}\right)\left(\frac{s^{-\gamma-\beta}}{s^{\alpha-\beta} + \frac{b}{a}}\right) + \left(\frac{m!}{a}\right)\left(\frac{s^{-\beta-m-1}}{s^{\alpha-\beta} + \frac{b}{a}}\right).$$

Using the inverse transform, we have

$$y(x) = \frac{cx_+^{\alpha+\gamma-1}}{a} E_{\alpha-\beta,\alpha+\gamma}\left(-\frac{b}{a}x_+^{\alpha-\beta}\right) + \frac{m!\, x_+^{\alpha+m}}{a} E_{\alpha-\beta,\alpha+m+1}\left(-\frac{b}{a}x_+^{\alpha-\beta}\right)$$

by the formula [46]

$$\int_0^\infty e^{-sx} x^{\beta-1} E_{\alpha,\beta}(-ax^\alpha) dx = \frac{s^{\alpha-\beta}}{s^\alpha + a}, \quad \operatorname{Re}(s) > |a|^{1/\alpha}.$$

Remark 3. *We must add that the following fractional differential equation*

$$ay^{(\alpha)}(x) + by^{(\beta)}(x) = c\Phi_\gamma(x) + x_+^{-3/2}$$

can also be solved by the same technique used in Example 3. Though it fails to do so by the Laplace transform, as the distribution $x_+^{-3/2}$ is singular.

Many applied problems from physical, engineering and chemical processes lead to integral equations, which at first glance have nothing in common with Abel's integral equations. Due to this perception, additional efforts are undertaken for the development of analytical or numerical procedure for solving these equations. However, their transformations to the form of Abel's integral equations will speed up the solution process [20], or, more significantly, lead to distributional solutions in cases where classical ones do not exist [40,41].

Example 4. *Let $0 \leq \theta < \pi/2$ and $\alpha < 1$. Then the following integral equation*

$$\int_\theta^{\pi/2} \frac{\sin\varphi\, y(\varphi)}{(\cos\theta - \cos\varphi)^\alpha} d\varphi = f(\theta) \tag{14}$$

has the solution

$$y(\arccos x) = \frac{1}{\Gamma(1-\alpha)} \frac{d^{1-\alpha}}{dx^{1-\alpha}} f(\arccos x)$$

where f is a differential function in $\mathcal{D}'(R^+)$.

Proof. Making the variable changes $\tau = \cos\varphi$ and $x = \cos\theta$. Then Equation (14) becomes

$$\int_0^x \frac{1}{(x-\tau)^\alpha} y(\arccos\tau) d\tau = \frac{\Gamma(1-\alpha)}{\Gamma(1-\alpha)} \int_0^x \frac{1}{(x-\tau)^\alpha} y(\arccos\tau) d\tau = f(\arccos x)$$

which is Abel's integral equation of the first kind. Therefore, we arrive at

$$\Phi_{1-\alpha}(\tau) * y(\arccos\tau) = \frac{1}{\Gamma(1-\alpha)} f(\arccos x)$$

which implies that

$$y(\arccos x) = \frac{1}{\Gamma(1-\alpha)} \Phi_{\alpha-1}(\tau) * f(\arccos\tau) = \frac{1}{\Gamma(1-\alpha)} \frac{d^{1-\alpha}}{dx^{1-\alpha}} f(\arccos x).$$

This completes the proof of Example 4. □

In particular, we have that for $\alpha = 1/2$ and $f(\theta) = \theta$

$$y(\arccos x) = \frac{1}{\Gamma(1/2)} \frac{d^{1/2}}{dx^{1/2}} \arccos x = \frac{1}{\sqrt{\pi}} \frac{d^{1/2}}{dx^{1/2}} \arccos x.$$

Using the Taylor series

$$\arccos x = \frac{1}{2}\pi - \sum_{n=0}^{\infty} \frac{(2n)!}{2^{2n}(n!)^2(2n+1)} x^{2n+1}$$

if $0 < x \leq 1$, and

$$\frac{d^{1/2}}{dx^{1/2}} x_+^{2n+1} = (2n+1)!\Phi_{-1/2}(x) * \frac{x_+^{2n+1}}{\Gamma(2n+2)} = (2n+1)!\Phi_{2n+3/2}(x)$$

we come to

$$y(\arccos x) = -\frac{1}{\sqrt{\pi}} \sum_{n=0}^{\infty} \frac{[(2n)!]^2}{2^{2n}(n!)^2} \Phi_{2n+3/2}(x)$$

which is obviously convergent. Furthermore, setting $t = \arccos x$ we finally infer that

$$y(t) = -\frac{1}{\sqrt{\pi}} \sum_{n=0}^{\infty} \frac{[(2n)!]^2}{2^{2n}(n!)^2} \Phi_{2n+3/2}(\cos t).$$

Remark 4. *Clearly, Equation (14) can be converted into*

$$\begin{aligned} \int_\theta^{\pi/2} \frac{\sin\varphi\, y(\varphi)}{(\cos\theta - \cos\varphi)^\alpha} d\varphi &= \int_{\pi/2}^{\theta} \frac{y(\varphi)}{(\cos\theta - \cos\varphi)^\alpha} d\cos\varphi \\ &= \int_0^{\cos\theta} \frac{y(\arccos\tau)}{(\cos\theta - \tau)^\alpha} d\tau \\ &= \int_0^{x} \frac{y(\arccos\tau)}{(x - \tau)^\alpha} d\tau \\ &= f(\arccos x). \end{aligned}$$

Setting

$$y(\varphi) = \frac{1}{\sin\varphi} Y(\varphi),$$

then the integral equation

$$\int_\theta^{\pi/2} \frac{Y(\varphi)}{(\cos\theta - \cos\varphi)^\alpha} d\varphi = f(\theta)$$

has the solution

$$Y(\arccos x) = \frac{\sqrt{1-x^2}}{\Gamma(1-\alpha)} \frac{d^{1-\alpha}}{dx^{1-\alpha}} f(\arccos x)$$

since

$$\sin(\arccos x) = \sqrt{1-x^2}.$$

Further, setting

$$y(\varphi) = \frac{1}{\sin\varphi \cos^\beta \varphi} Y(\varphi) \quad \textit{for } \beta < 1,$$

then the integral equation

$$\int_\theta^{\pi/2} \frac{Y(\varphi)}{\cos^\beta \varphi (\cos\theta - \cos\varphi)^\alpha} d\varphi = f(\theta)$$

has the solution

$$Y(\arccos x) = \frac{x^\beta \sqrt{1-x^2}}{\Gamma(1-\alpha)} \frac{d^{1-\alpha}}{dx^{1-\alpha}} f(\arccos x).$$

Example 5. *Assume that the functions g and f are given and g is a nonzero function satisfying the condition*

$$g(x+t) = g(x)g(t)$$

for all $x,\ t \in R$. Then the integral equation

$$\int_0^\infty x^{1/2} g(x) y(x+t) dx = f(t) \tag{15}$$

has the solution

$$y(1/s) = \frac{2s^{2.5}}{\sqrt{\pi}\, g\left(\frac{1}{s}\right)} \left(\frac{d^{1.5}}{ds^{1.5}} \frac{f(1/s) s^{0.5}}{g(-1/s)} \right).$$

Proof. Making the substitution

$$\tau = \frac{1}{x+t},$$

Equation (15) becomes

$$\int_0^{1/t} \left(\frac{1}{\tau} - t\right)^{1/2} g\left(\frac{1}{\tau} - t\right) \frac{y\left(\frac{1}{\tau}\right)}{\tau^2} d\tau = f(t)$$

which infers that

$$\int_0^{1/t} \left(\frac{1}{\tau} - t\right)^{1/2} g\left(\frac{1}{\tau}\right) \frac{y\left(\frac{1}{\tau}\right)}{\tau^2} d\tau = \frac{f(t)}{g(-t)}$$

since g is a nonzero function. Further, setting $s = 1/t$ we come to

$$\frac{\Gamma(1.5)}{\Gamma(1.5)} \int_0^s (s-\tau)^{1/2} \frac{g\left(\frac{1}{\tau}\right) y\left(\frac{1}{\tau}\right)}{\tau^{2.5}} d\tau = \frac{s^{0.5} f(1/s)}{g(-1/s)}$$

which is Abel's integral equation. Hence, we get the solution

$$y(1/s) = \frac{2s^{2.5}}{\sqrt{\pi}\, g\left(\frac{1}{s}\right)} \left(\frac{d^{1.5}}{ds^{1.5}} \frac{f(1/s)s^{0.5}}{g(-1/s)}\right)$$

using

$$\Gamma(1.5) = \sqrt{\pi}/2.$$

This completes the proof of Example 5. □

A particular example can be derived from setting $g(x) = e^{-x}$. We leave this to interested readers. We should point out that the term

$$\frac{d^{1.5}}{ds^{1.5}} \left(\frac{f(1/s)s^{0.5}}{g(-1/s)}\right)$$

is in the distributional sense. Otherwise, it is undefined if we let $g(x) \equiv 1$ and f be chosen in $\mathcal{D}'(R^+)$ such that $f(1/s)s^{0.5} = s_+^{-1.4}$.

3. The Applications in Viscoelastic Systems

A modeling is a cognitive activity which we use to describe how devices, or objects of interest, behave.

Elasticity is the ability of a material to resist a distortion or a deforming force and return to its original form when the force is removed. According to the classical theory in the infinitesimal deformation, the most elastic materials, based on Hooke's Law, can be described by a linear relation between the strain ϵ and stress σ and

$$\epsilon(t) = \frac{1}{E}\sigma(t)$$

where E is a constant, known as the elastic or Young's modulus.

However, in a more complicated fractional viscoelastic model, one [49,50] constructs the following integral equation

$$\epsilon(t) = \sigma(t)J(0^+) + \frac{1}{E\tau^\alpha}\left[\frac{1}{\Gamma(\alpha)} \int_0^t (t-\zeta)^{\alpha-1}\sigma(\zeta)d\zeta\right] \tag{16}$$

where

$$J\alpha(t) = \frac{1}{E}\frac{(t/\tau)^{\alpha}}{\Gamma(1+\alpha)},$$

and $\tau = E/\eta$, η being the shear modulus.

According to the kernel function $t^{\alpha-1}/\Gamma(\alpha)$ in Equation (16), $\alpha = 0$ and $\alpha = 1$ are equal with a memory-less system and full-memory system in creeping state respectively owing to $\Gamma(0) = \infty$. Clearly, we can derive that for $\alpha = 0$

$$\sigma(t) = \frac{\epsilon(t)}{J(0^+) + \frac{1}{E}} = \frac{E\epsilon(t)}{1 + EJ(0^+)}$$

using Equation (2) in distribution.

When $0 < \alpha \leq 1$, we convert Equation (16) into

$$\frac{\epsilon(t)}{J(0^+)} = \sigma(t) + \frac{1}{E\tau^{\alpha}J(0^+)}(\Phi_\alpha * \sigma)(t) = \left(\delta + \frac{1}{E\tau^{\alpha}J(0^+)}\Phi_\alpha\right) * \sigma(t).$$

By Babenko's approach, we imply that

$$\begin{aligned}\sigma(t) &= \frac{1}{J(0^+)}\left(\delta + \frac{1}{E\tau^{\alpha}J(0^+)}\Phi_\alpha\right)^{-1} * \epsilon(t)\\ &= \frac{1}{J(0^+)}\sum_{n=0}^{\infty}\frac{(-1)^n}{E^n\tau^{\alpha n}J^n(0^+)}\Phi_{\alpha n} * \epsilon(t)\end{aligned}$$

which is the relation between the stress $\sigma(t)$ and strain $\epsilon(t)$.

In particular, we derive that

$$\sigma(t) = \frac{1}{J(0^+)}\sum_{n=0}^{\infty}\frac{(-1)^n}{E^n\tau^{\alpha n}J^n(0^+)}\Phi_{\alpha n+1}$$

if the strain $\epsilon(t) = \theta(t)$.

4. Conclusions

With Babenko's approach, we have studied and solved several Abel's integral and fractional differential equations based on fractional calculus and convolutions of distributions in the space $\mathcal{D}'(R^+)$. Some of the results obtained are not achievable in the classical sense, such as numerical analysis methods or the Laplace transform, since the equations involve generalized functions which are not locally integrable, and undefined at points in R. Generally speaking, these equations can be expressed in terms of series or the Mittag–Leffer functions using inverse convolution operators in distribution. At the end, we demonstrate applications of Abel's integral equations in viscoelastic systems, and for solving other different types of integral equations with potential demands in physical problems.

Author Contributions: The order of the author list reflects contributions to the paper.

Acknowledgments: The authors are grateful to the reviewers and academic editor for their careful reading of the paper with very productive suggestions and corrections, which certainly improved its quality.

References

1. Kilbas, A.A.; Srivastava, H.M.; Trujillo, J.J. *Theory and Applications of Fractional Differential Equations*; Elsevier: Amsterdam, The Netherlands, 2006.
2. Ding, J.F.; Zhang, Y. Noether symmetries for the El-Nabulsi-Pfaff variational problem from extended exponentially fractional integral. *Acta Sci. Nat. Univ. Sunyatseni* **2014**, *53*, 150–154.
3. El-Wakil, S.A.; Abulwafa, E.M.; El-Shewy, E.K.; Mahmoud, A.A. Time-fractional KdV equation for plasma of two different temperature electrons and stationary ion. *Phys. Plasmas* **2011**, *18*, 092116. [CrossRef]
4. Neirameh, A. Soliton solutions of the time fractional generalized Hirota Satsuma coupled KdV equations. *Appl. Math. Inf. Sci.* **2015**, *9*, 1847–1853.
5. Zhang, X.; Zhang, Y. Lie symmetry and conserved quantity based on El-Nabulsi models in phase space. *J. Jiangxi Norm. Univ. Nat. Sci.* **2016**, *1*, 65–70.
6. Chen, J.; Zhang, Y. Perturbation to Noether symmetries and adiabatic invariants for disturbed Hamiltonian systems based on El-Nabulsi nonconservative dynamics model. *Nonlinear Dyn.* **2014**, *77*, 353–360. [CrossRef]
7. Golmankhaneh, A.K.; Lambert, L. *Investigations in Dynamics: With Focus on Fractional Dynamics*; Academic Publishing: Cambridge, MA, USA, 2012.
8. Monje, C.A.; Chen, Y.Q.; Vinagre, B.M.; Xue, D.; Feliu, V. *Fractional-Order Systems and Controls*; Series: Advances in Industrial Control; Springer: Berlin, Germany, 2010.
9. El-Nabulsi, R.A. The fractional Boltzmann transport equation. *Comp. Math. Appl.* **2011**, *62*, 1568–1575. [CrossRef]
10. Pu, Y.; Wang, W.X.; Zhou, J.L.; Wand, Y.Y.; Jia, H.D. Fractional differential approach to detecting textural features of digital image and its fractional differential filter implementation. *Sci. China Ser. F Inf. Sci.* **2008**, *51*, 1319–1339. [CrossRef]
11. El-Nabulsi, R.A. Spectrum of Schrodinger Hamiltonian operator with singular inverted complex and Kratzer's molecular potentials in fractional dimensions. *Eur. Phys. J. Plus* **2018**, *133*, 277. [CrossRef]
12. Chuanjing, S.; Zhang, Y. Conserved quantities and adiabatic invariants for El-Nabulsi's fractional Birkhoff system. *Int. J. Theor. Phys.* **2015**, *54*, 2481–2493.
13. El-Nabulsi, R.A. Glaeske-Kilbas-Saigo fractional integration and fractional Dixmier traces. *Acta Math. Vietnam.* **2012**, *37*, 149–160.
14. El-Nabulsi, R.A. Path integral formulation of fractionally perturbed Lagrangian oscillators on fractal. *J. Stat. Phys.* **2018**. [CrossRef]
15. Magin, R.L. *Fractional Calculus in Bioengineering*; Begell House: Redding, CT, USA, 2006.
16. West, B.J.; Bologna, M.; Grigolini, P. *Physics of Fractal Operators, Institute for Nonlinear Science*; Springer: New York, NY, USA, 2003.
17. Cafagna, D. Fractional calculus: A mathematical toll from the past for present engineering. *IEEE Ind. Electron. Mag.* **2007**, 1, 35–40. [CrossRef]
18. El-Nabulsi, R.A.; Wu, G.C. Fractional complexified field theory from Saxena-Kumbhat fraction integral, fractional derivative of order (α, β) and dynamical fractional integral exponent. *Afr. Diaspora J. Math. New Ser.* **2012**, *13*, 45–61.
19. Srivastava, T.; Singh, A.P.; Agarwal, H. Modeling the Under-Actuated Mechanical System with Fractional Order Derivative. *Progr. Fract. Differ. Appl.* **2015**, *1*, 57–64.
20. Srivastava, H.M.; Buschman, R.G. *Theory and Applications of Convolution Integral Equations*; Kluwer Academic Publishers: Dordrecht, The Netherlands; Boston, MA, USA; London, UK, 1992.
21. Gorenflo, R.; Mainardi, F. *Fractional Calculus: Integral and Differential Equations of Fractional Order*; Carpinteri, A., Mainardi, F., Eds.; Fractals and Fractional Calculus in Continuum Mechanics; Springer: New York, NY, USA, 1997; pp. 223–276.
22. Hatcher, J.R. A nonlinear boundary problem. *Proc. Am. Math. Soc.* **1985**, *95*, 441–448. [CrossRef]
23. Hilfer, R. *Applications of Fractional Calculus in Physics*; World Scientific Publishing: River Edge, NJ, USA, 2000.
24. Caponetto, R.; Dongola, G.; Fortuna, L.; Petras, I. *Fractional Order Systems: Modeling and Control Applications*; World Scientific: Singapore, 2010.
25. Yang, X.-J.; Baleanu, D.; Srivastava, H.M. *Local Fractional Integral Transforms and Their Applications*; Elsevier: Amsterdam, The Netherlands; Heidelberg, Germany; London, UK; New York, NY, USA, 2016.

26. Herrmann, R. *Fractional Calculus: An Introduction for Physicists*; World Scientific Publishing Company: Singapore, 2011.
27. El-Nabulsi, R.A. Fractional elliptic operator of order 2/3 from Glaeske-Kilbas-Saigo fractional integral transform. *Funct. Anal. Approx. Comput.* **2015**, *7*, 29–33.
28. Wang, J.R.; Zhu, C.; Fečkan, M. Analysis of Abel-type nonlinear integral equations with weakly singular kernels. *Bound. Value Probl.* **2014**, *2014*, *20*. [CrossRef]
29. Atkinson, K.E. An existence theorem for Abel integral equations. *SIAM J. Math. Anal.* **1974**, *5*, 729–736. [CrossRef]
30. Bushell, P.J.; Okrasinski, W. Nonlinear Volterra integral equations with convolution kernel. *J. Lond. Math. Soc.* **1990**, *41*, 503–510. [CrossRef]
31. Gorenflo, R.; Vessella, S. *Abel Integral Equations: Analysis and Applications*; Lect. Notes Math 1461; Springer: Berlin, Germany, 1991.
32. Okrasinski, W. Nontrivial solutions to nonlinear Volterra integral equations. *SIAM J. Math. Anal.* **1991**, *22*, 1007–1015. [CrossRef]
33. Gripenberg, G. On the uniqueness of solutions of Volterra equations. *J. Integral Equ. Appl.* **1990**, 2, 421–430. [CrossRef]
34. Mydlarczyk, W. The existence of nontrivial solutions of Volterra equations. *Math. Scand.* **1991**, *68*, 83–88. [CrossRef]
35. Kilbas, A.A.; Saigo, M. On solution of nonlinear Abel-Volterra integral equation. *J. Math. Anal. Appl.* **1999**, *229*, 41–60. [CrossRef]
36. Karapetyants, N.K.; Kilbas, A.A.; Saigo, M. Upper and lower bounds for solutions of nonlinear Volterra convolution integral equations with power nonlinearity. *J. Integral Equ. Appl.* **2000**, *12*, 421–448. [CrossRef]
37. Lima, P.; Diogo, T. Numerical solution of nonuniquely solvable Volterra integral equation using extrapolation methods. *J. Comput. App. Math.* **2002**, *140*, 537–557. [CrossRef]
38. Rahimy, M. Applications of fractional differential equations. *Appl. Math. Sci.* **2010**, *4*, 2453–2461.
39. Li, C.; Li, C.P.; Kacsmar, B.; Lacroix, R.; Tilbury, K. The Abel integral equations in distribution. *Adv. Anal.* **2017**, 2, 88–104. [CrossRef]
40. Li, C.; Clarkson, K. Babenko's approach to Abel's integral equations. *Mathematics* **2018**, *6*, 32. [CrossRef]
41. Li, C.; Li, C.P.; Clarkson, K. Several results of fractional differential and integral equations in distribution. *Mathematics* **2018**, *6*, 97. [CrossRef]
42. Mainardi, F. Fractional relaxation-oscillation and fractional diffusion-wave phenomena. *Chaos Solitons Fractals* **1996**, *7*, 1461–1477. [CrossRef]
43. Kilbas, A.A.; Raina, R.K.; Saigo, M.; Srivastava, H.M. Solution of multidimensional hypergeometric integral equations of the Abel type. *Dokl. Natl. Acad. Sci. Belarus* **1999**, *43*, 23–26. (In Russian)
44. Raina, R.K.; Srivastava, H.M.; Kilbas, A.A.; Saigo, M. Solvability of some Abel-type integral equations involving the Gauss hypergeometric function as kernels in the spaces of summable functions. *ANZJAM J.* **2001**, *43*, 291–320
45. Gel'fand, I.M.; Shilov, G.E. *Generalized Functions*; Academic Press: New York, NY, USA, 1964; Volume I.
46. Podlubny, I. *Fractional Differential Equations*; Academic Press: New York, NY, USA, 1999.
47. Matignon, D. Stability results for fractional differential equations with applications to control processing. In Proceedings of the IEEE-SMC Computational Engineering in System Applications, Lille, France, 10 February 1996; Volume 2, pp. 963–968.
48. Babenko, Y.I. *Heat and Mass Transfer*; Khimiya: Leningrad, Russian, 1986. (In Russian)
49. Amirian, M.; Jamali, Y. The Concepts and Applications of Fractional Order Differential Calculus in Modelling of Viscoelastic Systems: A primer. *arXiv* **2017**, arXiv:1706.06446.
50. Stiassnie, M. On the application of fractional calculus for the formulation of viscoelastic models. *Appl. Math. Model.* **1979**, *3*, 300–302. [CrossRef]

18

Fixed Point Results in Partial Symmetric Spaces with an Application

Mohammad Asim [1,*], A. Rauf Khan [2] and Mohammad Imdad [1]

1 Department of Mathematics, Aligarh Muslim University, Aligarh-202002, India; mhimdad@gmail.com
2 Department of Applied sciences, University Polytechnic, Aligarh Muslim University, Aligarh-202002, India; abdurraufamu@gmail.com
* Correspondence: mohdasim.rs@amu.ac.in

Abstract: In this paper, we first introduce the class of partial symmetric spaces and then prove some fixed point theorems in such spaces. We use one of the our main results to examine the existence and uniqueness of a solution for a system of Fredholm integral equations. Furthermore, we introduce an analogue of the Hausdorff metric in the context of partial symmetric spaces and utilize the same to prove an analogue of the Nadler contraction principle in such spaces. Our results extend and improve many results in the existing literature. We also give some examples exhibiting the utility of our newly established results.

Keywords: partial symmetric; fixed point; contraction and weak contraction; Nadler's theorem

---

1. Introduction

The classical Banach contraction principle is one of the most powerful and effective results in analysis established by Banach [1], which guarantees the existence and uniqueness of fixed points in complete metric spaces. This principle has been extended and generalized in many different directions. One of these ways is to enlarge the class of spaces, such as partial metric spaces [2], metric-like spaces [3], b-metric spaces [4], rectangular metric spaces [5], cone metric spaces [6], and several others. Sometimes, one may come across situations wherein all the metric conditions are not needed (see [7–11]). Motivated by this reality, several authors established fixed point and common fixed point results in symmetric spaces (or semi-metric spaces).

A symmetric d on a non-empty set X is a function $d : X \times X \to \mathbb{R}_+$ which satisfies $d(x,y) = d(y,x)$ and $d(x,y) = 0$ if and only if $x = y$, for all $x, y \in X$. Unlike the metric, the symmetric is not generally continuous. Due to the absence of a triangular inequality, the uniqueness of the limit of a sequence is no longer ensured. To have a workable setting, Wilson [12] suggested several related weaker conditions to overcome the earlier mentioned difficulties, which we will adopt to our setting. Such weaker conditions will be stated in the preliminaries.

In 1969, Nadler [13] initiated the study of fixed points for multi-valued contractions using the Hausdorff metric, and extended the Banach fixed point theorem to set-valued contractive maps. The theory of multi-valued maps has applications in control theory, convex optimization, differential equations, economics, and so on.

On the other hand, Matthews [2] introduced the concept of partial metric spaces as a part of the study of denotational semantics of dataflow networks, and proved an analogue of the Banach contraction theorem, and Kannan-Ćirić and Ćirić quasi-type fixed point results.

Combining the ideas involved in the concepts of partial metric spaces and symmetric spaces, we introduce the class of partial symmetric spaces, wherein we prove existence and uniqueness fixed point results for certain types of contractions in partial symmetric spaces. Furthermore, with a view to prove a multivalued analogue of Nadler's fixed point theorem, we adopt the idea of the Hausdorff metric in the setting of partial symmetric spaces. Finally, we use one of the our main results to examine the existence and uniqueness of a solution for a system of Fredholm integral equations.

2. Preliminaries

In this section, we collect some relevant definitions and examples which are needed in our subsequent discussions.

Now, we introduce the partial symmetric space as follows:

Definition 1. *Let X be a non-empty set. A mapping $\mathcal{P} : X \times X \to \mathbb{R}_+$ is said to be a partial symmetric if, for all $x, y, z \in X$:*

(1$\mathcal{P}$) $x = y$ if and only if $\mathcal{P}(x,x) = \mathcal{P}(y,y) = \mathcal{P}(x,y)$;
(2$\mathcal{P}$) $\mathcal{P}(x,x) \leq \mathcal{P}(x,y)$;
(3$\mathcal{P}$) $\mathcal{P}(x,y) = \mathcal{P}(y,x)$.

Then, the pair $(X,\mathcal{P})$ is said to be a partial symmetric space.

A partial symmetric space $(X,\mathcal{P})$ reduces to a symmetric space if $\mathcal{P}(x,x) = 0$, for all $x \in X$. Obviously, every symmetric space is a partial symmetric space, but not conversely.

Example 1. *Let $X = \mathbb{R}$ and define a mapping $\mathcal{P} : X \times X \to \mathbb{R}_+$ for all $x, y \in X$ and $p, q > 1$, as follows:*

$$\mathcal{P}(x,y) = |x-y|^p + |x-y|^q.$$

Then, the pair $(X,\mathcal{P})$ is a partial symmetric space.

Example 2. *Let $X = \mathbb{R}_+$ and define a mapping $\mathcal{P} : X \times X \to \mathbb{R}_+$ for all $x, y \in X$ and $p, q > 1$, as follows:*

$$\mathcal{P}(x,y) = (\max\{x,y\})^p + (\max\{x,y\})^q.$$

Then, the pair $(X,\mathcal{P})$ is a partial symmetric space.

Example 3. *Let $X = [0,\pi)$ and define a mapping $\mathcal{P} : X \times X \to \mathbb{R}_+$ for all $x, y \in X$ and $\alpha > 0$, as follows:*

$$\mathcal{P}(x,y) = \sin|x-y| + \alpha.$$

Then, the pair $(X,\mathcal{P})$ is a partial symmetric space.

Let $(X,\mathcal{P})$ be a partial symmetric space. Then, the $\mathcal{P}$-open ball, with center $x \in X$ and radius $\epsilon > 0$, is defined by:

$$B_{\mathcal{P}}(x,\epsilon) = \{y \in X : \mathcal{P}(x,y) < \mathcal{P}(x,x) + \epsilon\}.$$

Similarly, the $\mathcal{P}$-closed ball, with center $x \in X$ and radius $\epsilon > 0$, is defined by:

$$B_{\mathcal{P}}[x,\epsilon] = \{y \in X : \mathcal{P}(x,y) \leq \mathcal{P}(x,x) + \epsilon\}.$$

The family of $\mathcal{P}$-open balls for all $x \in X$ and $\epsilon > 0$,

$$\mathcal{U}_{\mathcal{P}} = \{B_{\mathcal{P}}(x,\epsilon) : x \in X, \epsilon > 0\},$$

forms a basis of some topology $\tau_{\mathcal{P}}$ on X.

Lemma 1. *Let $(X, \tau_{\mathcal{P}})$ be a topological space and $f : X \to X$. If f is continuous then, for every convergent sequence $x_n \to x$ in X, the sequence fx_n converges to fx. The converse holds if X is metrizable.*

In subsequent future discussions, we need some more basic definitions, namely: Convergent sequences, Cauchy sequences, and complete partial symmetric spaces, which are outlined in the following:

Definition 2. *A sequence $\{x_n\}$ in $(X, \mathcal{P})$ is said to be $\mathcal{P}$-convergent to $x \in X$, with respect to $\tau_{\mathcal{P}}$, if*

$$\mathcal{P}(x, x) = \lim_{n\to\infty} \mathcal{P}(x_n, x).$$

Definition 3. *A sequence $\{x_n\}$ in $(X, \mathcal{P})$ is said to be $\mathcal{P}$-Cauchy if and only if $\lim_{n,m\to\infty} \mathcal{P}(x_n, x_m)$ exists and is finite.*

Definition 4. *A partial symmetric space $(X, \mathcal{P})$ is said to be $\mathcal{P}$-complete if every $\mathcal{P}$-Cauchy sequence $\{x_n\}$ in X is $\mathcal{P}$-convergent, with respect to $\tau_{\mathcal{P}}$ to a point in $x \in X$, such that*

$$\mathcal{P}(x, x) = \lim_{n\to\infty} \mathcal{P}(x_n, x) = \lim_{n,m\to\infty} \mathcal{P}(x_n, x_m).$$

Now, we adopt some definitions from symmetric spaces in the setting of partial symmetric spaces:

Definition 5. *Let $(X, \mathcal{P})$ be a partial symmetric. Then*

(A1) $\lim_{n\to\infty} \mathcal{P}(x_n, x) = \mathcal{P}(x, x)$ *and* $\lim_{n\to\infty} \mathcal{P}(x_n, y) = \mathcal{P}(x, y)$ *imply that $x = y$, for a sequence $\{x_n\}$, x, and y in X.*

(A2) *A partial symmetric $\mathcal{P}$ is said to be 1-continuous if $\lim_{n\to\infty} \mathcal{P}(x_n, x) = \mathcal{P}(x, x)$ implies that $\lim_{n\to\infty} \mathcal{P}(x_n, y) = \mathcal{P}(x, y)$, where $\{x_n\}$ is a sequence in X and $x, y \in X$.*

(A3) *A partial symmetric $\mathcal{P}$ is said to be continuous if $\lim_{n\to\infty} \mathcal{P}(x_n, x) = \mathcal{P}(x, x)$ and $\lim_{n\to\infty} \mathcal{P}(x_n, y) = \mathcal{P}(x, y)$ imply that $\lim_{n\to\infty} \mathcal{P}(x_n, y_n) = \mathcal{P}(x, y)$ where $\{x_n\}$ and $\{y_n\}$ are sequences in X and $x, y \in X$.*

(A4) $\lim_{n\to\infty} \mathcal{P}(x_n, x) = \mathcal{P}(x, x)$ *and* $\lim_{n\to\infty} \mathcal{P}(x_n, y_n) = \mathcal{P}(x, x)$ *imply* $\lim_{n\to\infty} \mathcal{P}(y_n, x) = \mathcal{P}(x, x)$, *for sequences $\{x_n\}$, $\{y_n\}$, and x in X.*

(A5) $\lim_{n\to\infty} \mathcal{P}(x_n, y_n) = \mathcal{P}(x, x)$ *and* $\lim_{n\to\infty} \mathcal{P}(y_n, z_n) = \mathcal{P}(x, x)$ *imply* $\lim_{n\to\infty} \mathcal{P}(x_n, z_n) = \mathcal{P}(x, x)$, *for sequences $\{x_n\}$, $\{y_n\}$, $\{z_n\}$, and x in X.*

Remark 1. *From the Definition 5, it is observed that $(A3) \Rightarrow (A2)$, $(A4) \Rightarrow (A1)$, and $(A2) \Rightarrow (A1)$ but, in general, the converse implications are not true.*

3. Fixed Point Results

Let $(X, \mathcal{P})$ be a partial symmetric space and $f : X \to X$. Then, for every $x \in X$ and for all $i, j \in \mathbb{N}$, we define

$$\mathfrak{S}(\mathcal{P}, f, x) = \sup\{\mathcal{P}(f^i x, f^j x) : i, j \in \mathbb{N}\}. \tag{1}$$

Definition 6. *Let $(X, \mathcal{P})$ be a partial symmetric space. A mapping $f : X \to X$ is said to be a κ-contraction if*

$$\mathcal{P}(fx, fy) \leq \kappa \mathcal{P}(x, y), \ \ \forall\, x, y \in X, \tag{2}$$

where $\kappa \in (0, 1)$.

Now, we prove an analogue of the Banach contraction principle in the setting of partial symmetric spaces:

Theorem 1. *Let $(X, \mathcal{P})$ be a complete partial symmetric space and $f : X \to X$. Assume that the following conditions are satisfied:*

(i) f *is a* κ*-contraction, for some* $\kappa \in [0, 1)$;
(ii) *there exists* $x_0 \in X$ *such that* $\mathfrak{S}(\mathcal{P}, f, x_0) < \infty$; *and*
(iii) *either*
  (a) f *is continuous, or*
  (b) $(X, \mathcal{P})$ *satisfies the* $(A1)$ *property.*

Then, f *has a unique fixed point* $x \in X$ *such that* $\mathcal{P}(x, x) = 0$.

Proof. Choose $x_0 \in X$ and construct an iterative sequence $\{x_n\}$ by:

$$x_1 = fx_0,\ x_2 = f^2x_0,\ x_3 = f^3x_0, \ldots, x_n = f^nx_0, \ldots$$

Now, from (2) (for all $i, j \in \mathbb{N}$), we have

$$\mathcal{P}(f^{n+i}x_0, f^{n+j}x_0) \leq \kappa\mathcal{P}(f^{n-1+i}x_0, f^{n-1+j}x_0).$$

The above inequality holds for all $i, j \in \mathbb{N}$; therefore, by conditions (ii) and (1), we have

$$\mathfrak{S}(\mathcal{P}, f, f^nx_0) \leq \kappa\mathfrak{S}(\mathcal{P}, f, f^{n-1}x_0).$$

Repeating this procedure indefinitely, we have (for every $n \in \mathbb{N}$)

$$\mathfrak{S}(\mathcal{P}, f, f^nx_0) \leq \kappa^n\mathfrak{S}(\mathcal{P}, f, x_0). \tag{3}$$

Let $n, m \in \mathbb{N}$, such that $m = n + p$ for some $p \in \mathbb{N}$. Using (3), we have

$$\mathcal{P}(f^nx_0, f^{n+p}x_0) \leq \mathfrak{S}(\mathcal{P}, f, f^nx_0) \leq \kappa^n\mathfrak{S}(\mathcal{P}, f, x_0).$$

As $\mathfrak{S}(\mathcal{P}, f, x_0) < \infty$ and $\kappa \in (0, 1)$, we have

$$\lim_{n,m\to\infty} \mathcal{P}(x_n, x_m) = 0,$$

so that $\{x_n\}$ is a $\mathcal{P}$-Cauchy sequence in X. In light of the $\mathcal{P}$-completeness of X, there exists $x \in X$ such that $\{x_n\}$ $\mathcal{P}$-converges to x. Now, we show that $x \in X$ is a fixed point of f.

Assume that f is continuous. Then,

$$x = \lim_{n\to\infty} x_{n+1} = f(\lim_{n\to\infty} x_n) = fx.$$

Alternately, assume that $(X, \mathcal{P})$ satisfies the $(A1)$ property. Now, we have

$$\mathcal{P}(fx_n, fx) \leq \mathcal{P}(x_n, x),$$

which, on taking $n \to \infty$, implies that $\lim_{n\to\infty} \mathcal{P}(x_{n+1}, fx) = 0$. Thus, from the $(A1)$ property, $fx = x$. Therefore, x is a fixed point of f. To prove the uniqueness of the fixed point, let on contrary that there exist $x, y \in X$ such that $fx = x$ and $fy = y$. Then, by the definition of κ-contraction, we have

$$\mathcal{P}(x, y) = \mathcal{P}(fx, fy) \leq \kappa\mathcal{P}(x, y),$$

a contradiction. Hence, $x = y$; that is, x is a unique fixed point of f. Finally, we show that $\mathcal{P}(x,x) = 0$. Since, f is κ-contraction mapping, we have

$$\mathcal{P}(x,x) = \mathcal{P}(fx,fx) \leq \kappa\mathcal{P}(x,x).$$

This implies that $\mathcal{P}(x,x) < 0$, implying thereby that $\mathcal{P}(x,x) = 0$. This completes the proof. □

Now, we recall the definition of the Kannan-Ćirić contraction condition [14]:

Definition 7. *Let $(X,\mathcal{P})$ be a partial symmetric space. A mapping $f : X \to X$ is said to be a Kannan-Ćirić type κ-contraction if, for all $x,y \in X$,*

$$\mathcal{P}(fx,fy) \leq \kappa\max\{\mathcal{P}(x,fx),\mathcal{P}(y,fy)\}, \tag{4}$$

where $\kappa \in [0,1)$.

Next, we prove a fixed point result via Kannan-Ćirić type κ-contractions in the setting of partial symmetric spaces:

Theorem 2. *Let $(X,\mathcal{P})$ be a complete partial symmetric space and $f : X \to X$. Assume that the following conditions are satisfied:*

(i) f is a Kannan-Ćirić type κ-contraction mapping,
(ii) f is continuous.

Then, f has a unique fixed point $x \in X$ such that $\mathcal{P}(x,x) = 0$.

Proof. Take $x_0 \in X$, and construct an iterative sequence $\{x_n\}$ by:

$$x_1 = fx_0,\ x_2 = f^2x_0,\ x_3 = f^3x_0,\ldots,x_n = f^nx_0,\ldots$$

Now, we assert that $\lim_{n\to\infty}\mathcal{P}(x_n,x_{n+1}) = 0$. On setting $x = x_n$ and $y = x_{n+1}$ in (4), we get

$$\begin{aligned}\mathcal{P}(x_n,x_{n+1}) &= \mathcal{P}(fx_{n-1},fx_n)\\ &\leq \kappa\max\{\mathcal{P}(x_{n-1},fx_{n-1}),\mathcal{P}(x_n,fx_n)\}\\ &\leq \kappa\max\{\mathcal{P}(x_{n-1},x_n),\mathcal{P}(x_n,x_{n+1})\}.\end{aligned} \tag{5}$$

Assume that $\max\{\mathcal{P}(x_{n-1},x_n),\mathcal{P}(x_n,x_{n+1})\} = \mathcal{P}(x_n,x_{n+1})$, then from (5), we have

$$\mathcal{P}(x_n,x_{n+1}) \leq \kappa\mathcal{P}(x_n,x_{n+1}),$$

a contradiction (since $\kappa \in (0,1)$). Thus, $\max\{\mathcal{P}(x_{n-1},x_n),\mathcal{P}(x_n,x_{n+1})\} = \mathcal{P}(x_{n-1},x_n)$. Therefore, (5) gives rise

$$\mathcal{P}(x_n,x_{n+1}) = \kappa\mathcal{P}(x_{n-1},x_n) \text{ for all } n \in \mathbb{N}.$$

Thus, inductively, we have

$$\mathcal{P}(x_n,x_{n+1}) = \kappa^n\mathcal{P}(x_0,x_1) \text{ for all } n \in \mathbb{N}.$$

On taking the limit as $n \to \infty$, we get

$$\lim_{n\to\infty}\mathcal{P}(x_n,x_{n+1}) = 0. \tag{6}$$

Now, we assert that $\{x_n\}$ is a $\mathcal{P}$-Cauchy sequence. From (4), we have, for $n, m \in \mathbb{N}$,

$$\begin{aligned}\mathcal{P}(x_n, x_m) &= \mathcal{P}(fx_{n-1}, fx_{m-1}) \\ &\leq \kappa \max\{\mathcal{P}(x_{n-1}, fx_{n-1}), \mathcal{P}(x_{m-1}, fx_{m-1})\} \\ &\leq \kappa \max\{\mathcal{P}(x_{n-1}, x_n), \mathcal{P}(x_{m-1}, x_m)\}.\end{aligned}$$

By taking the limit as $n, m \to \infty$ and using (6), we have

$$\lim_{n,m\to\infty} \mathcal{P}(x_n, x_m) = 0. \tag{7}$$

Hence, $\{x_n\}$ is a $\mathcal{P}$-Cauchy sequence. Since $(X, \mathcal{P})$ is $\mathcal{P}$-complete, there exists $x \in X$ such that $\lim_{n\to\infty} \mathcal{P}(x_n, x) = 0$. Now, we show that $x \in X$ is a fixed point of f. By the continuity of f, we have

$$x = \lim_{n\to\infty} x_{n+1} = f(\lim_{n\to\infty} x_n) = fx.$$

Therefore, x is a fixed point of f. For the uniqueness part, let on contrary that there exist $x, y \in X$ such that $fx = x$ and $fy = y$. Then, from (4), we have

$$\begin{aligned}\mathcal{P}(x, y) &= \mathcal{P}(fx, fy) \\ &\leq \kappa \max\{\mathcal{P}(x, fx), \mathcal{P}(y, fy)\}, \\ &\leq \kappa \max\{\mathcal{P}(x, x), \mathcal{P}(y, y)\}.\end{aligned}$$

So, either $\mathcal{P}(x, y) \leq \kappa\mathcal{P}(x, x)$ or $\mathcal{P}(x, y) \leq \kappa\mathcal{P}(y, y)$, which is a contradiction. Therefore, x is a unique fixed point of f. Finally, we show that $\mathcal{P}(x, x) = 0$. From (4), we have

$$\begin{aligned}\mathcal{P}(x, x) &= \kappa\mathcal{P}(fx, fx) \\ &\leq \kappa \max\{\mathcal{P}(x, fx), \mathcal{P}(x, fx)\}, \\ &\leq \kappa \max\{\mathcal{P}(x, x), \mathcal{P}(x, x)\}, \\ &\leq \kappa\mathcal{P}(x, x),\end{aligned}$$

this implies that $\mathcal{P}(x, x) < 0$, implying thereby that $\mathcal{P}(x, x) = 0$. This completes the proof. □

Now, we present some fixed point results for Ćirić quasi contractions in the setting of partial symmetric spaces. We start with the following definition.

Definition 8. *Let $(X, \mathcal{P})$ be a partial symmetric space and $f : X \to X$. Then f is said to be κ-weak contraction if, for all $x, y \in X$, and $\kappa \in (0, 1)$*

$$\mathcal{P}(fx, fy) \leq \kappa \max\{\mathcal{P}(x, y), \mathcal{P}(x, fx), \mathcal{P}(y, fy), \mathcal{P}(x, fy), \mathcal{P}(y, fx)\}. \tag{8}$$

Proposition 1. *Let f be a κ-weak contraction for any $\kappa \in (0, 1)$. If x is a fixed point of f, then $\mathcal{P}(x, x) = 0$.*

Proof. Suppose $x \in X$ is a fixed point of f. Since f is a κ-weak contraction, we have that

$$\begin{aligned}\mathcal{P}(x, x) &= \mathcal{P}(fx, fx) \\ &\leq \kappa \max\{\mathcal{P}(x, x), \mathcal{P}(x, fx), \mathcal{P}(x, fx), \mathcal{P}(x, fx), \mathcal{P}(x, fx)\} \\ &= \kappa \max\{\mathcal{P}(x, x), \mathcal{P}(x, x), \mathcal{P}(x, x), \mathcal{P}(x, x), \mathcal{P}(x, x)\} \\ &= \kappa\mathcal{P}(x, x),\end{aligned}$$

this implies that $\mathcal{P}(x, x) < 0$, implying thereby $\mathcal{P}(x, x) = 0$. □

Theorem 3. *Let $(X, \mathcal{P})$ be a $\mathcal{P}$-complete partial symmetric space and $f : X \to X$. Suppose that the following conditions hold:*

(i) *f is a κ-weak contraction for some $\kappa \in [0, 1)$;*
(ii) *there exists $x_0 \in X$ such that $\mathfrak{S}(\mathcal{P}, f, x) < \infty$; and*
(iii) *f is continuous.*

Then, f has a unique fixed point.

Proof. Assume $x_0 \in X$, and construct an iterative sequence $\{x_n\}$ by:

$$x_1 = fx_0,\ x_2 = f^2x_0,\ x_3 = f^3x_0, \ldots, x_n = f^nx_0, \ldots$$

Let n be an arbitrary positive integer. Since f is a κ-weak contraction, for all $i, j \in \mathbb{N}$, we have

$$\begin{aligned}\mathcal{P}(f^{n+i}x_0, f^{n+j}x_0) \leq\ & \kappa \max\{\mathcal{P}(f^{n-1+i}x_0, f^{n-1+j}x_0), \mathcal{P}(f^{n-1+i}x_0, f^{n+i}x_0),\\ & \mathcal{P}(f^{n-1+j}x_0, f^{n+j}x_0), \mathcal{P}(f^{n-1+i}x_0, f^{n+j}x_0), \mathcal{P}(f^{n-1+j}x_0, f^{n+i}x_0)\}.\end{aligned}$$

Since the above inequality is true for all $i, j \in \mathbb{N}$, therefore by conditions (ii) and (1), we have

$$\mathfrak{S}(\mathcal{P}, f, f^nx_0) \leq \kappa\mathfrak{S}(\mathcal{P}, f, f^{n-1}x_0).$$

Continuing this process indefinitely, we have, for all $n \geq 1$,

$$\mathfrak{S}(\mathcal{P}, f, f^nx_0) \leq \kappa^n\mathfrak{S}(\mathcal{P}, f, x_0). \tag{9}$$

Now, for each $n, m \in \mathbb{N}$, such that $m = n + p$ for some $p \in \mathbb{N}$, we have, due to (9), that

$$\mathcal{P}(f^nx_0, f^{n+p}x_0) \leq \mathfrak{S}(\mathcal{P}, f, f^nx_0) \leq \kappa^n\mathfrak{S}(\mathcal{P}, f, x_0). \tag{10}$$

Since $\mathfrak{S}(\mathcal{P}, f, x_0) < \infty$ and $\kappa \in (0, 1)$, we have

$$\lim_{n,m\to\infty} \mathcal{P}(x_n, x_m) = 0,$$

so $\{x_n\}$ is a $\mathcal{P}$-Cauchy sequence in X. In view of the $\mathcal{P}$-completeness of X, there exists $x \in X$ such that $\{x_n\}$ $\mathcal{P}$-converges to x. Now, we show that x is a fixed point of f. By the continuity of f, we have

$$x = \lim_{n\to\infty} x_{n+1} = f(\lim_{n\to\infty} x_n) = fx.$$

Therefore, x is a fixed point of f. For the uniqueness part, let on contrary that there exist $x, y \in X$ such that $fx = x$ and $fy = y$. Thus, by using the condition (8), we have

$$\begin{aligned}\mathcal{P}(x, y) &= \mathcal{P}(fx, fy)\\ &\leq \kappa \max\{\mathcal{P}(x, y), \mathcal{P}(x, fx), \mathcal{P}(y, fy), \mathcal{P}(x, fy), \mathcal{P}(y, fx)\}\\ &= \kappa \max\{\mathcal{P}(x, y), \mathcal{P}(x, x), \mathcal{P}(y, y), \mathcal{P}(x, y), \mathcal{P}(y, x)\}.\end{aligned}$$

By using the property $(2\mathcal{P})$, we have

$$\mathcal{P}(x, y) \leq \kappa\mathcal{P}(x, y) < \mathcal{P}(x, y),$$

a contradiction, and so $\mathcal{P}(x, y) = 0$; which implies that $x = y$. Thus, f has a unique fixed point. This completes the proof. □

Now, we furnish the following example, which illustrates Theorem 3.

Example 4. *Consider $X = [0,1]$ and a partial symmetric $\mathcal{P} : X \times X \to \mathbb{R}_+$ defined by $\mathcal{P}(x,y) = \max\{x,y\}$, for all $x, y \in X$. Define a self-mapping f on X by*

$$fx = \frac{2x^2}{5}, \text{ for all } x \in X.$$

Observe that

$$\begin{aligned}
\mathcal{P}(fx, fy) &= \max\{fx, fy\} \\
&= \max\left\{\frac{2x^2}{5}, \frac{2y^2}{5}\right\} \\
&\leq \frac{2}{5}\max\{x,y\} = \frac{2}{5}\mathcal{P}(x,y) \\
&\leq \frac{2}{5}\max\left\{\mathcal{P}(x,y), \mathcal{P}(x,fx), \mathcal{P}(y,fy), \mathcal{P}(x,fy), \mathcal{P}(y,fx)\right\},
\end{aligned}$$

for all $x, y \in X$. Observe that f is continuous and condition (ii) holds. Thus, all the conditions of Theorem 3 are satisfied and so f has a unique fixed point (i.e., $x = 0$).

Notice that this example can not be covered by metrical fixed point theorems.

Corollary 1. *The conclusions of Theorem 3 remain true, if the contractive condition (8) is replaced by any one of the following:*

(i) $\mathcal{P}(fx, fy) \leq \frac{\kappa}{2}[\mathcal{P}(x,fy) + \mathcal{P}(y,fx)]$;
(ii) $\mathcal{P}(fx, fy) \leq \kappa \max\{\mathcal{P}(x,y), \mathcal{P}(x,fx), \mathcal{P}(y,fy)\}$;
(iii) $\mathcal{P}(fx, fy) \leq \kappa \max\{\mathcal{P}(x,y), \mathcal{P}(x,fy), \mathcal{P}(y,fx)\}$;
(iv) $\mathcal{P}(fx, fy) \leq \kappa \max\left\{\mathcal{P}(x,y), \frac{\mathcal{P}(x,fx)+\mathcal{P}(y,fy)}{2}, \frac{\mathcal{P}(x,fy)+\mathcal{P}(y,fx)}{2}\right\}$;
(v) $\mathcal{P}(fx, fy) \leq \kappa \max\left\{\mathcal{P}(x,y), \frac{\mathcal{P}(x,fx)+\mathcal{P}(y,fy)}{2}, \mathcal{P}(x,fy), \mathcal{P}(y,fx)\right\}$; *or*
(vi) $\mathcal{P}(fx, fy) \leq \kappa \max\left\{\mathcal{P}(x,y), \mathcal{P}(x,fx), \mathcal{P}(y,fy), \frac{\mathcal{P}(x,fy)+\mathcal{P}(y,fx)}{2}\right\}$.

4. Application

In this section, we endeavor to apply Theorem 1 to prove the existence and uniqueness of a solution of the following integral equation of Fredholm type:

$$x(t) = \int_a^b G(t,s,x(s))ds + h(t) \text{ for all } t,s \in [a,b], \tag{11}$$

where $G, h \in C([a,b], \mathbb{R})$ (say, $X = C([a,b], \mathbb{R})$. Define a partial symmetric space $\mathcal{P}$ on X:

$$\mathcal{P}(x,y) = \sup_{t\in[a,b]} |x(t) - y(t)|^p + \sup_{t\in[a,b]} |x(t) - y(t)|^q, \text{ for all } x, y \in X, \text{ and } p, q > 1.$$

Then, $(X, \mathcal{P})$ is a complete partial symmetric space.
Now we are equipped to state and prove our result, as follows:

Theorem 4. *Assume that, for all $x, y \in C([a,b], \mathbb{R})$,*

$$|G(t,s,x(s)) - G(t,s,y(s))| \leq \frac{1}{2(b-a)}|x(s) - y(s)|, \tag{12}$$

for all $t, s \in [a, b]$. *Then, Equation* (11) *has a unique solution.*

Proof. Define $f : X \to X$ by

$$fx(t) = \int_a^b G(t,s,x(s))ds + h(t) \text{ for all } t,s \in [a,b].$$

It is clear that x is a fixed point of the operator f if and only if it is a solution of Equation (11). Now, for all $x, y \in X$, we have

$$\begin{aligned}
|fx(t) - fy(t)|^p + |fx(t) - fy(t)|^q &\leq \Big(\int_a^b |G(t,s,x(s)) - G(t,s,y(s))|ds\Big)^p \\
&\quad + \Big(\int_a^b |G(t,s,x(s)) - G(t,s,y(s))|ds\Big)^q \\
&\leq \Big(\int_a^b \frac{1}{2(b-a)}|x(s) - y(s)|ds\Big)^p \\
&\quad + \Big(\int_a^b \frac{1}{2(b-a)}|x(s) - y(s)|ds\Big)^q \\
&\leq \frac{1}{2^p(b-a)^p} \sup_{t\in[a,b]} |x(t) - y(t)|^p \Big(\int_a^b ds\Big)^p \\
&\quad + \frac{1}{2^q(b-a)^q} \sup_{t\in[a,b]} |x(t) - y(t)|^q \Big(\int_a^b ds\Big)^q \\
&\leq \lambda\big(\sup_{t\in[a,b]} |x(t) - y(t)|^p + \sup_{t\in[a,b]} |x(t) - y(t)|^q\big).
\end{aligned}$$

Thus, condition (12) is satisfied, with $\lambda = \max\{\frac{1}{2^p}, \frac{1}{2^q}\} \in [0, 1)$. Hence, the operator f has a unique fixed point; that is, the Fredholm integral Equation (11) has a unique solution. □

5. Results Involving Set-Valued Map

In this section, first we extend the idea of Hausdorff distance to partial symmetric spaces. Let $(X, \mathcal{P})$ be a partial symmetric space and $\mathcal{CB}^{\mathcal{P}}(X)$ be the family of all nonempty, $\tau_{\mathcal{P}}$-closed, and bounded subsets of $(X, \mathcal{P})$. Observe that A will be bounded if there exist $x_0 \in X$ and $M \geq 0$ such that, for all $a \in A$, $\mathcal{P}(x_0, a) \leq \mathcal{P}(a, a) + M$.

Moreover, for $A, B \in \mathcal{CB}^{\mathcal{P}}(X)$ and $x \in X$, we define:

$$dist_{\mathcal{P}}(x, A) = \inf\{\mathcal{P}(x, a) : a \in A\};$$

$$\delta_{\mathcal{P}}(A, B) = \sup\{dist_{\mathcal{P}}(a, B) : a \in A\}; \text{ and}$$

$$\delta_{\mathcal{P}}(B, A) = \sup\{dist_{\mathcal{P}}(b, A) : b \in B\}.$$

Remark 2. *Let* $(X, \mathcal{P})$ *be a partial symmetric space and* A *a non-empty subset of* X, *then*

$$a \in \overline{A} \text{ if and only if } dist_{\mathcal{P}}(a, A) = \mathcal{P}(a, a),$$

where $\overline{A}$ *denotes the closure of* A, *with respect to the partial symmetric* $\mathcal{P}$. *Also,* A *is* $\mathcal{P}$-*closed in* $(X, \mathcal{P})$ *if and only if* $A = \overline{A}$.

Proposition 2. *Let* $(X, \mathcal{P})$ *be a partial symmetric space. For* $A, B, C \in \mathcal{CB}^{\mathcal{P}}X$, *we have the following:*

(i) $\delta_{\mathcal{P}}(A, A) = \sup\{\mathcal{P}(a, a) : a \in A\}$;
(ii) $\delta_{\mathcal{P}}(A, A) \leq \delta_{\mathcal{P}}(A, B)$;
(iii) $B \subset C \Rightarrow \delta_{\mathcal{P}}(A, C) \leq \delta_{\mathcal{P}}(A, B)$;

(*iv*) $\delta_{\mathcal{P}}(A,B) = 0 \Rightarrow A \subseteq B$; *and*
(*v*) $\delta_{\mathcal{P}}(A \cup B, C) = \max\{\delta_{\mathcal{P}}(A,C), \delta_{\mathcal{P}}(B,C)\}$.

Proof. (*i*) Suppose $A \in \mathcal{CB}^{\mathcal{P}}(X)$. Since $a \in \overline{A}$ if and only if $\mathcal{P}(a,A) = \mathcal{P}(a,a)$,

$$\delta_{\mathcal{P}}(A,A) = \sup\{dist_{\mathcal{P}}(a,A) : a \in A\} = \sup\{\mathcal{P}(a,a) : a \in A\}.$$

(*ii*) Suppose $a \in A$. By definition of the partial symmetric space, we know that $\mathcal{P}(a,a) \leq \mathcal{P}(a,b)$, which implies that

$$\mathcal{P}(a,a) \leq dist_{\mathcal{P}}(a,B) \leq \delta_{\mathcal{P}}(A,B).$$

Hence, condition (*i*) gives rise to

$$\delta_{\mathcal{P}}(A,A) \leq \delta_{\mathcal{P}}(A,B).$$

(*iii*) Suppose $A, B, C \in \mathcal{CB}^{\mathcal{P}}(X)$, such that $B \subseteq C$. Then,

$$dist_{\mathcal{P}}(x,B) \leq dist_{\mathcal{P}}(x,C) \text{ for all } a \in X.$$

Thus,

$$B \subset C \Rightarrow \delta_{\mathcal{P}}(A,C) \leq \delta_{\mathcal{P}}(A,B).$$

(*iv*) Suppose $A, B \in \mathcal{CB}^{\mathcal{P}}(X)$, such that $\delta_{\mathcal{P}}(A,B) = 0$. Then,

$$\sup\{dist_{\mathcal{P}}(a,A) : a \in A\} = 0 \Rightarrow dist_{\mathcal{P}}(a,B) = 0 \text{ for all } a \in A.$$

In view of the above conditions (*i*) and (*ii*), we have

$$\mathcal{P}(a,a) \leq \delta_{\mathcal{P}}(A,B) = 0 \Rightarrow \mathcal{P}(a,a) = 0, \text{ for all } a \in A.$$

Therefore, $dist_{\mathcal{P}}(a,B) = \mathcal{P}(a,a)$ for all $a \in A$ implies that 'a' is in the closure of B for all $a \in A$. Since B is $\mathcal{P}$-closed, we have $A \subseteq B$.

(*v*) Suppose $A, B, C \in \mathcal{CB}^{\mathcal{P}}(X)$. Then,

$$\begin{aligned}
\delta_{\mathcal{P}}(A \cup B, C) &= \sup\{dist_{\mathcal{P}}(x,C) : x \in A \cup B\} \\
&= \max\left\{\sup\{dist_{\mathcal{P}}(x,C) : x \in A\}, \sup\{dist_{\mathcal{P}}(x,C) : x \in B\}\right\} \\
&= \max\{\delta_{\mathcal{P}}(A,C), \delta_{\mathcal{P}}(B,C)\}.
\end{aligned}$$

□

Next, let $(X, \mathcal{P})$ be a partial symmetric space. Define

$$H_{\mathcal{P}}(A,B) = \max\{\delta_{\mathcal{P}}(A,B), \delta_{\mathcal{P}}(B,A)\}.$$

Proposition 3. *Let $(X, \mathcal{P})$ be a partial symmetric space. For $A, B, C, D \in \mathcal{CB}^{\mathcal{P}}(X)$, we have the following:*

(1H) $H_{\mathcal{P}}(A,A) \leq H_{\mathcal{P}}(A,B)$;
(2H) $H_{\mathcal{P}}(A,B) = H_{\mathcal{P}}(B,A)$; *and*
(3H) $H_{\mathcal{P}}(A \cup B, C \cup D) = \max\{H_{\mathcal{P}}(A,C), H_{\mathcal{P}}(B,D)\}$.

Proof. $(1H)$ By condition (*ii*) of Proposition 2, we have

$$H_{\mathcal{P}}(A,A) = \delta_{\mathcal{P}}(A,A) \leq \delta_{\mathcal{P}}(A,B) \leq H_{\mathcal{P}}(A,B).$$

$(2H)$ By the definition of $H_{\mathcal{P}}$, we have

$$H_{\mathcal{P}}(A,B) = \max\{\delta_{\mathcal{P}}(A,B), \delta_{\mathcal{P}}(B,A)\} = \max\{\delta_{\mathcal{P}}(B,A), \delta_{\mathcal{P}}(A,B)\} = H_{\mathcal{P}}(B,A).$$

$(3H)$ By condition (v) of Proposition 2, we have

$$\begin{aligned} \delta_{\mathcal{P}}(A\cup B, C\cup D) &= \max\{\delta_{\mathcal{P}}(A, C\cup D), \delta_{\mathcal{P}}(B, C\cup D)\} \\ &\leq \max\{\delta_{\mathcal{P}}(A,C), \delta_{\mathcal{P}}(B,D)\} \\ &\leq \max\{H_{\mathcal{P}}(A,C), H_{\mathcal{P}}(B,D)\}. \end{aligned}$$

Similarly, we obtain

$$\delta_{\mathcal{P}}(C\cup D, A\cup B) \leq \max\{H_{\mathcal{P}}(A,C), H_{\mathcal{P}}(B,D)\}.$$

Hence, by the definition of $H_{\mathcal{P}}$, we have, for all $A,B,C,D \in \mathcal{CB}^{\mathcal{P}}(X)$, that

$$H_{\mathcal{P}}(A\cup B, C\cup D) = \max\{H_{\mathcal{P}}(A,C), H_{\mathcal{P}}(B,D)\}.$$

□

Proposition 4. *Let $(X,\mathcal{P})$ be a partial symmetric space. For $A,B \in \mathcal{CB}^{\mathcal{P}}(X)$, we have*

$$H_{\mathcal{P}}(A,B) = 0 \Rightarrow A = B.$$

Proof. Let $H_{\mathcal{P}}(A,B) = 0$. Then, by the definition of $H_{\mathcal{P}}$, we have

$$\delta_{\mathcal{P}}(A,B) = \delta_{\mathcal{P}}(B,A) = 0.$$

Thus, by condition (iii) of Proposition 2, we get $A \subseteq B$ and $B \subseteq A$, which implies $A = B$. □

Now, we prove the following lemma which is needed in the sequel:

Lemma 2. *Let $(X,\mathcal{P})$ be partial symmetric space and $A,B \in \mathcal{CB}^{\mathcal{P}}(X)$. Then, for any $h > 1$ and $a \in A$, there exists $b \in B$ such that*

$$\mathcal{P}(a,b) \leq hH_{\mathcal{P}}(A,B). \tag{13}$$

Proof. First, we consider $A = B$. From (i) of Proposition 2,

$$H_{\mathcal{P}}(A,B) = H_{\mathcal{P}}(A,A) = \delta_{\mathcal{P}}(A,A) = \sup_{a\in A} \mathcal{P}(a,a).$$

Observe that, for any $a \in A$ and any $h > 1$, we have

$$\mathcal{P}(a,a) \leq \sup_{a\in A} \mathcal{P}(a,a) = H_{\mathcal{P}}(A,B) \leq hH_{\mathcal{P}}(A,B).$$

Consequently, $b = a$ satisfies the inequality (13). Now, let $A \neq B$. Then, there exists $a \in A$ such that

$$\mathcal{P}(a,b) > hH_{\mathcal{P}}(A,B) \text{ for all } b \in B.$$

Then,

$$\inf\{\mathcal{P}(a,b) : b \in B\} \geq hH_{\mathcal{P}}(A,B),$$

so that

$$dist_{\mathcal{P}}(a,B) \geq hH_{\mathcal{P}}(A,B).$$

Hence,

$$H_{\mathcal{P}}(A,B) \geq \delta_{\mathcal{P}}(A,B) = \sup_{a\in A} dist_{\mathcal{P}}(a,B) \geq dist_{\mathcal{P}}(a,B) \geq hH_{\mathcal{P}}(A,B),$$

a contradiction, since $h > 1$. □

Recall that, if $f : X \to \mathcal{CB}^{\mathcal{P}}(X)$ is a mapping, then an element $x \in X$ is said to be a fixed point of f if $x \in fx$.

Now, we state and prove our main result in this section:

Theorem 5. *Let $(X,\mathcal{P})$ be a complete partial symmetric space and $f : X \to \mathcal{CB}^{\mathcal{P}}(X)$. Assume that the following conditions are satisfied:*

(i) there exists $\kappa \in [0,1)$ such that

$$H_{\mathcal{P}}(fx,fy) \leq \kappa\mathcal{P}(x,y) \text{ for all } x,y \in X;$$

(ii) there exists $x_0 \in X$ such that $\mathfrak{S}(\mathcal{P},f,x_0) < \infty$; and

(iii) f is continuous.

Then, f has a unique fixed point $x \in X$, such that $\mathcal{P}(x,x) = 0$.

Proof. Suppose $x_0 \in X$ and $x_1 \in fx_0$. From Lemma 2 with $h = \frac{1}{\sqrt{\kappa}}$, there exists $x_2 \in fx_1$ such that $\mathcal{P}(x_1,x_2) \leq \frac{1}{\sqrt{\kappa}} H_{\mathcal{P}}(fx_0,fx_1)$. Since $H_{\mathcal{P}}(fx_0,fx_1) \leq \kappa\mathcal{P}(x_0,x_1)$, then $\mathcal{P}(x_1,x_2) \leq \sqrt{\kappa}\mathcal{P}(x_0,x_1)$. Similarly, for $x_2 \in fx_1$ there exists $x_3 \in fx_2$ such that

$$\mathcal{P}(x_2,x_3) \leq \frac{1}{\sqrt{\kappa}} H_{\mathcal{P}}(fx_1,fx_2) \leq \sqrt{\kappa}\mathcal{P}(x_1,x_2).$$

Inductively, we obtain a sequence $\{x_n\}$ in X, such that

$$x_{n+1} \in fx_n \text{ and } \mathcal{P}(x_{n+1},x_n) \leq \sqrt{\kappa}\mathcal{P}(x_n,x_{n-1}), \text{ for all } n \in \mathbb{N}.$$

By condition (i), for all $i,j \in \mathbb{N}$ we have

$$\mathcal{P}(f^{n+i}x_0, f^{n+j}x_0) \leq \sqrt{\kappa}\mathcal{P}(f^{n-1+i}x_0, f^{n-1+j}x_0).$$

Therefore, by condition (ii) and (1), we have

$$\mathfrak{S}(\mathcal{P},f,f^n x_0) \leq \sqrt{\kappa}\mathfrak{S}(\mathcal{P},f,f^{n-1}x_0).$$

Continuing this process, we have, for every $n \in \mathbb{N}$,

$$\mathfrak{S}(\mathcal{P},f,f^n x_0) \leq (\sqrt{\kappa})^n \mathfrak{S}(\mathcal{P},f,x_0). \tag{14}$$

By using (3), we have, for $n,m,p \in \mathbb{N}$ such that $m = n + p$,

$$\mathcal{P}(f^n x_0, f^{n+p}x_0) \leq \mathfrak{S}(\mathcal{P},f,f^n x_0) \leq (\sqrt{\kappa})^n \mathfrak{S}(\mathcal{P},f,x_0).$$

Since $\mathfrak{S}(\mathcal{P},f,x_0) < \infty$ and $\kappa \in (0,1)$, then

$$\lim_{n,m\to\infty} \mathcal{P}(x_n,x_m) = 0,$$

so that $\{x_n\}$ is a $\mathcal{P}$-Cauchy sequence in X. In view of the $\mathcal{P}$-completeness of X, there exists $x \in X$ such that $\{x_n\}$ $\mathcal{P}$-converges to x. Therefore,

$$\mathcal{P}(x,x) = \lim_{n\to\infty} \mathcal{P}(x_n, x) = \lim_{n,m\to\infty} \mathcal{P}(x_n, x_m) = 0.$$

As $H_{\mathcal{P}}(fx_n, fx) \leq \kappa \mathcal{P}(x_n, x)$, implies that

$$\lim_{n\to\infty} H_{\mathcal{P}}(fx_n, fx) = 0.$$

Hence, $x_{n+1} \in fx_n$. Therefore,

$$dist_{\mathcal{P}}(x_{n+1}, fx) \leq \delta_{\mathcal{P}}(fx_n, fx) \leq H_{\mathcal{P}}(fx_n, fx).$$

Hence,

$$\lim_{n\to\infty} dist_{\mathcal{P}}(x_{n+1}, fx) = 0.$$

By the continuity of f, we obtain

$$dist_{\mathcal{P}}(x, fx) = \lim_{n\to\infty} dist_{\mathcal{P}}(x_{n+1}, fx) = 0.$$

Thus, we have $\mathcal{P}(x,x) = dist_{\mathcal{P}}(x, fx) = 0$. As fx is $\mathcal{P}$-closed, then we have $x \in fx$. Hence, x is a fixed point of f in X. This completes the proof. □

Next, we adopt the following example to demonstrate Theorem 5.

Example 5. *Consider* $X = \{0,1,2\}$ *equipped with the partial symmetric* $\mathcal{P} : X \times X \to \mathbb{R}_+$ *defined by*

$$\mathcal{P}(x,y) = \frac{1}{2}|x-y|^2 + \frac{1}{4}(\max\{x,y\})^2, \text{ for all } x, y \in X.$$

Then $(X, \mathcal{P})$ *is a* $\mathcal{P}$*-complete symmetric space. Note that* $\{0\}$ *and* $\{0,1\}$ *are bounded sets in* $(X,\mathcal{P})$*. In fact, if* $x \in \{0,1,2\}$ *then*

$$\begin{aligned} x \in \overline{\{0\}} \quad &\Leftrightarrow \quad dist_{\mathcal{P}}(x, \{0\}) = \mathcal{P}(x,x) \\ &\Leftrightarrow \quad \frac{3x^2}{4} = \frac{x^2}{4} \\ &\Leftrightarrow \quad x = 0 \Leftrightarrow x \in \{0\}. \end{aligned}$$

Hence, $\{0\}$ *is closed with respect to the partial symmetric* $\mathcal{P}$*. Next,*

$$\begin{aligned} x \in \overline{\{0,1\}} \quad &\Leftrightarrow \quad dist_{\mathcal{P}}(x, \{0,1\}) = \mathcal{P}(x,x) \\ &\Leftrightarrow \quad \min\left\{\frac{3x^2}{2}, \frac{1}{2}|x-1|^2 + \frac{1}{4}(\max\{x,1\})^2\right\} = \frac{x^2}{4} \\ &\Leftrightarrow \quad x \in \{0,1\}. \end{aligned}$$

Hence, $\{0,1\}$ *is also closed with respect to the partial symmetric* $\mathcal{P}$*.*
Now, define $f : X \to \mathcal{CB}^{\mathcal{P}}(X)$ *by:*

$$f0 = f1 = \{0\} \text{ and } f2 = \{0,1\}.$$

Notice that f *is continuous under the partial symmetric* $\mathcal{P}$*. Now, to show that the contractive condition* (i) *of Theorem 5 is satisfied, we distinguish the following cases:*
Case 1. *Let* $x, y \in \{0,1\}$*. Then,*

$$H_{\mathcal{P}}(fx, fy) = H_{\mathcal{P}}(0,0) = 0,$$

so that the contractive condition (i) satisfied.
Case 2. *Let $x \in \{0,1\}$ and $y = 2$. Then, with $k = \frac{1}{2}$, we have*

$$\begin{aligned} H_{\mathcal{P}}(fx, f2) &= H_{\mathcal{P}}(\{0\}, \{0,1\}) \\ &= \max\left\{\delta_{\mathcal{P}}(\{0\}, \{0,1\}), \delta_{\mathcal{P}}(\{0,1\}, \{0\})\right\} \\ &= \max\left\{dist_{\mathcal{P}}(0, \{0,1\}), \max\{\mathcal{P}(0,0), \mathcal{P}(1,0)\}\right\} \\ &= \frac{3}{4} \leq kH_{\mathcal{P}}(x,2). \end{aligned}$$

Case 3. *Let $x = y = 2$. Then, with $k = \frac{1}{4}$, we have*

$$\begin{aligned} H_{\mathcal{P}}(f2, f2) &= H_{\mathcal{P}}(\{0,1\}, \{0,1\}) \\ &= \delta_{\mathcal{P}}(\{0,1\}, \{0,1\}) \\ &= \max\left\{dist_{\mathcal{P}}(0, \{0,1\}), dist_{\mathcal{P}}(1, \{0,1\})\right\} \\ &= \max\left\{0, \min\{\mathcal{P}(1,0), \mathcal{P}(1,1)\}\right\} \\ &= \frac{1}{4} \leq kH_{\mathcal{P}}(2,2). \end{aligned}$$

Hence, the contractive condition (i) of Theorem 5 is satisfied for $k = \frac{1}{2}$.

By routine calculation, one can verify the other conditions of Theorem 5. Observe that f has a unique fixed point (namely, $x = 0$).

6. Conclusions

First, we enlarged the class of symmetric spaces to the class of partial symmetric spaces, wherein we proved several results which included analogues of the Banach contraction principle, the Kannan-Ćirić fixed theorem, and the Ćirić quasi-fixed point theorem, in such spaces. We also furnished some examples, exhibiting the utility of our newly established results. Furthermore, we used one of the our main results to examine the existence and uniqueness of a solution for a system of Fredholm integral equations. Moreover, we extended the idea of Hausdorff distance to partial symmetric spaces, and proved an analogue of Nadler's fixed point theorem and some related results.

Author Contributions: All authors contributed equally in writing this article. All authors read and approved the final manuscript.

References

1. Banach, S. Sur les operations dans les ensembles abstraits et leur application aux equations integrals. *Fund. Math.* **1922**, *3*, 133–181. [CrossRef]
2. Matthews, S.G. Partial metric topology. *Ann. N. Y. Acad. Sci.* **1994**, *728*, 183–197. [CrossRef]
3. Amini-Harandi, A. Metric-like spaces, partial metric spaces and fixed points. *Fixed Point Theory Appl.* **2012**, *2012*, 204. [CrossRef]
4. Czerwik, S. Contraction mappings in b-metric spaces. *Acta Math. Inf. Univ. Ostrav.* **1993**, *1*, 5–11.
5. Branciari, A. A fixed point theorem of Banach–Caccioppoli type on a class of generalized metric spaces. *Publ. Math.* **2000**, *57*, 31–37.
6. Huang, L.G.; Zhang, X. Cone metric spaces and fixed point theorems of contractive mappings. *J. Math. Anal. Appl.* **2007**, *332*, 1468–1476. [CrossRef]
7. Imdad, M.; Asim, M.; Gubran, R. Common fixed point theorems for g-Generalized contractive mappings in b-metric spaces. *Indian J. Math.* **2018**, *60*, 85–105.
8. Ciric, L.B. A generalization of Banach's contraction principle. *Proc. Am. Math. Soc.* **1974**, *45*, 267–273. [CrossRef]
9. Mustafa, Z.; Roshan, J.R.; Parvaneh, V.; Kadelburg, Z. Some common fixed point results in ordered partial b-metric spaces. *J. Inequal. Appl.* **2013**, *2013*, 562. [CrossRef]

10. Imdad, M.; Ali, J. Common fixed point theorems in symmetric spaces employing a new implicit function and common property (EA). *Bull. Belg. Math. Soc. Simon Stev.* **2009**, *16*, 421–433.
11. Jleli, M.; Samet, B. A generalized metric space and related fixed point theorems. *Fixed Point Theory Appl.* **2015**, *2015*, 61. [CrossRef]
12. Wilson, W.A. On semi-metric spaces. *Am. J. Math.* **1931**, *53*, 361–373. [CrossRef]
13. Nadler, S.B. Multivalued contraction mappings. *Pac. J. Math.* **1969**, *30*, 475–488. [CrossRef]
14. Villa-Morales, J. Subordinate Semimetric Spaces and Fixed Point Theorems. *J. Math.* **2018**, *2018*, 7856594. [CrossRef]

Permissions

All chapters in this book were first published by MDPI; hereby published with permission under the Creative Commons Attribution License or equivalent. Every chapter published in this book has been scrutinized by our experts. Their significance has been extensively debated. The topics covered herein carry significant findings which will fuel the growth of the discipline. They may even be implemented as practical applications or may be referred to as a beginning point for another development.

The contributors of this book come from diverse backgrounds, making this book a truly international effort. This book will bring forth new frontiers with its revolutionizing research information and detailed analysis of the nascent developments around the world.

We would like to thank all the contributing authors for lending their expertise to make the book truly unique. They have played a crucial role in the development of this book. Without their invaluable contributions this book wouldn't have been possible. They have made vital efforts to compile up to date information on the varied aspects of this subject to make this book a valuable addition to the collection of many professionals and students.

This book was conceptualized with the vision of imparting up-to-date information and advanced data in this field. To ensure the same, a matchless editorial board was set up. Every individual on the board went through rigorous rounds of assessment to prove their worth. After which they invested a large part of their time researching and compiling the most relevant data for our readers.

The editorial board has been involved in producing this book since its inception. They have spent rigorous hours researching and exploring the diverse topics which have resulted in the successful publishing of this book. They have passed on their knowledge of decades through this book. To expedite this challenging task, the publisher supported the team at every step. A small team of assistant editors was also appointed to further simplify the editing procedure and attain best results for the readers.

Apart from the editorial board, the designing team has also invested a significant amount of their time in understanding the subject and creating the most relevant covers. They scrutinized every image to scout for the most suitable representation of the subject and create an appropriate cover for the book.

The publishing team has been an ardent support to the editorial, designing and production team. Their endless efforts to recruit the best for this project, has resulted in the accomplishment of this book. They are a veteran in the field of academics and their pool of knowledge is as vast as their experience in printing. Their expertise and guidance has proved useful at every step. Their uncompromising quality standards have made this book an exceptional effort. Their encouragement from time to time has been an inspiration for everyone.

The publisher and the editorial board hope that this book will prove to be a valuable piece of knowledge for researchers, students, practitioners and scholars across the globe.

List of Contributors

Margarita Besova and Vasiliy Kachalov
Department of Mathematical Sciences, National Research University "MPEI", ul. Krasnokazarmennaya 14,11250 Moscow, Russia

P. Njionou Sadjang
Department of Common Core, National Higher Polytechnic School Douala, Douala, Cameroon

S. Mboutngam
Higher Teachers' Training College, University of Maroua, Maroua, Cameroon

Taekyun Kim
Department of Mathematics, Kwangwoon University, Seoul 139-701, Korea

Dae San Kim
Department of Mathematics, Sogang University, Seoul 121-742, Korea

Bruna Germano and Maria Renata Martinelli
Department of Methods and Mathematic Models for Applied Sciences, University of Rome, La Sapienza, Via A. Scarpa, 14, 00161 Rome, Italy

Yang-Hi Lee
Department of Mathematics Education, Gongju National University of Education, Gongju 32553, Korea

Gwang Hui Kim
Department of Mathematics, Kangnam University, Yongin 16979, Korea

Hsien-Chung Wu
Department of Mathematics, National Kaohsiung Normal University, Kaohsiung 802, Taiwan

Julalak Prabseang and Kamsing Nonlaopon
Department of Mathematics, Faculty of Science, Khon Kaen University, Khon Kaen 40002, Thailand

Jessada Tariboon
Department of Mathematics, Faculty of Applied Science, King Mongkut's University of Technology North Bangkok, Bangkok 10800, Thailand

Hanaa M. Zayed and Adela O. Mostafa
Department of Mathematics, Faculty of Science, Menofia University, Shebin Elkom 32511, Egypt

Mohamed Kamal Aouf
Department of Mathematics, Faculty of Science, Mansoura University, Mansoura 35516, Egypt

Giuseppe Dattoli, Silvia Licciardi and Elio Sabia
ENEA – Frascati Research Center, Via Enrico Fermi 45, 00044 Frascati (Rome), Italy

Rosa Maria Pidatella
Department of Mathematics and Computer Sciences, University of Catania, Viale A. Doria 6, 95125 Catania, Italy

Yilmaz Simsek
Department of Mathematics, Faculty of Science University of Akdeniz, TR-07070 Antalya, Turkey

Andriy Bandura
Department of Advanced Mathematics, Ivano-Frankivsk National Technical University of Oil and Gas, 76019 Ivano-Frankivsk, Ukraine

Oleh Skaskiv
Department of Mechanics and Mathematics, Ivan Franko National University of Lviv, 79000 Lviv, Ukraine

Cheon Seoung Ryoo
Department of Mathematics, Hannam University, Daejeon 306-791, Korea

Haitham Qawaqneh, Mohd Noorani and Habes Alsamir
Department of Mathematics, University Kebangsaan Malaysia, Bangi 43600, Malaysia

Wasfi Shatanawi
Department of Mathematics, Hashemite University, Zarqa 1315, Jordan
Department of Mathematics and General Courses, Prince Sultan University, Riyadh 11586, Saudi Arabia

Konstantin Zhukovsky, Dmitrii Oskolkov and Nadezhda Gubina
Faculty of Physics, M.V. Lomonosov Moscow State University, Leninskie Gory, 119991 Moscow, Russia

Pierpaolo Natalini
Dipartimento di Matematica e Fisica, Università degli Studi Roma Tre, Largo San Leonardo Murialdo, 1, 00146 Roma, Italy

Paolo Emilio Ricci
Sezione di Matematica, International Telematic University UniNettuno, Corso Vittorio Emanuele II, 39, 00186 Roma, Italy

Maxie D. Schmidt
School of Mathematics, Georgia Institute of Technology, Atlanta, GA 30318, USA

Chenkuan Li, Thomas Humphries and Hunter Plowman
Department of Mathematics and Computer Science, Brandon University, Brandon, MB R7A 6A9, Canada

Mohammad Asim and Mohammad Imdad
Department of Mathematics, Aligarh Muslim University, Aligarh-202002, India

A. Rauf Khan
Department of Applied Sciences, University Polytechnic, Aligarh Muslim University, Aligarh-202002, India

Index

www.ingramcontent.com/pod-product-compliance
Lightning Source LLC
LaVergne TN
LVHW060525210926
848485LV00005B/1150